Computational Statistics in Climatology

Computational Statistics in Climatology

Ilya Polyak

New York   Oxford
OXFORD UNIVERSITY PRESS
1996

Oxford University Press

Oxford   New York
Athens   Auckland   Bangkok   Bogota   Bombay
Buenos Aires   Calcutta   Cape Town   Dar es Salaam
Delhi   Florence   Hong Kong   Istanbul   Karachi
Kuala Lumpur   Madras   Madrid   Melbourne
Mexico City   Nairobi   Paris   Singapore
Taipei   Tokyo   Toronto

and associated companies in
Berlin   Ibadan

Copyright © 1996 Oxford University Press, Inc.

Published by Oxford University Press, Inc.,
198 Madison Avenue, New York, New York 10016

Oxford is a registered trademark of Oxford University Press

All rights reserved. No part of this publication may be reproduced,
stored in a retrieval system, or transmitted, in any form or by any means
electronic, mechanical, photocopying, recording, or otherwise,
without the prior permission of Oxford University Press.

Library of Congress Cataloging-in-Publication Data
Polyak, Ilya.
Computational statistics in climatology / Ilya Polyak.
p.   cm.
Includes bibliographical references and index.
ISBN 0-19-509999-0
1. Climatology—Statistical methods.   I. Title.
QC981.P63   1996
551.5'01'176—dc20   95-35132

9 8 7 6 5 4 3 2 1

Printed in the United States of America
on acid-free paper

Preface

The range of statistical applications in the atmospheric sciences is so wide and diverse that it is difficult to mention even the most important areas of study. The proceedings of the Conferences on Probability and Statistics in the Atmospheric Sciences and of the Meetings on Statistical Climatology number in the dozens, and several books were published more than a decade ago (Gandin, 1965; Panofsky and Brier, 1968; Kagan, 1979; Epstein, 1985). Recently published textbooks (Thiebaux, 1994; Wilks, 1995) basically contain an introduction to the statistical methods, and an interesting monograph by R. Daley (1991) shows the possibility of the mutual physical and statistical analysis and modeling of atmospheric data. There are also a monograph, *Applied Statistics in Atmospheric Sciences*, by Essenwanger (1989) and a book by Preisendorfer and Mobley (1988) devoted solely to the principal component analysis in meteorology and oceanography.

Although much work has been done in developing statistical climatology, many important topics remain to be studied—for instance, topics in the field of computational statistics and its climatological applications.

This book presents the advanced course on computational statistics that I have taught for some years to Ph.D. students and scientists in the Main Geophysical Observatory, Hydrological Institute, and Civil Engineering Institute at St. Petersburg, Russia. The main objective of the course is to progress from univariate modeling and spectral and correlation analysis to multivariate modeling and multidimensional spectral and correlation analysis. The course developed as a result of my many years of involvement in a variety of projects dealing with design and development of software for climate change studies by using statistic and random functions methods. The elaboration of the corresponding algorithms, their application to analyzing and modeling climatic data, and consulting in the widespread area of such methodologies have made it possible to gain an understanding of the importance of various statistical disciplines for climate study. The two major areas of application presented in this book are multidimensional spectral and correlation analysis and multivariate autoregressive modeling. Because the spectrum estimation methodology is based on the choice of smoothing window, the book opens with a consideration of digital filters (Chapter 1). Questions of averaging (as the dominant statistical procedure in climatology) and fitting simple linear models are given principal attention in Chapter 2. Estimations of spectral and correlation functions of random processes and fields are discussed in Chapter 3. Algorithms of univariate (Chapter 4) and multivariate (Chapter 5) modeling, potentially the most important methodologies for climate study, conclude the consideration of statistics.

Chapters 6, 7, and 8 are devoted to the analysis and modeling of climatological data. (The climatological results given are limited to those that were obtained with my participation.)

The goal of the methodological sections is to present numerical algorithms in a form that can immediately be used for software development. For example, the formulas for several classes of one- and two-dimensional smoothing and differentiating digital filters, each of which depends on several parameters (width, length, order), can be easily programmed and presented as a system of corresponding routines. The numerical schemes for one-, two-, and multidimensional spectral and correlation analyses, gathered in Tables 3.1 and 3.2, make it possible to design corresponding software with a general focus on the application of some classes of filters of different dimensionality. In many respects this work can serve as a handy reference book for software development.

One of the features of the approach used is a numerical analysis of the algorithms to gain insight into the theoretical accuracy (and limitations) of different methods and to obtain maximum preliminary information for the particular application. Such an approach is especially effective for linear numerical procedures when the influence of data transformation, which determines the standard error value to within some constant factor, on the estimate accuracy can be analyzed before dealing with real observations.

The computational orientation of the text has necessitated that theoretical-statistical questions are only briefly outlined. Thus this specialized book cannot substitute for the study of more general theoretical courses. Moreover, a certain level of knowledge in statistics and random function theory has been assumed, and the material presented can help the reader design software applications and efficiently use them. For many of the algorithms considered, I have developed software in Fortran and the Interactive Data Language (IDL) and have used them for climate studies.

Multiple examples with illustrations developed while teaching have a special methodological meaning, not only as appropriate examples, but also as general results that represent the accuracy of some classes of methods (see, for instance, the figures in Chapter 4 with normalized standard deviations of the forecast errors for the entire class of first- and second-order ARMA processes). Actually, most of the book consists of methodological or climatological examples.

Statistical descriptions of climate represent an important aspect of multipurpose scientific studies and provide a means for finding solutions to many climatological problems. Historically, descriptions of climate have been conducted by studying the average values (norms) of meteorological observations from different points on the Earth. As climatology has developed, it has become necessary to consider not only averages but also numerous other statistics that characterize spatial and temporal climate variability. The most important topics in statistical climatology are problems that are (or can be) stated within the strict frameworks of theoretically founded statistical methods. With this in mind, the following group of studies should be mentioned.

1. Climate modeling presents the very substantial problem of verifying and confirming GCM (General Circulation Model) experiments as part of the statistical identification of complicated stochastic systems. The obvious requirement (to have identical probability distributions of the observed and simulated data) led to the necessity of the estimation and multiple comparison of the corresponding first- and second-order moments. The second moments can be presented in the form of the multidimensional spectra and correlation functions or multivariate stochastic models. Subsequent hypothesis testing and statistical inferences can help in evaluating the closeness of the observed and simulated climate and reveal the validity of GCMs. The subsequent physical analysis and interpretation of the parameter matrices of the fitted models can become a major diagnostic tool and can help solve some of the inverse problems for hydrodynamic equations. In other words, at this stage of study, the interpretation of the model parameter matrices as a system of interactions and feedbacks of the analyzed processes is as important as the forecast. In principle, according to system identification theory, statistical and physical modeling must result in a general model of the climate system.

2. Some engineering and practical projects require an analogous comparison of the probability distributions of data produced either by different types of measurements (such as point gauges, radar, and satellite) or different data collecting and averaging methodologies. For instance, the ground truth problem for the Tropical Rainfall Measuring Mission (TRMM) from satellite or the evaluation of the validity of different surface air temperature data bases necessitates the estimation and comparison of second-moment statistics. Generally, most of the questions sur-

rounding retrieval, calibration, beam filling error analysis, and so on, are the nonlinear statistical problems.

3. Another important direction of statistical modeling in climatology that is an inseparable part of the GCM development arises in a wide variety of parametrization problems. It seems that the replacement of the existing, pure empirical parametrization schemes by the multivariate and multidimensional stochastic models can help to create scientifically sound, more accurate and reliable physico-stochastic GCMs corresponding to the real physico-stochastic character of the climate system.

4. Statistical (as well as physical) modeling of the climatic system is necessary for developing climatology as a scientific discipline and, in particular, for describing the causes of climate change. The formulation of such problems is based on the study of the Earth's history, which shows that climate changes have taken place on global and regional scales: ice-ages followed by warming periods, damp regions turned into deserts and semi-deserts, significantly fluctuating levels of spacious internal reservoirs, growth and decay of glaciers, and so on. The dependence of every sphere of human life on the climate makes it obvious that statistical descriptions of climate constitute some of the most important scientific work of our time.

5. There are some other interesting global and climate change problems; for example, verification of the astronomical (Milankovitch) climate theory, and the paleoclimate and tree-ring data analysis, which require application of different statistical procedures.

Estimating the statistical moments of atmospheric processes (as the components of a stochastic climate system) provides a means for developing stochastic models for any of their subsystems. The problem of such modeling is to describe (theoretically and experimentally) the temporal evolution of the climatic system (or its statistical structure). But the methods of computational statistics used for such modeling are based, as a rule, on the assumptions that the variability of the corresponding stochastic system is spatially and temporally stable (stationary and homogeneous) and that it possesses a number of other theoretic properties (such as ergodicity and normality). Of course, real processes of nature are far more complicated than the above theoretical abstractions. Consequently, the results of the estimation (for example, of the second moments) obtained within the framework of any mathematical condition are only approximate, which might not necessarily reflect reality in its details and complexity. Absolutely accurate stochastic (as well as physical) models of natural processes are impossible in principle. We can only design and fit some images, the goodness of fit of which is, among other things, determined by the applied technique.

Nevertheless, conducting investigations using standard statistical methodologies remains an important step in the development of climatology. It also allows an orientation among multiple empirical approaches and nonscientific interpretations of the results of estimation. Generally, the formulation of climatological problems within the strict framework of fundamental statistical methods promotes the natural progress of climatology along with the simultaneous theoretical development of such methods.

I do not pretend to present the entire range of statistical methods used in climatology. Several very interesting, and, in some cases, more advanced, branches of applied statistics (for example, nonlinear modeling, principal component analysis, methods of objective analysis of meteorological fields, and others) are not considered here at all.

Of course, the reliability of climate statistics depends primarily on the quality and volume of data. The climatological results of this book were obtained using the most complete climate data bases available. These include the surface air temperature of the Northern Hemisphere for 100 years of observations, special kinds of precipitation measurements such as GATE rain rate data and PRE-STORM precipitation, historical records (unique in duration) of temperature (315 years of observations) and precipitation (250 years of observations) in Central England, atmospheric pressure in Basel (205 years of observations), the simulation data of one of the best GCMs, and many others.

The second half of the book presents wide-ranging statistical results, which describe the temporal and spatial statistical structure of the large-scale climate fluctuations for both point gauges and spatially averaged data. In many cases, the closeness of climate fluctuations to white noise or to first-order multivariate AR models is discussed. We expect that limiting the order of models (as well as their types) roughens our understanding of climate, but this stage of climatology development cannot be bypassed. Such an approach may either manifest the most general regularities of climate or give the first very rough approximation of climate as a stochastic system with the possibility of analyzing the interactions and feedbacks of the diverse climatological processes.

Presentation of the spectral and correlation estimates of different climatological fields is restricted to the two-dimensional case (three-dimensional figures) because pictorial representation of the four- (and more) dimensional structures is too technically complex, given the book's framework. The best medium for multidimensional illustrations is film, which enables us to demonstrate variations of climate in the spaces of time, lags, and frequencies (wave numbers) simultaneously.

In presenting the illustrations, I also tried to show a little bit of beauty, as statistics sometimes seems to the uninitiated a dry and tedious science. In this aspect an excellent IDL software developed by Research Systems, Inc., was very helpful in drawing three-dimensional graphics.

This book is intended for specialists in statistical software development and application and for climatologists. I also hope that this book (and especially its inevitable errors and mistakes) will attract the attention of specialists in theoretical statistics to climatological problems, the solutions of which, in many cases, are a continuing challenge to modern science.

I am grateful to Dr. Gerald North for his support, which made this work possible. I thank Phyllis McBride for help with technical editing of the first version of the book, and Stephen Cronin for help with the LaTex representation of the formulas and the typing of several chapters.

I also appreciate the permission of the American Meteorological Society to reproduce materials from the papers Polyak et al., 1994; Polyak and North, 1995; and Polyak, 1996.

This work was made possible in part by NASA grants.

College Station, Texas I. P.
December 1995

Contents

1. **DIGITAL FILTERS** — 3
   1.1 Gauss-Markov Theory — 3
   1.2 Polynomial Approximations — 6
   1.3 Variance of the Point Estimates — 15
   1.4 The Fourier Set — 21
   1.5 Examples — 23
   1.6 Smoothing Digital Filters — 24
   1.7 Regressive Filters — 28
   1.8 Harmonic Filters — 39
   1.9 Applications of Digital Filters — 40
   1.10 Differentiating Filters — 44
   1.11 Two-Dimensional Filters — 51
   1.12 Multidimensional Filters — 58

2. **AVERAGING AND SIMPLE MODELS** — 61
   2.1 Correlated Observations — 61
   2.2 The Mean and the Linear Trend — 66
   2.3 Nonoptimal Estimation — 75
   2.4 Spatial Averaging — 79
   2.5 Smoothing of Correlated Observations — 87
   2.6 Filters of Finite Differences — 91
   2.7 Regression and Instrumental Variable — 99
   2.8 Nonlinear Processes — 107

3. **RANDOM PROCESSES AND FIELDS** — 110
   3.1 Stationary Process — 111
   3.2 Cross-Statistical Analysis — 117
   3.3 Nonstationary Processes — 122
   3.4 Nonergodic Stationary Process — 127

3.5 Time Series with Missing Data	129
3.6 Two-Dimensional Fields	131
3.7 Multidimensional Fields	138
3.8 Examples of Climatological Fields	147
3.9 Anisotropy of Climatological Fields	160
4. VARIABILITY OF ARMA PROCESSES	**162**
4.1 Fundamental ARMA Processes	163
4.2 AR Processes	165
4.3 AR(1) and AR(2) Processes	167
4.4 Order of the AR Process	174
4.5 MA(1) and MA(2) Processes	179
4.6 ARMA(1,1) Process	183
4.7 Comments	186
4.8 Signal-Plus-White-Noise Type Processes	190
4.9 Process with Stationary Increments	198
4.10 Modeling the Five-Year Mean Surface Air Temperature	201
4.11 Nonstationary and Nonlinear Models	204
5. MULTIVARIATE AR PROCESSES	**208**
5.1 Fundamental Multivariate AR Processes	208
5.2 Multivariate AR(1) Process	212
5.3 Algorithm for Multivariate Model	218
5.4 AR(2) Process	223
5.5 Examples of Climate Models	225
5.6 Climate System Identification	234
6. HISTORICAL RECORDS	**239**
6.1 Linear Trends	239
6.2 Climate Trends over Russia	245
6.3 Periodograms	247
6.4 Spectral and Correlation Analysis	251
6.5 Univariate Modeling	259
6.6 Statistics and Climate Change	263
7. THE GCM VALIDATION	**266**
7.1 Objectives	267
7.2 Data	268
7.3 Zonal Time Series	271
7.4 Multivariate Models	279
7.5 The Diffusion Process	285

7.6	Latitude-Temporal Fields	291
7.7	Conclusion	299
8.	SECOND MOMENTS OF RAIN	302
8.1	GATE Observations	303
8.2	PRE-STORM Precipitation	331
8.3	Final Remarks	344
	References	348
	Index	355

Computational Statistics in Climatology

1
Digital Filters

In this chapter, several systems of digital filters are presented. The first system consists of regressive smoothing filters, which are a direct consequence of the least squares polynomial approximation to equally spaced observations. Descriptions of some particular univariate cases of these filters have been published and applied (see, for example, Anderson, 1971; Berezin and Zhidkov, 1965; Kendall and Stuart, 1963; Lanczos, 1956), but the study presented in this chapter is more general, more elaborate in detail, and more fully illustrated. It gives exhaustive information about classical smoothing, differentiating, one- and two-dimensional filtering schemes with their representation in the spaces of time, lags, and frequencies. The results are presented in the form of algorithms, which can be directly used for software development as well as for theoretical analysis of their accuracy in the design of an experiment. The second system consists of harmonic filters, which are a direct consequence of a Fourier approximation of the observations. These filters are widely used in the spectral and correlation analysis of time series.

1.1 Gauss-Markov Theory

The foundation for developing regressive filters is the least squares polynomial approximation (of equally spaced observations), a principal notion that will be considered briefly. We look first at the independent and equally accurate observations

$$Y_0, Y_1, \ldots, Y_k \qquad (1.1)$$

with conditions

$$E(Y_i) = x_{i0}\beta_0 + x_{i1}\beta_1 + \ldots + x_{im}\beta_m$$

$$(i = 0, 1, \ldots, k; \quad k > m), \tag{1.2}$$

where β_i are unknown parameters which must be estimated and the x_{ij} are determined by the structure of the approximation function.

Independence and equal accuracy mean that

$$V(\mathbf{Y}) = \sigma^2 \mathbf{I}, \tag{1.3}$$

where V denotes variance, $\mathbf{Y} = \{Y_i\}_{i=0}^{k}$, $\mathbf{I}$ is the identity matrix of size $(k+1) \times (k+1)$, and σ^2 is the variance of the observations. In matrix notation,

$$\mathbf{X} = \{x_{ij}\}_{i,j=0}^{k,m}, \quad \boldsymbol{\beta} = \{\beta_j\}_{j=0}^{m}, \tag{1.4}$$

and the system of conditions can be written

$$\mathbf{E}(\mathbf{Y}) - \mathbf{X}\boldsymbol{\beta} = 0. \tag{1.5}$$

To estimate the parameters β_j by the least squares method, the quadratic form

$$(\mathbf{Y} - \mathbf{X}\boldsymbol{\beta})^{\mathrm{T}}(\mathbf{Y} - \mathbf{X}\boldsymbol{\beta}) = \sum_{i=0}^{k}\left[Y_i - \sum_{j=0}^{m} x_{ij}\beta_j\right]^2 \tag{1.6}$$

must be minimized over β_j.

Differentiating (1.6) with respect to β_j yields a system of equations for finding the unknown parameters:

$$\mathbf{X}^{\mathrm{T}}\mathbf{X}\boldsymbol{\beta} = \mathbf{X}^{\mathrm{T}}\mathbf{Y}. \tag{1.7}$$

This system is called the normal system (or the normal equations), and the matrix

$$\mathbf{C} = \mathbf{X}^{\mathrm{T}}\mathbf{X} \tag{1.8}$$

is called the matrix of the normal system. $\mathbf{C}$ is a symmetric matrix of size $(m+1) \times (m+1)$. Let us assume that matrix $\mathbf{C}$ is not singular and that the solution of system (1.7) is

$$\hat{\boldsymbol{\beta}} = (\mathbf{X}^{\mathrm{T}}\mathbf{X})^{-1}\mathbf{X}^{\mathrm{T}}\mathbf{Y}, \tag{1.9}$$

where

1.1 Gauss-Markov Theory

$$\hat{\boldsymbol{\beta}} = \{\hat{\beta}_j\}_{j=0}^{m} . \qquad (1.10)$$

Vector $\hat{\boldsymbol{\beta}}$ is a random vector (as a linear function of $\mathbf{Y}$) with a covariance matrix $\mathbf{M}_\beta$. Introducing the notation $\hat{\boldsymbol{\beta}} = \mathbf{FY}$, where $\mathbf{F} = (\mathbf{X}^T\mathbf{X})^{-1}\mathbf{X}^T$, it is possible to obtain

$$\begin{aligned}\mathbf{M}_\beta &= \mathbf{F}\sigma^2\mathbf{I}\mathbf{F}^T = \sigma^2(\mathbf{X}^T\mathbf{X})^{-1}\mathbf{X}^T\mathbf{X}(\mathbf{X}^T\mathbf{X})^{-1} \\ &= \sigma^2(\mathbf{X}^T\mathbf{X})^{-1} = \sigma^2\mathbf{C}^{-1}.\end{aligned} \qquad (1.11)$$

If observations (1.1) are from a normal distribution, then an estimator $\hat{\sigma}^2$ of variance σ^2 (1.3) is given by the expression

$$\hat{\sigma}^2 = \frac{1}{k-m}(\mathbf{Y} - \mathbf{X}\hat{\boldsymbol{\beta}})^T(\mathbf{Y} - \mathbf{X}\hat{\boldsymbol{\beta}}). \qquad (1.12)$$

The vector $\hat{\mathbf{Y}} = \{\hat{Y}_i\}_{i=0}^{k} = \mathbf{Y} - \mathbf{X}\hat{\boldsymbol{\beta}}$ is called the vector of the point estimates.

The covariance matrix $\mathbf{M}_{\hat{Y}}$ of the random vector $\hat{\mathbf{Y}}$ is

$$\mathbf{M}_{\hat{Y}} = \mathbf{X}\mathbf{M}_\beta\mathbf{X}^T = \sigma^2\mathbf{X}(\mathbf{X}^T\mathbf{X})^{-1}\mathbf{X}^T = \sigma^2\mathbf{X}\mathbf{C}^{-1}\mathbf{X}^T. \qquad (1.13)$$

Although observations (1.1) are independent and equally accurate, the point estimates are statistically dependent with covariance matrix (1.13). Consider, for instance, the normalized (standardized) matrix

$$\frac{1}{\sigma^2}\mathbf{M}_{\hat{Y}} = \mathbf{X}(\mathbf{X}^T\mathbf{X})^{-1}\mathbf{X}^T \qquad (1.14)$$

and its diagonal elements

$$\sigma_i^2 = \{\mathbf{X}(\mathbf{X}^T\mathbf{X})^{-1}\mathbf{X}^T\}_{ii} \quad (i = 0, 1, \ldots, k). \qquad (1.15)$$

The elements of matrix (1.14) [which are proportional to the elements of covariance matrix (1.13)] do not depend on either the observations or their statistical characteristics. The elements are determined only by the structure of the system of conditions (i.e., by the elements of matrix $\mathbf{X}$). For example, the polynomial approximations produce a special class of conditional matrices. Therefore, during the design stage of an experiment it is possible to analyze the matrices $(\mathbf{X}^T\mathbf{X})^{-1}$ and $\mathbf{X}(\mathbf{X}^T\mathbf{X})^{-1}\mathbf{X}^T$, which determine the statistical characteristics of the parameters and point estimates (to within a factor σ^2). This preliminary information enables us to make the necessary computations more efficiently. Such an analysis can even provide enough information to make a decision about the value of the suggested model without the need for calculations.

The variances of the point estimates [the diagonal elements of (1.13)] look like $\sigma^2 \sigma_i^2$, where σ^2 is a constant for given observations but σ_i^2 depends on i and is determined by the matrix $\mathbf{X}$. Experimenting with different systems of conditions allows us to analyze the dependence of σ_i^2 on i and to obtain information about the accuracy of the point estimates $\widehat{\mathbf{Y}} = X\widehat{\boldsymbol{\beta}}$ before any calculations are conducted. The values of σ_i^2 are called normalized (or standardized) variances of the point estimates; they differ from the variances of the point estimates only by the factor σ^2.

Sometimes, we need to consider the statistical characteristics of the residuals $\mathbf{R} = \mathbf{Y} - \mathbf{X}\widehat{\boldsymbol{\beta}}$. The covariance matrix of $\mathbf{R}$ is

$$\mathbf{M}_R = \sigma^2 \mathbf{I} - \mathbf{M}_{\widehat{Y}}. \tag{1.16}$$

To check the accuracy of the computations, we can use the following expression:

$$S_{tr}(\mathbf{M}_{\widehat{Y}}) = (m+1)\sigma^2, \tag{1.17}$$

where $S_{tr}(\mathbf{M}_{\widehat{Y}})$ is a trace of the matrix $\mathbf{M}_{\widehat{Y}}$.

If observations Y_i are normal, the statistical significance of the estimates of the parameters β_i is determined by the Student statistic (Rao, 1973)

$$t(k-m) = \frac{\beta_i - \widehat{\beta}_i}{\widehat{\sigma}\sqrt{(C^{-1})_{ii}}}, \tag{1.18}$$

which follows a t distribution with $k - m$ degrees of freedom.

1.2 Polynomial Approximations

In this section the least squares polynomial approximations of the observations, given on the equally spaced grid of points, are considered. One convenient way to construct such polynomials is based on a system of orthogonal polynomials (see Bolshev and Smirnov, 1965). But for some practical purposes, such as designing digital filters, the direct construction of the polynomials is preferred.

Consider the observations given on the sequence of equally spaced points

$$t'_0, t'_1, \ldots, t'_k; \quad (t'_{i+1} - t'_i = h = \text{const}).$$

1.2 Polynomial Approximations

The transformation

$$t = \frac{1}{h}\left(t'_i - \frac{1}{k+1}\sum_{j=0}^{k} t'_j\right) = i - \frac{k}{2} = i - r, \qquad (1.19)$$

where $r = k/2$, $(k+1 = 2r+1)$, changes this sequence to the symmetrical grid points

$$-r, -(r-1), \ldots, r-1, r. \qquad (1.20)$$

If the number of observations is odd (i.e., k is even), the grid points are

$$-r, \ldots, -2, -1, 0, 1, 2, \ldots, r.$$

If the number of points is even (i.e., k is odd), the grid points are

$$-r, \ldots, -\frac{3}{2}, -\frac{1}{2}, \frac{1}{2}, \frac{3}{2}, \ldots, r.$$

The notation $f_{i-r} = Y_i$ $(i = 0, 1, \ldots, k)$ leads to the symmetrical indexing of the observations:

$$f_{-r}, f_{-(r-1)}, \ldots, f_{r-1}, f_r. \qquad (1.21)$$

If k is odd, the values f_{i-r} have fractional subscripts, but these will not cause any difficulties in our presentation.

The coefficients of the polynomial

$$f(t) = \beta_0 + \beta_1 t + \ldots + \beta_m t^m \qquad (1.22)$$

under the condition

$$\min \sum_{i=0}^{k} [f_{i-r} - f(i-r)]^2 \qquad (1.23)$$

must then be estimated.

It is clear that

$$x_{ij} = (i-r)^j. \qquad (1.24)$$

The normal equations in this case are

$$\beta_0(k+1) + \beta_1 \sum_{i=0}^{k}(i-r) + \beta_2 \sum_{i=0}^{k}(i-r)^2 + \ldots$$

$$+\beta_m \sum_{i=0}^{k}(i-r)^m = \sum_{i=0}^{k} f_{i-r},$$

$$\beta_0 \sum_{i=0}^{k}(i-r) + \beta_1 \sum_{i=0}^{k}(i-r)^2 + \beta_2 \sum_{i=0}^{k}(i-r)^3 + \ldots$$

$$+\beta_m \sum_{i=0}^{k}(i-r)^{m+1} = \sum_{i=0}^{k} f_{i-r}(i-r),$$

$$\cdots\cdots\cdots\cdots$$

$$\beta_0 \sum_{i=0}^{k}(i-r)^m + \beta_1 \sum_{i=0}^{k}(i-r)^{m+1} + \beta_2 \sum_{i=0}^{k}(i-r)^{m+2} + \ldots$$

$$+\beta_m \sum_{i=0}^{k}(i-r)^{2m} = \sum_{i=0}^{k} f_{i-r}(i-r)^m. \qquad (1.25)$$

In matrix notation, this system is

$$\mathbf{C}\boldsymbol{\beta} = \mathbf{Q}, \qquad (1.26)$$

where

$$\mathbf{C} = \mathbf{X}^T\mathbf{X} = \begin{pmatrix} k+1 & \sum_{i=0}^{k}(i-r) & \ldots & \sum_{i=0}^{k}(i-r)^m \\ \vdots & \vdots & & \vdots \\ \sum_{i=0}^{k}(i-r)^m & \sum_{i=0}^{k}(i-r)^{m+1} & \ldots & \sum_{i=0}^{k}(i-r)^{2m} \end{pmatrix}$$
$$(1.27)$$

and

1.2 Polynomial Approximations

$$\mathbf{Q} = \mathbf{X}^T \mathbf{Y} = \begin{pmatrix} \sum_{i=0}^{k} f_{i-r} \\ \sum_{i=0}^{k} f_{i-r}(i-r) \\ \vdots \\ \sum_{i=0}^{k} f_{i-r}(i-r)^m \end{pmatrix}. \qquad (1.28)$$

For the chosen grid points, if p is odd then

$$\sum_{i=0}^{k}(i-r)^p = 0 \qquad (1.29)$$

and the normal system is separated into two independent systems of the following orders: $INT\,[(m+2)/2]$ and $INT\,[(m+1)/2]$. [1] For the same observations, polynomials of degrees m and $m+1$ have $INT\,[(m+2)/2]$ identical coefficients. Polynomial approximations (to the same observations) with subsequent increasing (by one) degree lead to the necessity of finding the solution to two independent systems, of orders $INT\,[(m+2)/2]$ and $INT\,[(m+1)/2]$, instead of one system of order $m+1$, as is necessary in general cases. In other words, such polynomials with degrees $2p$ and $2p+1$ ($p = 0, 1, \ldots$) have identical coefficients with even numbers. Polynomials with degrees $2p-1$ and $2p$ ($p = 1, 2, \ldots$) have identical coefficients with odd numbers. This is one of the advantages of the symmetrical system of points (1.20). To compute the elements of the normal system's matrix (1.27), it is necessary to find the sums

$$\sum_{i=0}^{k}(i-r)^p \qquad (1.30)$$

if p is even. These sums can be calculated by the formula

$$\frac{1}{2}\sum_{i=0}^{k}(i-r)^p = \frac{1}{p+1}B_{p+1}(r+1), \qquad (1.31)$$

---

[1] $INT(a)$ means the integer part of a. For example, $INT(5/2) = 2$.

where $B_{p+1}(r+1)$ are the Bernoulli polynomials.[2]

The definition of the Bernoulli polynomials produces the equalities

$$\sum_{i=0}^{k}(i-r)^2 = \frac{1}{3}r(r+1)(2r+1),$$

$$\sum_{i=0}^{k}(i-r)^4 = \frac{1}{3}r(r+1)(2r+1)\frac{1}{5}(3r^2+3r-1),$$

---

[2]The Bernoulli polynomials $B_p(x)$ are determined by

$$B_p(x) = x^p + \binom{p}{1}B_1 x^{p-1} + \ldots + \binom{p}{p-1}B_p x + \binom{p}{p}B_p,$$

where B_p are the Bernoulli numbers, defined by the formula

$$\frac{z}{e^z-1} = \sum_{i=0}^{\infty} B_i \frac{z^i}{i!}.$$

Values for the first few Bernoulli numbers can be found, for instance, in Berezin and Zhidkov (1965).

Note the following properties of the Bernoulli polynomials:

$$\int_x^{x+1} B_p(t)dt = x^p, \qquad (*)$$

$$\int_x^y B_p(t)dt = \frac{B_{p+1}(y) - B_{p+1}(x)}{p+1}, \qquad (**)$$

$$B_p(1-x) = (-1)^p B_p(x). \qquad (***)$$

Now we can find a sum (if p is even)

$$\sum_{i=0}^{2r}(i-r)^p$$

$$= \sum_{i=0}^{2r} \int_{i-r}^{i-r+1} B_p(t)dt = \int_{-r}^{r+1} B_p(t)dt = \frac{B_{p+1}(r+1) - B_{p+1}(-r)}{p+1}$$

$$= \frac{2B_{p+1}(r+1)}{p+1},$$

where the first equality is conditioned by (*), the third by (**), and the fourth by (***).

1.2 Polynomial Approximations

$$\sum_{i=0}^{k}(i-r)^6 = \frac{1}{3}r(r+1)(2r+1)\frac{1}{7}(3r^4 + 6r^3 - 3r + 1),$$

$$\sum_{i=0}^{k}(i-r)^8 = \frac{1}{3}r(r+1)(2r+1)\frac{1}{15}(5r^6 + 15r^5$$
$$+5r^4 - 15r^3 - r^2 + 9r - 3),$$

$$\sum_{i=0}^{k}(i-r)^{10} = \frac{1}{3}r(r+1)(2r+1)\frac{1}{11}(3r^8 + 12r^7 + 8r^6$$
$$-18r^5 - 10r^4 + 24r^3 + 2r^2 - 15r + 5), \quad (1.32)$$

which determine the matrix elements of the normal systems if $m = 0, 1, 2, 3, 4, 5$.

Therefore, the elements of the normal system's matrices are the Bernoulli polynomials with argument $r + 1 = k/2 + 1$. This enables us to find the analytical expressions for the elements of the inverse matrix $\mathbf{C}^{-1}$ as functions of r. The inversion of the matrix $\mathbf{C}$ amounts to the inversion of the two matrices

$$\begin{pmatrix} B_1(r+1) & \frac{1}{3}B_3(r+1) & \frac{1}{5}B_5(r+1) \ldots \\ \frac{1}{3}B_3(r+1) & \frac{1}{5}B_5(r+1) & \frac{1}{7}B_7(r+1) \ldots \\ \frac{1}{5}B_5(r+1) & \frac{1}{7}B_7(r+1) & \frac{1}{9}B_9(r+1) \ldots \\ \vdots & \vdots & \vdots \end{pmatrix} \quad (1.33)$$

and

$$\begin{pmatrix} \frac{1}{3}B_3(r+1) & \frac{1}{5}B_5(r+1) & \frac{1}{7}B_7(r+1) \ldots \\ \frac{1}{5}B_5(r+1) & \frac{1}{7}B_7(r+1) & \frac{1}{9}B_9(r+1) \ldots \\ \frac{1}{7}B_7(r+1) & \frac{1}{9}B_9(r+1) & \frac{1}{11}B_{11}(r+1) \ldots \\ \vdots & \vdots & \vdots \end{pmatrix}. \quad (1.34)$$

Note the formulas for the elements of the matrices

$$\mathbf{C}^{-1} = \{c_{ij}^m\}_{i,j=0}^m$$

($m = 0, 1, 2, 3, 4, 5$; m is the superscript, not the power) that satisfy the requirements of practical applications:

$$\mathbf{C}^{-1} = \begin{pmatrix} c_{00}^1 & 0 \\ 0 & c_{11}^1 \end{pmatrix}, \tag{1.35}$$

where

$$c_{00}^1 = \frac{1}{2r+1}, \quad c_{11}^1 = \frac{3}{r(r+1)(2r+1)};$$

$$\mathbf{C}^{-1} = \begin{pmatrix} c_{00}^3 & 0 & c_{02}^3 & 0 \\ 0 & c_{11}^3 & 0 & c_{13}^3 \\ c_{20}^3 & 0 & c_{22}^3 & 0 \\ 0 & c_{31}^3 & 0 & c_{33}^3 \end{pmatrix}, \tag{1.36}$$

where

$$c_{00}^3 = 3(3r^2 + 3r - 1)\,\lambda, \qquad c_{02}^3 = c_{20}^3 = -15\,\lambda,$$

$$c_{11}^3 = 25(3r^4 + 6r^3 - 3r + 1)\,\eta\,\lambda,$$

$$c_{13}^3 = c_{31}^3 = -35(3r^2 + 3r - 1)\,\eta\,\lambda,$$

$$c_{22}^3 = \tfrac{45}{r(r+1)}\,\lambda, \qquad c_{33}^3 = 175\,\eta\,\lambda,$$

$$\lambda = \frac{1}{(4r^2 - 1)(2r+3)}, \qquad \eta = \frac{1}{r(r^2 - 1)(r+2)};$$

and

$$\mathbf{C}^{-1} = \begin{pmatrix} c_{00}^5 & 0 & c_{02}^5 & 0 & c_{04}^5 & 0 \\ 0 & c_{11}^5 & 0 & c_{13}^5 & 0 & c_{15}^5 \\ c_{20}^5 & 0 & c_{22}^5 & 0 & c_{24}^5 & 0 \\ 0 & c_{31}^5 & 0 & c_{33}^5 & 0 & c_{35}^5 \\ c_{40}^5 & 0 & c_{42}^5 & 0 & c_{44}^5 & 0 \\ 0 & c_{51}^5 & 0 & c_{53}^5 & 0 & c_{55}^5 \end{pmatrix}, \tag{1.37}$$

where

$$c_{00}^5 = \frac{15}{4\gamma}(15r^4 + 30r^3 - 35r^2 - 50r + 12),$$

1.2 Polynomial Approximations

$$c_{02}^5 = c_{20}^5 = -\frac{75 \cdot 7}{4\gamma}(2r^2 + 2r - 3), \qquad c_{04}^5 = c_{40}^5 = \frac{27 \cdot 35}{4\gamma},$$

$$c_{22}^5 = \frac{7 \cdot 15 \cdot 21}{4\gamma} \frac{4r^4 + 8r^3 - 4r^2 - 8r + 5}{r(r^2-1)(r+2)},$$

$$c_{42}^5 = c_{24}^5 = -\frac{21 \cdot 75}{4\gamma} \frac{6r^2 + 6r - 5}{r(r^2-1)(r+2)},$$

$$c_{44}^5 = \frac{7 \cdot 21 \cdot 75}{4\gamma} \frac{1}{r(r^2-1)(r+2)},$$

$$c_{11}^5 = \frac{49 \cdot 3}{4\mu}(25r^8 + 100r^7 - 50r^6 - 500r^5 - 95r^4 + 760r^3$$
$$+ 180r^2 - 300r + 72),$$

$$c_{13}^5 = c_{31}^5 = -\frac{49 \cdot 45}{4\mu}(6r^6 + 18r^5 - 15r^4 - 60r^3 + 17r^2$$
$$+ 50r - 12),$$

$$c_{15}^5 = c_{51}^5 = \frac{63 \cdot 11}{4\mu}(15r^4 + 30r^3 - 35r^2 - 50r + 12),$$

$$c_{33}^5 = \frac{63 \cdot 75}{4\mu}(12r^4 + 24r^3 - 28r^2 - 40r + 39),$$

$$c_{53}^5 = c_{35}^5 = -\frac{11 \cdot 45 \cdot 49}{4\mu}(2r^2 + 2r - 3), \qquad c_{55}^5 = \frac{11 \cdot 81 \cdot 49}{4\mu},$$

$$\gamma = (4r^2 - 9)(4r^2 - 1)(2r + 5),$$

$$\mu = \gamma r(r^2-1)(r^2-4)(r+3). \tag{1.38}$$

If m is even,

$$c_{ij}^m = \begin{cases} c_{ij}^{m-1} & \text{if } i \text{ is odd,} \\ c_{ij}^{m+1} & \text{if } i \text{ is even.} \end{cases} \tag{1.39}$$

For example, for the second- and fourth-degree polynomials, the matrices $\mathbf{C}^{-1}$ can be composed from the above elements as follows:

$$\mathbf{C}^{-1} = \begin{pmatrix} c_{00}^3 & 0 & c_{02}^3 \\ 0 & c_{11}^1 & 0 \\ c_{20}^3 & 0 & c_{22}^3 \end{pmatrix}, \tag{1.40}$$

$$\mathbf{C}^{-1} = \begin{pmatrix} c_{00}^5 & 0 & c_{02}^5 & 0 & c_{04}^5 \\ 0 & c_{11}^3 & 0 & c_{13}^3 & 0 \\ c_{20}^5 & 0 & c_{22}^5 & 0 & c_{24}^5 \\ 0 & c_{13}^3 & 0 & c_{33}^3 & 0 \\ c_{40}^5 & 0 & c_{42}^5 & 0 & c_{44}^5 \end{pmatrix}. \quad (1.41)$$

For polynomials with $m > 5$, the matrices $\mathbf{C}^{-1}$ can be found by the method of dividing matrices into blocks [see Berezin and Zhidkov (1965)], using the inverse matrices that have already been found for the smaller values of m.

Sometimes, the least squares polynomial is used to estimate the derivatives of the observed function. In this case, the polynomial building can be carried out on the system of points

$$-rh, \ -(r-1)h, \ldots, (r-1)h, \ rh, \quad (1.42)$$

where h is the step length along the t axis. The elements of the inverse matrices of the normal equations are determined by the formula

$$c_{ijh}^m = h^{-(i+j)} c_{ij}^m, \quad (1.43)$$

where c_{ij}^m is given by (1.35)–(1.41).

Concluding this section, it is natural to ask: for what do we really need these simple, particular results?

First, when we work with a very long time series of observations, the values of the elements of the normal systems' matrices can be large enough either to cause overflow or to produce inaccuracy. This must be taken into account when developing computer programs. Second, the formulas save computer time whenever it is necessary to find a large number of polynomials that approximate many different time series. Third, the analytical expressions for the elements of the inverse matrices make it possible to investigate before the experiment the accuracy (and correlations) for both the polynomial coefficient and the point estimates. Finally, the availability of analytical expressions for the elements of the inverse matrices of the normal equations makes it possible to build classic smoothing and differentiating filters, which are widely applied.

Consider the expressions for the polynomial coefficients needed to obtain such digital filters. Because the polynomials of even and odd degrees have a set of identical coefficients, we need formulas for

1.3 Variance of the Point Estimates

estimating only parameters β_p, if m is odd. From (1.28) and (1.35)–(1.38) it follows that

$$\widehat{\beta}_p = \begin{cases} \sum_{v=0}^{(m-1)/2} c_{p,2v}^m \left[\sum_{i=0}^{k} f_{i-r}(i-r)^{2v} \right] & \text{if } p \text{ is even or } 0, \\ \sum_{v=0}^{(m-1)/2} c_{p,2v+1}^m \left[\sum_{i=0}^{k} f_{i-r}(i-r)^{2v+1} \right] & \text{if } p \text{ is odd,} \end{cases}$$

$$(m = 1, 3, 5, \ldots). \qquad (1.44)$$

Letting $i - r = j$, we get

$$\widehat{\beta}_p = \sum_{j=-r}^{r} \alpha_j^{m,r,p} f_j, \qquad (1.45)$$

where

$$\alpha_j^{m,r,p} = \begin{cases} \sum_{v=0}^{(m-1)/2} c_{p,2v}^m j^{2v} & \text{if } p \text{ is even or } 0, \\ \sum_{v=0}^{(m-1)/2} c_{p,2v+1}^m j^{2v+1} & \text{if } p \text{ is odd.} \end{cases} \qquad (1.46)$$

Replacing the c_{ij}^m in (1.46) by their expressions from (1.35)–(1.38) produces simple formulas for parameters $\alpha_j^{m,r,p}$, for $m = 0, 1, \ldots, 5$.

1.3 Variance of the Point Estimates

Expression (1.14) makes it possible to investigate the correlation of the point estimates. However, the practical interpretation of this correlation is often cumbersome. It is much simpler to determine only the variances (1.15) and to draw the corresponding graphics, which illustrate the accuracy of the point estimates.

Placing (1.24) in formula (1.15) yields the expression for computing the normalized variance as a function of the number of points.

For example, formulas

$$\sigma^2_{v-r} = c^1_{00} + c^1_{11}(v-r)^2,$$

$$\sigma^2_{v-r} = c^3_{00} + (c^1_{11} + 2c^3_{20})(v-r)^2 + c^3_{22}(v-r)^4,$$

$$\sigma^2_{v-r} = c^3_{00} + (c^3_{11} + 2c^3_{20})(v-r)^2 + (c^3_{22} + 2c^3_{13})(v-r)^4$$
$$+ c^3_{33}(v-r)^6, \qquad (v = 0, 1, \ldots, k), \qquad (1.47)$$

where the values c^m_{ij} are taken from (1.35) and (1.36), determine the diagonal elements of the covariance matrix (1.14) of the point estimates, if $m = 1, 2$, and 3 respectively.

For the degree m polynomial, the general formula is

$$\sigma^2_{v-r} = \sum_{l=0}^{m} \left[\left(\sum_{i+j=2l} c^m_{ij} \right) (v-r)^{2l} \right], \qquad (v = 0, 1, \ldots, k). \qquad (1.48)$$

Formula (1.48) was derived by replacing the elements of the matrices $\mathbf{C}^{-1}$ and $\mathbf{X}$ in (1.15) by (1.24), because the polynomials are built on the system of points (1.20). From (1.47) and (1.48), it follows that the normalized variance of any point estimate depends on this point value, on the number of observations, and on the polynomial degree. For different m and r, one can calculate the values of such variances for all the points, then draw the corresponding graphics.

Figure 1.1 illustrates the shape of the variance curve of the point estimates on the observational interval, when the degree is fixed ($m = 3$) and the number of points varies ($k = 10, 12, 14$). (For a visual demonstration, the discrete points in Fig. 1.1, as well as in other figures, are joined by the continuous curves.) Figure 1.1 shows that the variance curves are symmetrical relative to the vertical axis and that they have identical shapes. The fewer the number of observations, the higher the curve location. If $k \to m$, the curves converge into the straight line $\sigma^2_{v-r} = 1$.

Figure 1.2 illustrates the shapes of the curves of variance (1.48) for different polynomial degrees ($m = 0, 1, \ldots, 7$) and for the fixed number of points k (30). The curves have different shapes, and they are symmetric relative to the vertical axis. The curves show that the variance of the point estimates increases as the polynomial degree increases. The values of the variance of point estimates close to the ends of the interval are large (close to one); it is only by moving

1.3 Variance of the Point Estimates

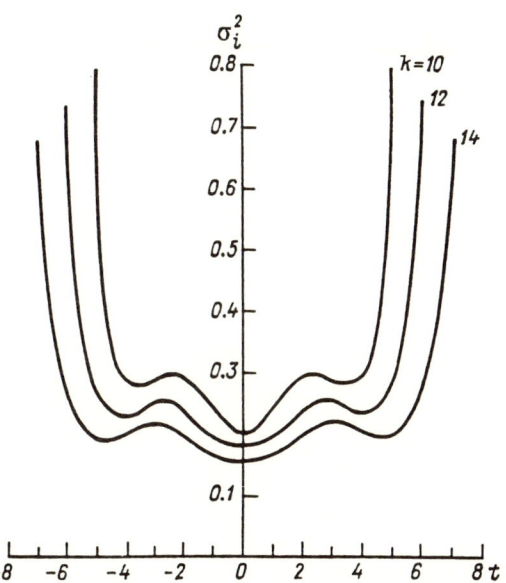

Figure 1.1: Normalized variances of the point estimates when the polynomial degree is fixed ($m = 3$) and the number of observations k varies.

toward the center of the interval that one can obtain estimates with a small variance.

For the first- and third-degree polynomials, it is possible to show [with the aid of formula (1.47) and the expressions for the corresponding elements c_{ij}^1 and c_{ij}^3] that the minimal variance corresponds to the center point of the interval. It is very likely that this assertion will prove correct for any odd m. If m is even, the minimal variance corresponds to some of the points that are close to the origin and that are located symmetrically relative to the center of the interval.

If $t = 0$, the points on the graphics (Figs. 1.1 and 1.2) present the value $c_{00}^m = \sigma_{\beta_0}^2/\sigma^2$ (the normalized variance of the estimate of β_0). If the polynomial degree m and the number of points $2r+1$ are known, one can find the value of c_{00}^m and draw a conclusion (with the aid of Fig. 1.2) about the shape and location of the variance curve of the point estimates.

The graph in Figure 1.1 resembles a menorah, and Figure 1.2 and equations (1.47)–(1.48) can be used for designing menorahs of

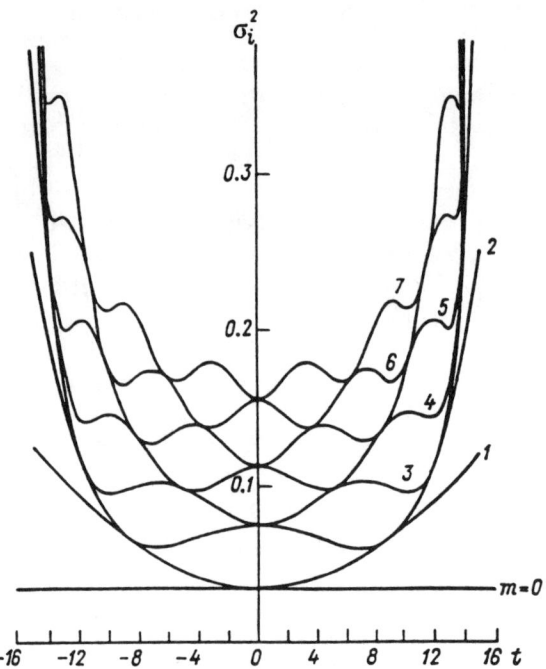

Figure 1.2: Normalized variances of the point estimates, when the number of observations is fixed ($k = 30$) and the polynomial degree m varies.

many types and shapes. Therefore, formula (1.48) can be called the menorah equation.

Figure 1.3 illustrates the dependence of c_{00}^m on the number of points k for different degrees m. As this illustration clearly shows, increasing the number of observations to more than 30 does not greatly decrease the point estimate variance (when $m \leq 5$). Having represented the observation variance value, one can use Figure 1.3 to determine the sample size, which is necessary to obtain estimate $\hat{\beta}_0$ with the required accuracy.

The results of this section enable us to evaluate (within a constant factor σ^2) the accuracy of the point estimates of the least squares polynomials before making any calculations and estimations. For example, if one wants to build a second-degree polynomial having 11 observations, one can refer to Figure 1.3 [or formula (1.36)], which gives $c_{00}^3 = \sigma_{\hat{\beta}_0}^3/\sigma^2 \approx 0.2$. Therefore, the point estimate variance is

1.3 Variance of the Point Estimates

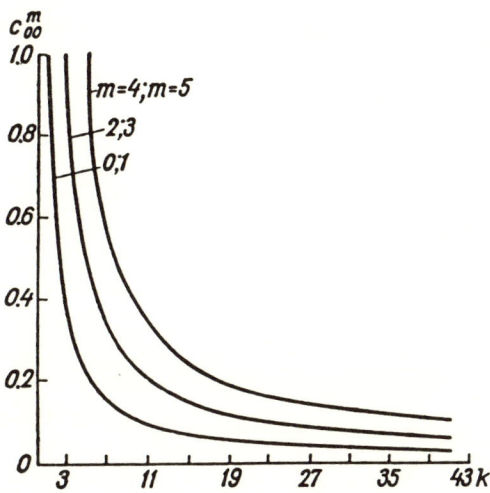

Figure 1.3: Dependence of the normalized variance of the β_0 estimate on the number of points and on the polynomial degree.

approximately one fifth of the observation variance. (Of course, the result for the end points is worse.) These calculations provide some quantitative results, which previously had been clear only intuitively:

1. The highest accuracy of the point estimates corresponds to the central point of the approximation interval (if $k+1$ is odd) or to some points close to the center (if $k+1$ is even).

2. Formula (1.48) reveals that, outside the approximation interval, the variance σ_t^2 of the extrapolated value of f_t increases as t^{2m} with increasing t. Therefore, from a purely statistical viewpoint, the least squares polynomial cannot be used for the extrapolation.

3. Within the framework of the conditions outlined in this chapter, for the variance $\sigma_i^2\,(k,m)$ (1.48) of any point i of the approximational interval, the following inequalities hold true:

$$\sigma_i^2(k, m_1) \leq \sigma_i^2(k, m_2), \text{ if } m_1 < m_2, \qquad (1.49)$$

$$\sigma_i^2(k_1, m) < \sigma_i^2(k_2, m), \text{ if } k_1 > k_2. \qquad (1.50)$$

If the inequality (1.50) is correct for at least one point, then it is correct for all of the other points.

4. In spite of the fact that the observations have equal accuracy, the point estimates have different variances that characterize the nonstationary dependence. If the point estimates are used for further processing, this fact must be taken into account, especially when there are few observations.

5. The estimates of the coefficients β_i with even and odd numbers are not correlated because the system of normal equations is separated into two independent systems.

Now let us consider the identification of the polynomial degree. The least squares estimators have been derived under the following condition:
$$E(\mathbf{Y}) = \mathbf{X}\boldsymbol{\beta}. \tag{1.51}$$
If (1.51) is not satisfied, the above least squares estimates will be biased. Condition (1.51) can be fulfilled by increasing the polynomial degree. However, as has been shown, the higher the degree, the greater the variance of the parameters and of the point estimates. Moreover, increasing the polynomial degree increases the volume of calculations. Finding the lowest polynomial degree for which condition (1.51) is satisfied is the basis for obtaining unbiased estimates. An elaboration of the theoretical approach to this question has been given by Anderson (1971) and Tutubalin (1972).

The test of the statistical significance of $\hat{\beta}_i$ can be carried out with the aid of (1.18). If one is interested in comparing $\hat{\beta}_i$ with zero, (1.18) and (1.35)–(1.38) yield

$$t(k-m) = \frac{\hat{\beta}_i}{\hat{\sigma}_{\beta_i}} = \frac{\hat{\beta}_i}{\hat{\sigma}\sqrt{c_{ii}^m}}. \tag{1.52}$$

If k is large ($k > 30$), the t-distribution is close to the normal distribution, and the inequality

$$|\hat{\beta}_i| > t\hat{\sigma}\sqrt{c_{ii}^m}, \tag{1.53}$$

(where $t \approx 2$ for the 5% or 3 for the 1% significance levels) must be tested.

1.4 The Fourier Set

In this section the trigonometrical polynomials (Fourier sets) are considered briefly. For the Fourier approximation, the principal least squares relationships, derived in Section 1.1, can be significantly simplified, and all the formulas can be presented by simple and clear expressions without applying the matrix apparatus. This simplicity is possible because of the orthogonal properties of the trigonometrical polynomials.

Consider the principal formulas for estimating the Fourier coefficients and their statistical characteristics.

Let us assume that the observations

$$Y_0, Y_1, \ldots, Y_{k-1} \tag{1.54}$$

are given on the equally spaced system of points

$$x_t = \frac{2\pi}{k} t = \omega t, \tag{1.55}$$

where

$$t = 0, 1, \ldots, k-1; \quad \omega = 2\pi/k. \tag{1.56}$$

The system of conditions is

$$E(Y_t) = A_0 + \sum_{p=1}^{v} (A_p \cos \omega_p t + B_p \sin \omega_p t), \tag{1.57}$$

where $\omega_p = \omega p$; v is an integer ($v < k/2$); and A_0, A_p, B_p ($p = 1, 2, \ldots, v$) are unknown parameters that must be estimated. The matrix $\mathbf{C}$ of the normal equations is

$$\mathbf{C} = \mathbf{X}^T \mathbf{X} = \begin{pmatrix} k & 0 & 0 & \cdots & 0 \\ 0 & k/2 & 0 & \cdots & 0 \\ 0 & 0 & k/2 & \cdots & 0 \\ \vdots & \vdots & \vdots & & \vdots \\ 0 & 0 & 0 & \cdots & k/2 \end{pmatrix}. \tag{1.58}$$

The trigonometric relationships that are necessary to prove this formula can be found in Berezin and Zhidkov (1965). From (1.9) one can obtain the formulas for the Fourier coefficients

$$\widehat{A}_0 = \frac{1}{k}\sum_{j=0}^{k-1} Y_j, \quad \widehat{A}_p = \frac{2}{k}\sum_{j=0}^{k-1} Y_j \cos\omega_p j,$$

$$\widehat{B}_p = \frac{2}{k}\sum_{j=0}^{k-1} Y_j \sin\omega_p j, \qquad (p = 1, 2, \ldots, v). \qquad (1.59)$$

The elements $\widehat{Y}_t$ ($t = 0, 1, \ldots, k-1$) of the vector of the point estimates are

$$\widehat{Y}_t = \widehat{A}_0 + \sum_{p=1}^{v}\left(\widehat{A}_p \cos\omega_p t + \widehat{B}_p \sin\omega_p t\right). \qquad (1.60)$$

The parameter covariance matrix $\sigma^2 \mathbf{C}^{-1}$ reveals that the estimates of the Fourier coefficients are statistically independent. From (1.11) and (1.58) we get

$$\sigma_{A_0}^2/\sigma^2 = 1/k,$$

$$\sigma_{A_p}^2/\sigma^2 = \sigma_{B_p}^2/\sigma^2 = 2/k \quad (p = 1, 2, \ldots, v). \qquad (1.61)$$

The normalized variances of the Fourier coefficients do not depend upon the polynomial degree. With the aid of (1.14), one can determine the elements of the covariance matrix of the point estimates (1.60) by introducing the notation

$$\mathbf{X}\mathbf{C}^{-1}\mathbf{X}^T = \{\mu_{tj}\}_{t,j=0}^{k-1}. \qquad (1.62)$$

Then

$$\frac{k}{2}\mu_{tj} = \frac{1}{2} + \sum_{i=1}^{v}(\cos\omega_i t \cos\omega_i j + \sin\omega_i t \sin\omega_i j)$$

$$= \frac{1}{2} + \sum_{i=1}^{v} \cos\omega_i(t-j) = \begin{cases} \dfrac{2v+1}{2} & \text{if } t = j, \\[2mm] \dfrac{\sin\left(\omega_{2v+1}\frac{t-j}{2}\right)}{2\sin\omega\frac{t-j}{2}} & \text{if } t \neq j. \end{cases} \qquad (1.63)$$

1.5 Examples

From (1.63) we get

$$\mu_{tj} = \mu_v(\tau) = \begin{cases} \dfrac{2v+1}{k} & \text{if } t = j, \\ \\ \dfrac{\sin\dfrac{\pi(2v+1)}{k}\tau}{k\sin\dfrac{\pi\tau}{k}} & \text{if } t \neq j, \end{cases} \quad (1.64)$$

where $\tau = t - j$.

Equation (1.64) shows that for the given k and v, the elements of the covariance matrix of the point estimates depend only upon the difference $t - j$. Therefore, expression (1.64) determines the covariance function of the stationary random process. This function is an even periodic function of $\tau = t - j$ with period k. Stationary processes will be examined in more detail in Chapters 3 and 4.

The normalized variances of the point estimates (1.60), equal to $\frac{2v+1}{k}$, are identical for the given grid (1.56) and value of v; they do not change from point to point as they do in the algebraic polynomial approximation. Formula (1.64) shows that, for a fixed number of observations k, an increase in the degree of v leads to an increase in the variance of the point estimates. If k becomes equal to the number of parameters, this variance equals the observed variance value.

1.5 Examples

Now let us consider several examples of the graphs of the normalized variances of the point estimates.

1. Figure 1.4 presents normalized variances of the point estimates for the polynomial approximation of the unequally spaced observations. The number of points is 31, and the polynomial degree m varies from zero to six. The number of observations to the right of zero is twice that of those to the left of zero. The minimum value of the normalized variance is located to the right of zero. The location and shape of the variance curves depend upon the location of the points with observations on the interval. It is apparent that the curves in Figure 1.4 are the deformed curves in Figure 1.2.

2. Figure 1.5a shows graphs of the normalized variances of the point estimates obtained for the approximation of equally spaced observations by the function

$$y(t) = \beta_0 + \beta_1 e^t + \beta_2 e^{2t} + \ldots + \beta_m e^{mt}. \quad (1.65)$$

Figure 1.5b shows the analogous variances for the function

$$y(t) = \beta_0 + \beta_1 t^{-1} + \beta_2 t^{-2} + \ldots + \beta_m t^{-m}. \qquad (1.66)$$

Assuming that $x = e^t$ in (1.65) and that $x = t^{-1}$ in (1.66), we come to the problem of the polynomial approximation of the observations given on the unequally spaced grid.

1.6 Smoothing Digital Filters

In general, when the observed curve has a complicated shape and the number of observations is large, the problem of finding the appropriate approximation function is not a simple one. Its solution is theoretically attainable but practically cumbersome. One must have some assumptions not only about statistical characteristics of observations but also about the deterministic (geometric) structure of the curve. The large diversity of observed natural processes and of techniques that have different geometrical properties can lead to a situation in which purely formal approximation cannot satisfy the physical content of the problem. Moreover, sometimes the necessity of smoothing arises in the process of computations (for example, smoothing of the periodogram for the spectrum estimation of the stationary random function) when the empirical curve shape is completely unknown. This circumstance requires setting a smoothing problem in a more practically suitable but theoretically less rigorous form.

First, let us not attempt to approximate the entire set of observations by one function, but rather consider its subsets. The sizes of such subsets can be chosen small enough that initially there would be no question about finding a specific type of approximation function. This assumption can be best explained with the aid of formulas.

Consider the time series of the independent, equally spaced, and equally accurate (with variance σ^2) observations

$$Y_0, Y_1, \ldots, Y_N, \qquad (1.67)$$

given at the points

$$t_0, t_1, \ldots, t_N; \ (t_{i+1} - t_i = \text{const.}).$$

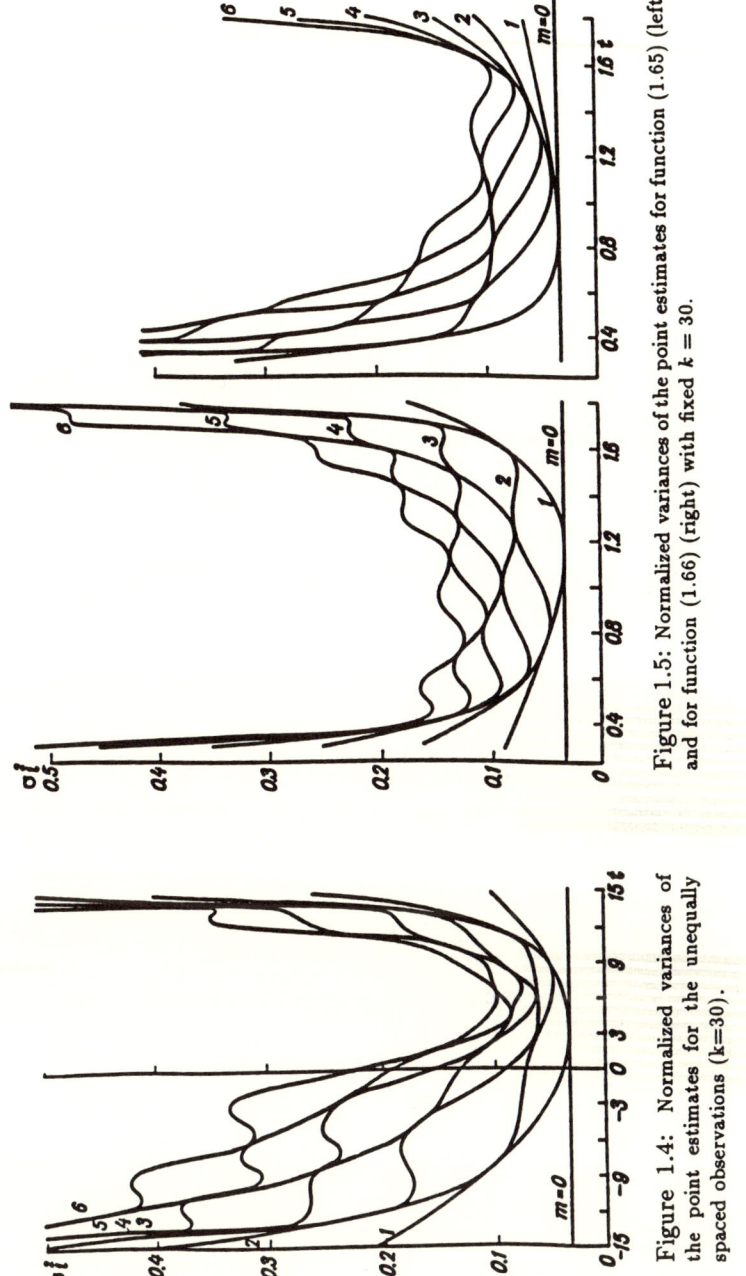

Figure 1.5: Normalized variances of the point estimates for function (1.65) (left) and for function (1.66) (right) with fixed $k = 30$.

Figure 1.4: Normalized variances of the point estimates for the unequally spaced observations ($k=30$).

The mean value of Y_t will be estimated by the formula

$$\bar{Y}_t = \sum_{j=-r}^{r} \alpha_j Y_{t+j}, \qquad (1.68)$$

where $t = r, r-1, \ldots, N-r$; $2r+1 = k$; $k << N$; $\alpha_{-j} = \alpha_j$.

The set of the coefficients (weights) $\{\alpha_j\}_{j=-r}^{r}$ will be called the *digital filter*. The graph of the dependence of α_j on j is referred to as the smoothing window (or the filter window), and the value $k = 2r+1$ is the window width (or the filter width). The smoothing procedure [the process of finding estimates by (1.68)] is referred to as the moving average procedure. The smoothing procedure (with the specific weights α_j) makes it possible to decrease or to filter out the high-frequency component of the fluctuations. This approach is the particular case of the process of filtration, which is widely used for modifying the Fourier components (of the Fourier expansion of the observations) at any frequencies. The filtration procedure is related to nonparametric methods of estimation. Its application differs from the least squares approximation, which has a precise and elegant theory. Smoothing does not (and possibly, cannot) have such a theory. It is a purely empirical approach to data processing, but the application of digital filters is one of the important elements in the methodology of computational statistics (see, for example, Anderson, 1971; Box and Jenkins, 1976; Kendall and Stuart, 1963; and Lanczos, 1956).

The filter parameters α_j can be obtained from different physical and statistical assumptions.

The scheme (1.68) cannot give the estimates for the first and last r points of the set of observations (1.67). But this disadvantage can be easily overcome, by, for example, the periodic extension of the data (1.67) in both directions. It is also possible to work out procedures, similar to (1.68) that compute estimates for the first and last r points of the interval.

Let us study the statistical and harmonic structure of the set of point estimates

$$\bar{Y}_r, \bar{Y}_{r+1}, \ldots, \bar{Y}_{N-r}, \qquad (1.69)$$

obtained by (1.68).

The moving average procedure (1.68) is a linear transformation of the random variables Y_t:

$$\bar{\mathbf{Y}} = \mathbf{F}\mathbf{Y}, \qquad (1.70)$$

where

1.6 Smoothing Digital Filters

$$\mathbf{Y} = \{Y_i\}_{i=0}^{N}, \quad \bar{\mathbf{Y}} = \{\bar{Y}_i\}_{i=r}^{N-r},$$

and

$$\mathbf{F} = \{F_{ij}\}_{i=r,j=0}^{N-r,N} = \begin{pmatrix} \alpha_{-r} & \alpha_{-(r-1)} & \cdots & \alpha_r & 0 & \cdots & 0 \\ 0 & \alpha_{-r} & \cdots & \alpha_{r-1} & \alpha_r & \cdots & 0 \\ \vdots & \vdots & & \vdots & \vdots & & \vdots \\ 0 & 0 & \cdots & 0 & 0 & \cdots & \alpha_r \end{pmatrix}. \quad (1.71)$$

The covariance matrix of the random vector $\bar{\mathbf{Y}}$ is $\mathbf{M}_{\bar{Y}} = \sigma^2 \mathbf{F}\mathbf{F}^T$. This matrix has identical elements located on the diagonals, which are parallel to the main diagonal.

By performing the necessary rearrangements in $\mathbf{M}_{\bar{Y}} = \sigma^2 \mathbf{F}\mathbf{F}^T$, the covariances M_τ of the point estimates (1.69) can be presented as

$$M_\tau = \begin{cases} \sigma^2 \sum_{j=-(r-|\tau|)}^{r} \alpha_j \alpha_{j-|\tau|} & \text{if } |\tau| < 2r+1 \\ 0 & \text{if } |\tau| \geq 2r+1, \end{cases} \quad (1.72)$$

where $\tau = q - s$ is the difference between the subscripts of the estimates $\bar{Y}_s$ and $\bar{Y}_q$, for which the covariance is computed.

Therefore, the moving average procedure changes the statistical structure of the observations. The values of point estimates $\bar{Y}_s$ and $\bar{Y}_q$ are statistically dependent (if $|q - s| = |\tau| < 2r+1$), with the covariance determined by (1.72).

The variance $\bar{\sigma}^2$ of the point estimates (1.69) is

$$\bar{\sigma}^2 = M_0 = \sigma^2 \sum_{j=-r}^{r} \alpha_j^2, \quad (1.73)$$

which is less than the variance of the original observations ($\bar{\sigma}^2 < \sigma^2$).

By combining (1.73) with the covariance function (1.72), it is possible to determine the correlation function of the point estimates:

$$\widetilde{M}_\tau = \frac{M_\tau}{\bar{\sigma}^2} = \begin{cases} \dfrac{\sum_{j=-(r-|\tau|)}^{r} \alpha_j \alpha_{j-|\tau|}}{\sum_{j=-r}^{r} \alpha^2} & \text{if } |\tau| < 2r+1 \\ 0 & \text{if } |\tau| \geq 2r+1. \end{cases} \quad (1.74)$$

The moving average procedure changes not only the statistical but also the harmonic structure of the observations.

Let $\varphi_t = e^{i\omega t}$ be any of the harmonics of the Fourier expansion of the observations (1.67). In accordance with (1.68), the point estimate $\bar{\varphi}_t$ is

$$\bar{\varphi}_t = \sum_{j=-r}^{r} \alpha_j \varphi_{t+j} = \sum_{j=-r}^{r} \alpha_j e^{i\omega(t+j)} = e^{i\omega t} \sum_{j=-r}^{r} \alpha_j e^{i\omega j}; \qquad (1.75)$$

that is, each harmonic with the frequency ω is multiplied by the factor

$$I(\omega) = \sum_{j=-r}^{r} \alpha_j e^{i\omega j} = \alpha_0 + 2\sum_{j=1}^{r} \alpha_j \cos\omega j, \qquad (1.76)$$

which does not depend on time t. The resulting factor of the fluctuation amplitude with frequency ω, which is equal to

$$I^2(\omega) = \left(\sum_{j=-r}^{r} \alpha_j e^{i\omega j}\right)^2 = \left(\alpha_0 + 2\sum_{j=1}^{r} \alpha_j \cos\omega j\right)^2, \qquad (1.77)$$

will be called the frequency characteristic of the filtering scheme (1.68). By analyzing functions (1.76) and (1.77) for the given values of α_j, it becomes possible to study the way in which the frequency composition of (1.69) changes as compared with that of (1.67).

The basis for constructing different classes of filters is presented by the character of the random and deterministic variations of the observations (or by different statistical and physical assumptions). The principal requirement is, of course, to provide the most accurate estimation (unbiased and with minimal variance). One of the classes of filters that approximately satisfied these conditions can be derived from the least squares polynomials, which were considered in Section 1.2. These filters have been used in applications so widely and so long that they can be called classical digital filters. They are analyzed in the next section.

1.7 Regressive Filters

The least squares point estimates, which were examined in Section 1.3, have a minimal variance; it is convenient to use them as a basis for constructing different filtering schemes. Such least squares estimates are linear functions of the observations, and the coefficients

1.7 Regressive Filters

(weights) of these functions can be calculated. The analysis of point estimates for polynomial approximations, which was made in Section 1.3, shows that the most accurate estimates are in the center of the observational interval. Therefore, it makes sense to construct the smoothing procedure for the central point of the interval.

Putting $t = 0$ and $p = 0$ in (1.22) and (1.45) yields

$$f(0) = \hat{\beta}_0 = \sum_{j=-r}^{r} \alpha_j^{m,r} f_j, \tag{1.78}$$

where

$$\alpha_j^{m,r} = \alpha_j^{m,r,0} = \sum_{v=0}^{(m-1)/2} c_{p,2v}^m j^{2v}, \tag{1.79}$$

and $c_{p,2v}^m$ are the elements (1.35)–(1.37) of the inverse matrix of the system (1.27), which was obtained for the least squares polynomial of degree m.

Filters (1.79) will be called *regressive filters*. For polynomials of $2p$ and $2p+1$ ($p = 0, 1, \ldots$) degrees, these filters have identical weights because such polynomials have identical corresponding coefficients $\hat{\beta}_0$ (see Section 1.2). For this reason, parameters $\alpha_j^{m,r}$ can be determined only for odd m:

$$\alpha_j^{m,r} = c_{00}^m + c_{02}^m j^2 + c_{04}^m j^4 + \ldots + c_{0m-3}^m j^{m-3} + c_{0m-1}^m j^{m-1}, \tag{1.80}$$

$$(m = 1, 3, 5, \ldots).$$

In Section 1.2, the formulas (1.35)–(1.37) for elements c_{ij}^m (for $m \leq 5$) were obtained. Using these formulas, the expressions for $\alpha_j^{m,r}$ are noted in the following way:

For $m = 0$ and 1

$$f(0) = \frac{1}{2r+1} \sum_{j=-r}^{r} f_j \tag{1.81}$$

and

$$\alpha_j^{1,r} = \frac{1}{2r+1}. \tag{1.82}$$

For $m = 2$ and 3

$$f(0) = \frac{3}{(4r^2-1)(2r+3)} \left[(3r^2 + 3r - 1) \sum_{j=-r}^{r} f_j - 5 \sum_{j=-r}^{r} j^2 f_j \right]$$

$$= \sum_{j=-r}^{r} \frac{3(3r^2 + 3r - 1 - 5j^2)}{(4r^2 - 1)(2r + 3)} f_j \qquad (1.83)$$

and α in (1.68) is

$$\alpha_j^{3,r} = \frac{3(3r^2 + 3r - 1 - 5j^2)}{(4r^2 - 1)(2r + 3)}. \qquad (1.84)$$

For $m = 4$ and 5

$$f(0) = \frac{[15\,(15r^4 + 30r^3 - 35r^2 - 50r + 12) \sum_{j=-r}^{r} f_j}{-75 \cdot 7\,(2r^2 + 2r - 3) \sum_{j=-r}^{r} j^2 f_j + 27 \cdot 35 \sum_{j=-r}^{r} j^4 f_j]}{4\,(4r^2 - 9)(4r^2 - 1)(2r + 5)}$$

$$= \sum_{j=-r}^{r} f_j \frac{[15\,(15r^4 + 30r^3 - 35r^2 - 50r + 12)}{-75 \cdot 7\,(2r^2 + 2r - 3)j^2 + 27 \cdot 35 j^4]}{4(4r^2 - 9)(4r^2 - 1)(2r + 5)} \qquad (1.85)$$

and α in (1.68) is

$$\alpha_j^{5,r} = \frac{[15\,(15r^4 + 30r^3 - 35r^2 - 50r + 12)}{-75 \cdot 7\,(2r^2 + 2r - 3)j^2 + 27 \cdot 35 j^4]}{4(4r^2 - 9)(4r^2 - 1)(2r + 5)} \qquad (1.86)$$

Expressions (1.84) and (1.86) can be used as recipes for software development. For instance, different smoothing numerical filters, which are described by numerous authors (Anderson, 1971; Berezin and Zhidkov, 1965; Box and Jenkins, 1976; Jenkins Watts, 1968; and Lanczos, 1956), are the particular examples of formula (1.84). Those filters can easily be obtained from expression (1.84) by substituting the appropriate numerical value for r.

Figure 1.6 shows the shape of the smoothing windows of the regressive filters as functions of m and r. Taking into consideration (1.73) and (1.78), the variance of point estimates (1.69) can be presented as

$$\bar{\sigma}^2 = \sigma_{\beta_0}^2 = \sigma^2 c_{00}^m. \qquad (1.87)$$

The graphs of the dependence of c_{00}^m on value $k = 2r$ for different m

1.7 Regressive Filters

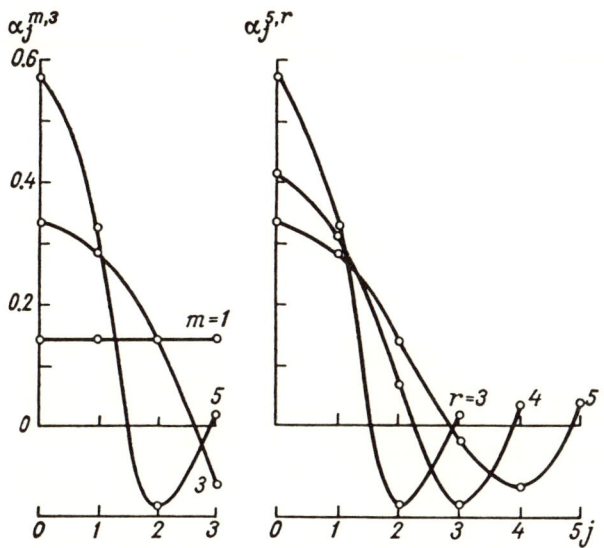

Figure 1.6: Dependence of the smoothing window shapes on the polynomial degree (left) and on the filter width (right).

are given in Figure 1.3.

The covariance function of the point estimates, obtained by applying the regressive filter, can be determined by substituting parameters $\alpha_j^{m,r}$ from (1.80) for α_j in (1.72).

Such substituting yields:

$$M_\tau^{m,r} = \sigma^2$$

$$\times \sum_{j=-(r-|\tau|)}^{r} (c_{00}^m + c_{02}^m j^2 + \ldots + c_{0m-3}^m j^{m-3} + c_{0m-1}^m j^{m-1})[c_{00}^m$$

$$+ c_{02}^m (j-|\tau|)^2 + \ldots + c_{0m-3}(j-|\tau|)^{m-3} + c_{0m-1}^m (j-|\tau|)^{m-1}]$$
(1.88)

if $|\tau| < 2r + 1$. And

$$M_\tau^{m,r} = 0 \quad \text{if } |\tau| \geq 2r + 1.$$

For example, for $m = 1$

$$M_\tau^{1,r} = \begin{cases} \dfrac{\sigma^2}{2r+1}\left(1 - \dfrac{|\tau|}{2r+1}\right) & \text{if } |\tau| < 2r+1, \\ 0 & \text{if } |\tau| \geq 2r+1. \end{cases} \quad (1.89)$$

Figures 1.7 and 1.8 show the graphs of the covariance functions (1.88) (to within factor σ^2) of the point estimates (1.69) (the results of smoothing by regressive filters) for different r and $m = 1, 3$, and 5.

The correlation function corresponding to (1.88) is

$$\widetilde{M}_\tau^{m,r} = \dfrac{M_\tau^{m,r}}{\bar{\sigma}^2} = \begin{cases} \dfrac{1}{c_{00}^m} \sum\limits_{j=-(r-|\tau|)}^{r} \alpha_j^{m,r}\alpha_{j-|\tau|}^{m,r} & \text{if } |\tau| < 2r+1, \\ 0 & \text{if } |\tau| \geq 2r+1. \end{cases} \quad (1.90)$$

For example, setting $m = 1$ in (1.90) yields

$$\widetilde{M}_\tau^{1,r} = \begin{cases} 1 - \dfrac{|\tau|}{2r+1} & \text{if } |\tau| < 2r+1, \\ 0 & \text{if } |\tau| \geq 2r+1. \end{cases} \quad (1.91)$$

The dependence of the shape of the correlation function on the parameters m and r is illustrated in Figures 1.9 and 1.10.

The natural property of the regressive filters is that for any m and r, the identity

$$\sum_{j=-r}^{r} \alpha_j^{m,r} \equiv 1 \quad (1.92)$$

is true. Really, substituting 1 for each f in (1.78) gives

$$1 = \hat{\beta}_0 = f(0) = \sum_{j=-r}^{r} \alpha_j^{m,r}. \quad (1.93)$$

Let us study the change of the observation Fourier transform amplitudes imposed by the smoothing procedure. For this purpose we

1.7 Regressive Filters

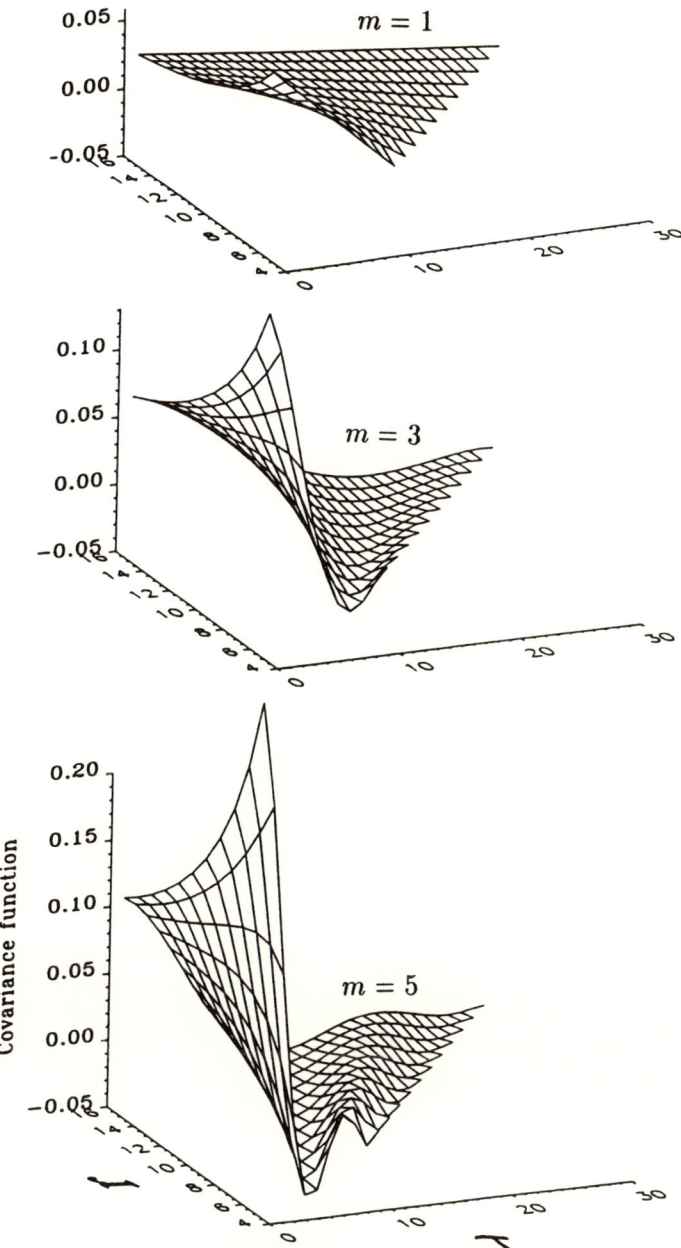

Figure 1.7: Dependence of the covariance function (1.88) of the estimates on filter width (r) and polynomial degree (m).

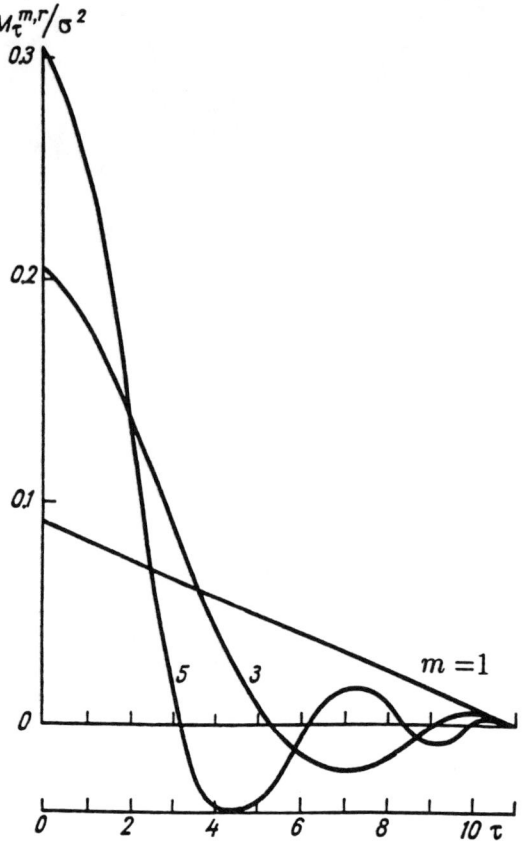

Figure 1.8: Dependence of the covariance function of the estimates on degree m when the filter width is fixed ($2r + 1 = 11$).

will deduce the analytical expressions for the frequency characteristics (1.77) of the regressive filters with $m = 1, 3$, and 5; then we will obtain a general scheme for any m.

Substituting the coefficients determined by (1.82), (1.84), and (1.86) for α_j in (1.76) yields for $m = 0$ and 1:

$$I_{1,r} = \frac{1}{2r+1} \sum_{j=-r}^{r} e^{i\omega j} = \frac{\sin \frac{2r+1}{2}\omega}{(2r+1)\sin \frac{\omega}{2}} = \frac{L(\omega)}{2r+1}, \qquad (1.94)$$

where

1.7 Regressive Filters

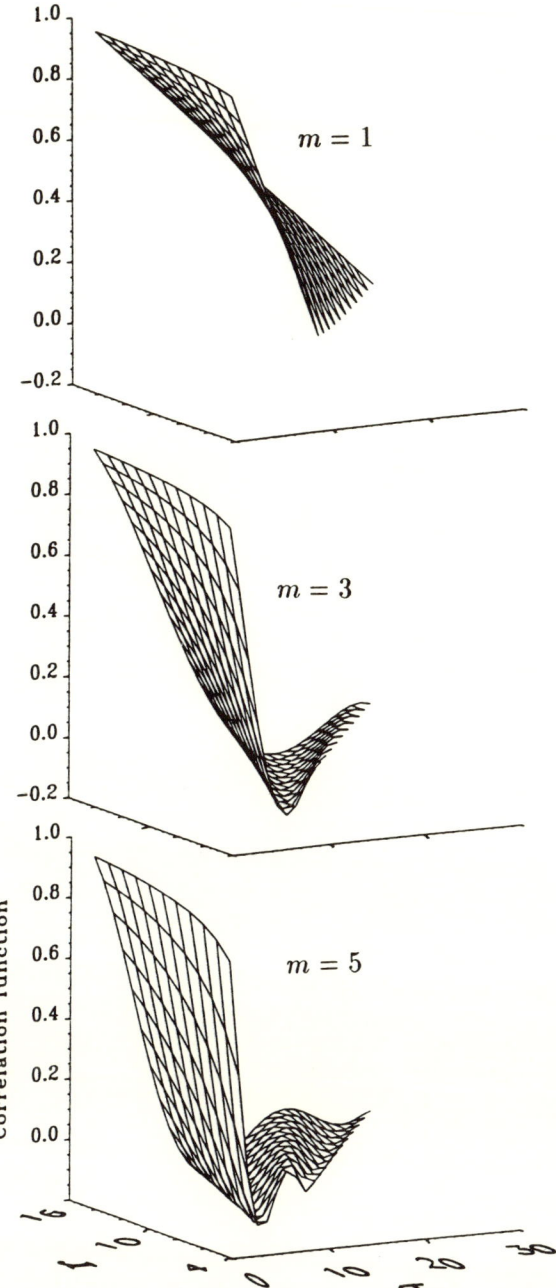

Figure 1.9: Dependence of the correlation function (1.90) of the estimates on filter width (r) and polynomial degree (m).

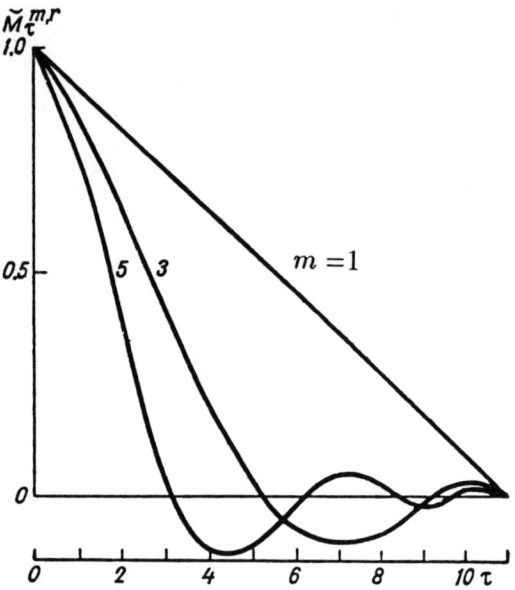

Figure 1.10: Dependence of the correlation function of the estimates on degree m, when the width is fixed $(2r+1=11)$.

$$L(\omega) = \frac{\sin\dfrac{2r+1}{2}\omega}{\sin\dfrac{\omega}{2}}, \qquad (1.95)$$

and the frequency characteristic is

$$I^2_{1,r}(\omega) = \left[\frac{\sin\dfrac{2r+1}{2}\omega}{(2r+1)\sin\dfrac{\omega}{2}}\right]^2 = \frac{L^2(\omega)}{(2r+1)^2}. \qquad (1.96)$$

This is, of course, a well-known formula (see, for example, Hamming, 1973).

To infer the other formulas, one must notice that the derivatives of the even orders of $L(\omega)$ are

1.7 Regressive Filters

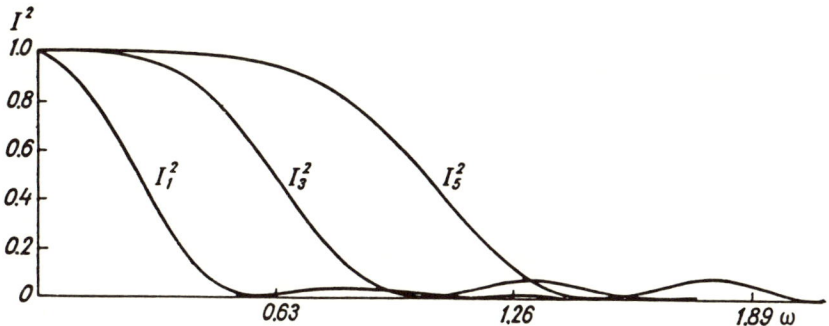

Figure 1.11: Frequency characteristics of the polynomial filters for different degrees and fixed filter width ($r = 5$).

$$L^{(2v)}(\omega) = (-1)^v \sum_{j=-r}^{r} j^{2v} e^{i\omega j}. \qquad (1.97)$$

So

$$I_{3,r}^2(\omega) = \left\{ \frac{3[(3r^2 + 3r - 1)L(\omega) + 5L''(\omega)]}{(4r^2 - 1)(2r + 3)} \right\}^2 \qquad (1.98)$$

and

$$I_{5,r}^2(\omega) = \left\{ \frac{\begin{array}{l}[15(15r^4 + 30r^3 - 35r^2 - 50r + 12)L(\omega) \\ + 75 \cdot 7(2r^2 + 2r - 3)L''(\omega) + 27 \cdot 35 L^{(4)}(\omega)]\end{array}}{4(4r^2 - 9)(4r^2 - 1)(2r + 5)} \right\}^2 \qquad (1.99)$$

The graphics of the frequency characteristics (1.96), (1.98), and (1.99) for different r and m are presented in Figures 1.11 and 1.12. These illustrations show that smoothing by the regressive filter suppresses the power of the high-frequency harmonics. Increasing m or decreasing r leads to an increase in the width of the corresponding spectral window. These illustrations make it possible to choose different filters depending on the required width of the low-frequency part of the spectrum.

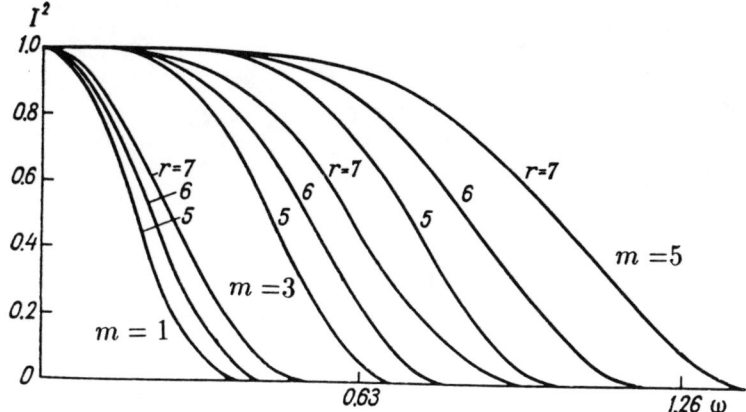

Figure 1.12: Frequency characteristics of the polynomial filters for different degrees (m) and filter widths ($2r+1$).

The amplitudes of the Fourier transform of estimates (1.69) can be obtained from the corresponding amplitudes of the Fourier transform of observations (1.67) with the aid of multiplication by the corresponding frequency characteristic.

To suppress the large power of the high frequencies (or retained power leaked out through the side lobes of the frequency characteristics), the set of observations can be smoothed several times. In this case, the amplitudes are multiplied (the same number of times) by the corresponding frequency characteristics. In such multiple smoothing, the filtering schemes must be chosen so that the points of maxima and minima of the side lobes of the frequency characteristics of the two consecutive smoothing procedures will coincide.

However, it is important to notice that multiple smoothing can lead to the misrepresentation of the real frequency composition of the set of observations. This can occur because

$$\sum_{j=-r}^{r} \alpha_j^{m,r} \equiv 1,$$

and

$$I_{m,n}(0) \equiv 1. \qquad (1.100)$$

So in the vicinity of the ω axis origin, amplitudes do not change markedly no matter what values of m and r are chosen. Formula (1.100) shows that the point estimates, obtained by multiple smoothing, converge to a constant because $\bar{\varphi}_t \to 0$, when $\omega \neq 0$.

The general expression (for arbitrary m) for the frequency characteristics of the regressive filters is obtained by substituting $\alpha_j^{m,r}$ (1.80) for α_j in (1.77) and taking into account (1.97). This yields

$$I_{m,r}^2 = [c_{00}^m L(\omega) - c_{02}^m L''(\omega) + c_{04}^m L^{(4)}(\omega) - \ldots$$

$$+ (-1)^{\frac{m-1}{2}} c_{0m-1}^m L^{(m-1)}(\omega)]^2 \qquad (m = 1, 3, 5, \ldots), \qquad (1.101)$$

where c_{ij}^m are the elements of the inverse matrix of the normal equations.

1.8 Harmonic Filters

In the previous sections the regressive filters, obtained on the basis of the least squares polynomial approximation, were analyzed in detail. Now the analogous class of the Fourier set filters will be considered.

First, let us rewrite (1.60) in the following way. Substituting expressions (1.59) for the parameters in (1.60) and rearranging yields

$$\begin{aligned}\hat{Y}_t &= \sum_{j=0}^{k-1} Y_j \frac{1}{k}[1 + 2\sum_{p=1}^{v}(\cos\omega_p t \cos\omega_p j + \sin\omega_p t \sin\omega_p j)] \\ &= \sum_{j=0}^{k-1} Y_j \frac{1}{k}[1 + 2\sum_{p=1}^{v}\cos\omega_p(t-j)] = \sum_{j=0}^{k-1} Y_j \alpha^{v,k}(t-j),\end{aligned}$$
(1.102)

where

$$\alpha^{v,k}(\tau) = \frac{1}{k}\left[1 + 2\sum_{p=1}^{v}\cos\omega_p(t-j)\right]$$

$$= \begin{cases} \dfrac{2v+1}{k} & \text{if } \tau = t-j = 0, \\[2mm] \dfrac{\sin\frac{\pi(2v+1)}{k}\tau}{k\sin\frac{\pi\tau}{k}} & \text{if } \tau = t-j \neq 0. \end{cases} \qquad (1.103)$$

This means that the point estimates (1.60) represent the smoothing values, obtained by the filter with the coefficients coinciding with the covariance function (1.64) of the point estimates (to within factor σ^2).

Expression (1.103) determines the class of the filters, which depends on the parameter (degree) v and number of observations k. These filters will be referred to as the *harmonic filters*. The change of the harmonic structure of the time series as the result of the harmonic filter application is the same as in the case of the truncated Fourier approximation: The terms with frequencies ω_p for numbers p greater than v are omitted (filtered out). The covariance of point estimates $\widehat{Y}_t$ and $\widehat{Y}_j$ is

$$M_\tau = \begin{cases} \sigma^2 \dfrac{2v+1}{k} & \text{if } t = j, \\[2ex] \sigma^2 \dfrac{\sin \frac{\pi(2v+1)}{k}\tau}{k \sin \frac{\pi\tau}{k}} & \text{if } t \neq j, \end{cases} \qquad (1.104)$$

where $\tau = t - j$.

1.9 Applications of Digital Filters

Designing, constructing, and developing filters, and then applying them to solve a variety of practical problems, is just as significant as identifying the function, which is used for approximating the observations. But applications of digital filters have some features and difficulties caused by the fact that identification of the order and the width depends on both statistical and deterministic (geometric) structures of observations. The process of smoothing by a digital filter is an empirical procedure with solid theoretical ground. General rigorous statistical inferences and hypothesis testings for evaluation of the correspondence between filter parameters and data is difficult to develop in this case. Identification of the regressive filter parameters is reduced to a compromise between the smallest possible m (order) and the greatest possible r (width) that provide approximate unbiasedness and sufficiently small variance to the estimates. Let us consider some questions of such identification.

1. For the most practical applications, the order of the filter $m \leq 5$ is sufficient, and the problem is to choose m (1, 3, or 5) and find an appropriately large width r, thus ensuring a small variance.

1.9 Applications of Digital Filters

The dependence of the estimate accuracy on m and r is determined by (1.87); its graphical representation is given in Figure 1.3. As a rule, for the final identification of the filter parameters, multiple consecutive smoothings are needed by varying m and r. For each set of m and r, a value

$$\sigma_s^2 = \frac{\sum_{i=r}^{N-r}(Y_i - \bar{Y}_i)^2}{N - 2r}$$

is computed, which is considered an approximate estimate of the variance of the observations. If this estimate does not change (becomes stable) much for several values of r, the largest r is accepted as an appropriate filter parameter. An approximate variance of the estimate $\bar{Y}_i$ can be computed with the aid of the following formula:

$$\bar{\sigma}^2 = \sigma_s^2 \sum_{j=-r}^{r} \alpha_j^2,$$

where α_j are the parameters of the fitted filter. For the regressive filters,

$$\sum_{j=-r}^{r} \alpha_j^2 = c_{00}^m.$$

2. In the statistical literature, several digital filters are recommended for a wide range of applications. But even with only one filter, it is possible to construct an infinite number of different filters by the superposition of its application. For example, if the time series is smoothed two times, and if each time the smoothing value is the average of the two adjacent terms, it yields

$$\bar{Y}_t = \frac{\frac{Y_{t-1}+Y_t}{2} + \frac{Y_t+Y_{t+1}}{2}}{2} = 0.25Y_{t-1} + 0.5Y_t + 0.25Y_{t+1}. \quad (1.105)$$

This filter is the Tukey filter, whose correlation and spectral characteristics are well-known (Jenkins and Watts, 1968).

3. Superposition of smoothing procedures can also be used by applying different filters. As an example of constructing another filter, let us apply any filter $\{\alpha_j\}_{j=-r}^{r}$ to point estimates (1.102):

$$\tilde{Y}_t = \sum_{j=-r}^{r} \alpha_j \hat{Y}_{t+j}. \quad (1.106)$$

The variance $\tilde\sigma^2$ of random variable $\tilde Y_t$ is

$$\tilde\sigma^2 = \sum_{i=-r}^{r}\sum_{j=-r}^{r}\alpha_i\alpha_j M_\tau, \qquad (1.107)$$

where $\tau = i - j$ and M_τ is determined by (1.104). By disregarding the correlations of values (1.102), one can obtain an approximate expression for the variance of estimates $\tilde Y_t$:

$$\tilde\sigma^2 \approx \sigma^2 \frac{2v+1}{k}\sum_{j=-r}^{r}\alpha_j^2. \qquad (1.108)$$

Assuming $\alpha_j = \alpha_j^{r,m}$, (1.108) yields

$$\tilde\sigma^2 \approx \sigma^2 \frac{2v+1}{k} c_{00}^m. \qquad (1.109)$$

If the Tukey filter (1.105) is used in (1.106), then (1.108) gives

$$\sigma_v^2 \approx \sigma^2 \frac{2v+1}{k}\left(\frac{1}{16}+\frac{1}{4}+\frac{1}{16}\right) = \frac{3(2v+1)}{8k}\sigma^2. \qquad (1.110)$$

4. In order to compare different smoothing schemes, let us introduce a notion for the equivalence of different filters. Two different filters can be considered equivalent if their point estimates have equal variances. As an example, we will find conditions for the equivalence between the regressive filters and scheme (1.106), when α_j is the Tukey window. Equating expressions (1.87) and (1.110) yields (for sufficiently large k)

$$v/k \approx 4/3c_{00}^m. \qquad (1.111)$$

For the commonly used regressive schemes, the relationships are

$$\frac{v}{k} = \begin{cases} \dfrac{4}{3(2r+1)} & \text{if } m = 1, \\[2mm] \dfrac{4(3r^2+3r-1)}{(4r^2-1)(2r+3)} & \text{if } m = 3, \\[2mm] \dfrac{5(15r^4+30r^3-35r^2-50r+12)}{(4r^2-9)(4r^2-1)(2r+5)} & \text{if } m = 5. \end{cases} \qquad (1.112)$$

1.9 Applications of Digital Filters

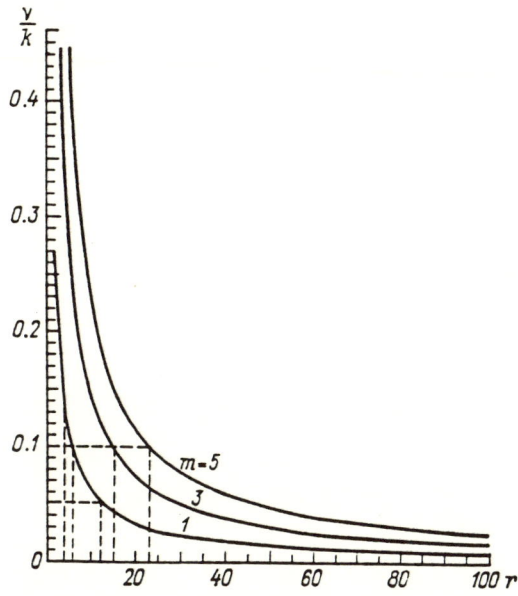

Figure 1.13: The equivalence of the filters.

The graphic representation of (1.112) is given in Figure 1.13, which can be used for choosing the width of the regressive filter, corresponding to a certain ratio v/k of the Fourier set in the scheme (1.106). For example, if v/k in (1.106) is equal to 0.1, then in order to obtain the estimator with the same variance by applying the regressive filter, it is necessary that $2r + 1 \approx 13$, if $m = 1$; or that $2r + 1 \approx 31$, if $m = 3$; and so on.

Conversely, smoothing by a regressive filter with width $2r+1 = 21$ is equivalent to the ratio value in (1.106) of $v/k \approx 0.05$, if $m = 1$, or to $v/k \approx 0.12$, if $m = 3$. Obviously, such an approach does not take into account any errors caused by smoothing that would lead to a bias.

5. It is impossible to list all of the smoothing filters that are used in applications. Identification of the filter parameters or construction of a new one is dictated by the statistical character of the observations as well as by the physical features of the problem under consideration.

It is important to notice that the unsubstantiated or careless use of the smoothing technique can distort the real information of the observations and result in incorrect conclusions. This is especially clear in the illustrations of the frequency characteristics (Figs. 1.11 and 1.12), which show the transformation of the frequency composition of data by the process of smoothing (which is not always desirable).

These figures show that smoothing eliminates or reduces the high-frequency part of the Fourier expansion of the observations. Therefore, its application is justified only when the high frequencies are not of any interest in the studies being performed.

1.10 Differentiating Filters

Now let us consider the procedures for the moving average differentiation of an equally spaced set of observations. The inference is based on the least squares polynomial approximation.

1.10.1 The First-Order Derivative

We proceed from (1.22). The estimation of the mean of the first-order derivative in the central point of an approximational interval yields

$$f'(0) = \widehat{\beta}_1. \tag{1.113}$$

The value of $\widehat{\beta}_1$ can be accepted as the derivative estimate by the moving average procedure for the different points of the observational interval. As in the case of smoothing, it is impossible to apply this procedure to the derivative estimation for the first and last r points of the interval. To overcome this difficulty, one can estimate all of the coefficients $\widehat{\beta}_i$ in order to differentiate the obtained polynomial and calculate the derivatives for these points. Another approach to finding a derivative for these points is the periodic expansion of the observations.

The general formula (based on the least squares polynomials) for estimating the first-order derivative is

$$f'(0) = \widehat{\beta}_1 = \sum_{j=-r}^{r} \gamma_j^{m,r} f_j, \tag{1.114}$$

where $\{\gamma_j^{m,r}\}_{j=-r}^{r}$ is the set of the coefficients, which depend on the polynomial degree m and number of points r,

$$\gamma_j^{m,r} = c_{11}^m j + c_{13}^m j^3 + c_{15}^m j^5 + \ldots + c_{1m-2}^m j^{m-2} + c_{1m}^m j^m,$$

1.10 Differentiating Filters

$$(m = 1, 3, 5, \ldots). \tag{1.115}$$

Formula (1.115) is obtained from (1.45) under the assumption that

$$\alpha_j^{m,r,1} = \gamma_j^{m,r}.$$

It is obvious that

$$\gamma_j^{m,r} = -\gamma_{-j}^{m,r} \tag{1.116}$$

and that the number of terms of the sum (1.114) is always even (for the odd $k + 1 = 2r + 1$, the coefficient for f_0 is 0).

Because the estimates of the coefficient $\hat{\beta}_1$ for degrees $2p - 1$ and $2p$ ($p = 1, 2, \ldots$) polynomials are identical, their differentiating schemes are identical, too.

Practical applied formulas for the parameters $\gamma_j^{m,r}$ (m is from 1 to 5) are easily derived from expressions (1.35), (1.36), and (1.37).

For $m = 1$ and 2

$$\gamma_j^{1,r} = \frac{3j}{r(r+1)(2r+1)}. \tag{1.117}$$

For $m = 3$ and 4

$$\gamma_j^{3,r} = \frac{5[5(3r^4 + 6r^3 - 3r + 1)j - 7(3r^2 + 3r - 1)j^3]}{r(r^2 - 1)(r+2)(4r^2 - 1)(2r+3)}. \tag{1.118}$$

For $m = 5$ and 6

$$\gamma_j^{5,r} = c_{11}^5 j + c_{13}^5 j^3 + c_{15}^5 j^5, \tag{1.119}$$

where values c_{ij}^5 are determined by (1.37). Formulas (1.117)–(1.119) can be used as recipes for software development.

Finding the variance σ_p^2 of the derivative point estimate obtained by these moving average differentiating filters is not difficult because

$$\sigma_p^2 = \sigma_{\hat{\beta}_1}^2 = \sigma^2 c_{11}^m = \sigma^2 \sum_{j=-r}^{r} (\gamma^{m,r})^2 \tag{1.120}$$

and expressions for the c_{11}^m (m is from 1 to 5) have already been found in Section 1.2.

Illustration of the dependencies of values $\sqrt{c_{11}^m}$ (normalized standard deviations of the point estimates of the derivatives) on m and r are given in Figure 1.14.

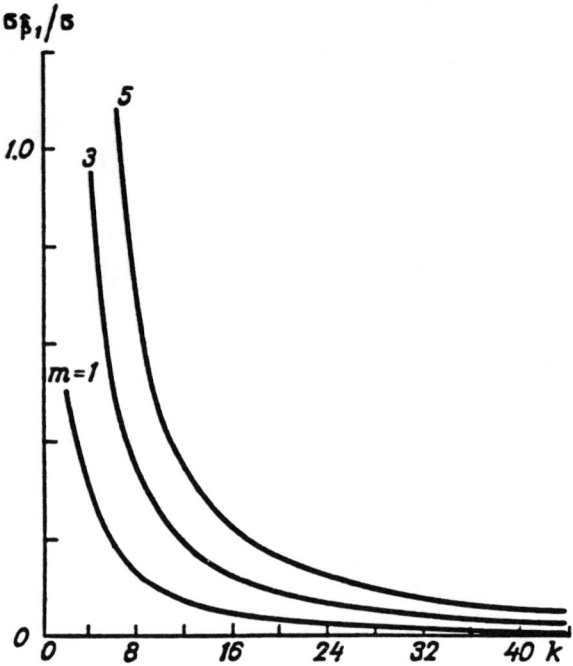

Figure 1.14: Dependence of the normalized standard deviation of the derivative estimate on the filter width.

It is possible to investigate (1.115) in detail in order to determine, for example, the covariance functions of the point estimates; but we will forego this investigation so as not to complicate our discussion.

Let us carry out the harmonic analysis of transformation (1.114), presenting it in the form

$$\bar{Y}'_t = \sum_{j=-r}^{r} \gamma_j^{m,r} Y_{t+j}, \qquad (1.121)$$

used in the applications. Let $\varphi(t) = e^{i\omega t}$ be any of the harmonics of the Fourier expansion of observations (1.67). According to (1.121),

$$\bar{\varphi}'(t) = e^{i\omega t} \sum_{j=-r}^{r} \gamma_j^{m,r} e^{i\omega j}. \qquad (1.122)$$

1.10 Differentiating Filters

The precise equality is obtained by immediately differentiating the $\varphi(t)$

$$\varphi'(t) = i\omega e^{i\omega t}. \tag{1.123}$$

The ratio

$$J_{m,r}(\omega) = \frac{\bar{\varphi}'(t)}{\varphi'(t)} = \frac{1}{i\omega} \sum_{j=-r}^{r} \gamma_j^{m,r} e^{i\omega j} \tag{1.124}$$

does not depend on t; it shows the effect of the moving average differentiating procedure (1.121) on each harmonic of the derivative.

Replacing coefficients $\gamma_j^{m,r}$ in (1.124) with formulas (1.117) to (1.119), based on the least squares polynomials, yields analytical expressions that show the transformation of each harmonic with a frequency ω by the differentiating regressive filter.

If $m = 1$ and 2

$$J_{1,r}(\omega) = \frac{3}{i\omega r(r+1)(2r+1)} \sum_{j=-r}^{r} j e^{i\omega j} = \frac{3}{\omega r(r+1)(2r+1)}$$

$$\times \left[-r \frac{\cos\left(\frac{2r+1}{2}\omega\right)}{\sin\frac{\omega}{2}} + \frac{\sin \omega r}{2\sin^2 \frac{\omega}{2}} \right] = \frac{3p(\omega)}{\omega r(r+1)(2r+1)}, \tag{1.125}$$

where

$$p(\omega) = \frac{1}{i} \sum_{j=-r}^{r} j e^{i\omega j} = -r \frac{\cos \frac{2r+1}{2}\omega}{\sin \frac{\omega}{2}} + \frac{\sin \omega r}{2\sin^2 \frac{\omega}{2}}. \tag{1.126}$$

To obtain other formulas, one must notice that the $2v$-order derivative of the function $p(\omega)$ is

$$p^{(2v)}(\omega) = \frac{(-1)^v}{i} \sum_{j=-r}^{r} j^{2v+1} e^{i\omega j}. \tag{1.127}$$

If $m = 3$, then from (1.124) and (1.118),

$$J_{3,r}(\omega) = \frac{5[5(3r^4 + 6r^3 - 3r + 1)p(\omega) + 7(3r^2 + 3r - 1)p''(\omega)]}{\omega r(r^2 - 1)(r+2)(4r^2 - 1)(2r+3)}. \tag{1.128}$$

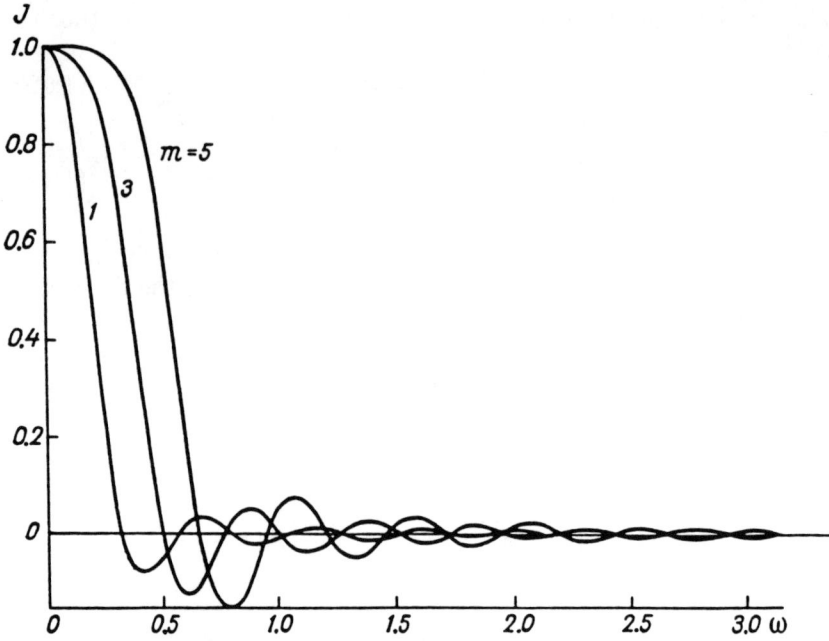

Figure 1.15: Function $J_{m,r}(\omega)$ for different m and the fixed filter width ($r = 5$).

If $m = 5$, formulas (1.124) and (1.119) yield

$$J_{5,r}(\omega) = \left[c_{11}^5 p(\omega) - c_{13}^5 p''(\omega) + c_{15}^5 p^{(4)}(\omega)\right]/\omega. \quad (1.129)$$

Illustrations of functions (1.125), (1.128), and (1.129) for different m and r are provided in Figures 1.15 and 1.16.

Notice that formula (1.124) can be presented through the elements c_{ij}^m of the inverse matrix of the normal equations for the least squares polynomials obtained above. By replacing $\gamma_j^{m,r}$ in (1.124) with (1.115) and by taking into account (1.127), one gets

$$J_{m,r}(\omega) = [c_{11}^m p(\omega) - c_{13}^m p''(\omega) + c_{15}^m p^{(4)}(\omega) - \ldots$$
$$+ (-1)^{\frac{m-1}{2}} c_{1m}^m p^{(m-1)}(\omega)]/\omega, \quad (m = 1, 3, 5, \ldots). \quad (1.130)$$

1.10 Differentiating Filters

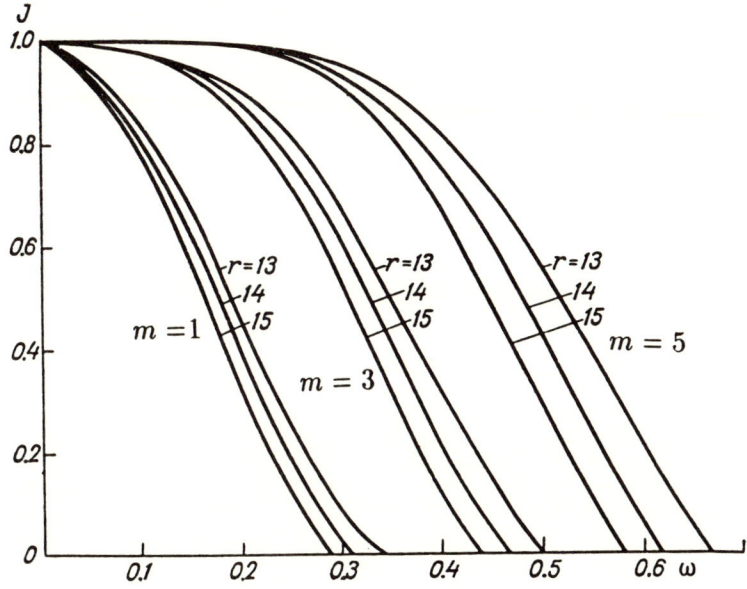

Figure 1.16: Function $J_{m,r}(\omega)$ for different r and m.

1.10.2 The Second-Order Derivative

Let us obtain a system of linear filters for estimation of the mean of the second-order derivative. From (1.22) we have

$$f''(0) = 2\widehat{\beta}_2, \tag{1.131}$$

where

$$\widehat{\beta}_2 = \sum_{j=-r}^{r} \left[c_{20}^m + c_{22}^m j^2 + \ldots + c_{2m-1}^m j^{(m-1)} \right] f_i, (m = 3, 5, \ldots). \tag{1.132}$$

A variance of the derivative estimate (1.131) is

$$\sigma_p^2 = 4c_{22}^m \sigma^2 = 4\sigma_{\beta_2}^2.$$

For $m = 3$, (1.131) and (1.36) yield

$$f''(0) = \frac{30}{(4r^2 - 1)(2r + 3)} \sum_{j=-r}^{r} f_j \left[\frac{3j^2}{r(r+1)} - 1 \right]. \tag{1.133}$$

For $m = 5$, (1.131) and (1.37) give

$$f''(0) = \frac{105}{2(4r^2 - 9)(4r^2 - 1)(2r + 5)} \sum_{j=-r}^{r} f_j[-5(2r^2 + 2r - 3)$$

$$+ 21\frac{4r^4 + 8r^3 - 4r^2 - 8r + 5}{r(r^2 - 1)(r + 2)} j^2 - 15\frac{6r^2 + 6r - 5}{r(r^2 - 1)(r + 2)} j^4]. \quad (1.134)$$

Formulas (1.133) and (1.134) can be used to develop software in problems connected with estimation of the second-order derivative of equally spaced observations.

1.10.3 Any Order Derivative

Estimation of the mean of any order derivative can be based on least squares polynomials as well. For example, for the order p derivative, computed at the central point of the observational interval, one has [from (1.22)]

$$f^{(p)}(0) = p!\hat{\beta}_p \quad (p \leq m), \quad (1.135)$$

where $\hat{\beta}_p$ is determined by expression (1.44).

Therefore, the order p derivative estimator is

$$\bar{Y}_t^{(p)} = p! \sum_{j=-r}^{r} \alpha_j^{m,r,p} Y_{t+j}, \quad (1.136)$$

where $\alpha_j^{m,r,p}$ is determined by (1.46).

The variance σ_c^2 of estimate (1.136) is equal to

$$\sigma_c^2 = \sigma^2 (p!)^2 c_{pp}^m. \quad (1.137)$$

Numerical differentiation schemes were derived for the grid point equal to one. If the grid step is equal to any number h, estimate (1.135) must be divided by h^p to get a derivative in the appropriate units of measurement.

Let us carry out a harmonic analysis of transformation (1.136). If $\varphi(t) = e^{i\omega t}$ is any of the harmonics of the Fourier expansion of observations Y_t, then (1.136) gives

$$\bar{\varphi}^{(p)}(t) = p! e^{i\omega t} \sum_{j=-r}^{r} \alpha_j^{m,r,p} e^{i\omega j}. \quad (1.138)$$

1.11 Two-Dimensional Filters

Because
$$\varphi^{(p)}(t) = (i\omega)^p e^{i\omega t},$$
the ratio
$$\frac{\bar{\varphi}^{(p)}(t)}{\varphi^{(p)}(t)} = \frac{p!}{(i\omega)^p} \sum_{j=-r}^{r} \alpha_j^{m,r,p} e^{i\omega j} \qquad (1.139)$$

reveals the influence of transformation (1.136) on the terms of the Fourier expansion of the observations. For example, (1.124) is a particular case of expression (1.139) for the first-order derivative.

The statistical analysis of estimates (1.136) can be carried out analogously to the approach that was developed in Section 1.8 for smoothing schemes, but such consideration is very formal and will be omitted. Notice that for the application of the various differentiating schemes, which can be obtained from (1.136) for the fixed numerical values of p, m, and r, such an analysis can help to reveal the transformation of a statistical structure of the observations and to evaluate its significance for the physical problem under consideration. The discussion (in Section 1.10) of some aspects of the application of smoothing filters is also valid for differentiating filters.

An accurate identification of the differentiating filter parameters is very important because the derivative estimate can have a very large error, exceeding a corresponding observation value by several times. As a rule, all the quantities used for the derivative estimation should have a double precision type.

1.11 Two-Dimensional Filters

Consider the two-dimensional smoothing and differentiating schemes that are analogous to the one-dimensional filters described in the previous sections.

1.11.1 Smoothing Filters

The smoothing (estimating the mean) of the two-dimensional field is necessary for solving many applied statistical problems connected with processing observations that are dependent on two variables. In particular, the two-dimensional filter is necessary for estimating a spectrum of a two-dimensional homogeneous random field, for conducting the spectral analysis of the nonstationary processes, and for many other problems.

Let us derive a formula for the point estimate $\bar{Y}_{lv}$ of the mean of an element Y_{lv} of the random field

$$\{Y_{lv}\}_{l=0,v=0}^{n,N} \tag{1.140}$$

under the assumption that its elements are independent, have equal variances σ^2, and are given on a system of two-dimensional equally spaced grid points, with the grid interval equal to one along each axis.

The point estimate (smoothing value) of the mean of Y_{lv} is presented as the weighted sum of the observations

$$\{Y_{lv}\}_{l-s,v-r}^{l+s,v+r} \tag{1.141}$$

in the vicinity of Y_{lv}. The optimum weights can be obtained with the aid of the least squares two-dimensional polynomials, which approximate observations (1.141), analogous to the one-dimensional schemes introduced in Section 1.2. Choosing a symmetric two-dimensional coordinate system [analogous to (1.20)] in the vicinity of Y_{lv}, one can ensure that half of the elements of the corresponding normal matrix (that is, the sums with odd powers) will be equal to zeroes [see (1.29)]. In other words, the least squares two-dimensional polynomials of degree q of two variables, τ and t, are built on the grid $(\tau = -s, -(s-1), \ldots, s-1, s; \; t = -r, -(r-1), \ldots, r-1, r)$. Omitting the formal (and cumbersome) algebraic transformations, we note the resulting formulas for $q = 1$ and $q = 3$.

The smoothing scheme is

$$\bar{Y}_{lv} = \sum_{\tau=-s}^{s} \sum_{t=-r}^{r} \alpha_{\tau,t}^{s,r} Y_{l+\tau,v+t}, \tag{1.142}$$

where weights

$$\alpha_{\tau,t}^{s,r} = \frac{1}{(2s+1)(2r+1)} \tag{1.143}$$

if $q = 1$.

If $q = 3$ the weights are

$$\alpha_{\tau,t}^{s,r} = c_0 + c_1 \tau^2 + c_2 t^2, \tag{1.144}$$

where

$$c_0 = \frac{56(s^2+s)(r^2+r) - 27(s^2+s+r^2+r) + 9}{(4s^2-1)(2s+3)(4r^2-1)(2r+3)}, \tag{1.145}$$

1.11 Two-Dimensional Filters

$$c_1 = \frac{-15}{(4s^2-1)(2s+3)(2r+1)}, \qquad (1.146)$$

$$c_2 = \frac{-15}{(4r^2-1)(2r+3)(2s+1)}, \qquad (1.147)$$

$$l = s, s+1, \ldots, n-s, \qquad (1.148)$$

$$v = r, r+1, \ldots, N-r. \qquad (1.149)$$

From (1.143) and (1.144), it follows that the filter weights are symmetric relative to the coordinate system origin. This relationship is also shown in Figure 1.17 for the two-dimensional window (1.144) for the fixed values of the widths s and r.

It is easy to show that for any s and r, the identity

$$\sum_{\tau=-s}^{s}\sum_{t=-r}^{r}\alpha_{\tau,t}^{s,r} \equiv 1 \qquad (1.150)$$

is true.

The variance σ_c^2 of point estimate $\overline{Y}_{lv}$ is

$$\sigma_c^2 = \sigma^2 c_0, \qquad (1.151)$$

where for $q = 1$

$$c_0 = \frac{1}{(2s+1)(2r+1)} \qquad (1.152)$$

and for $q = 3$, c_0 is determined by (1.145).

If the widths of filter (1.144) are identical ($s = r$) along both axes, then

$$\overline{Y}_{lv}$$

$$= \frac{[14(r^2+r)-3]\sum_{\tau=-r}^{r}\sum_{t=-r}^{r}Y_{l+\tau,v+t} - 15\sum_{\tau=-r}^{r}\sum_{t=-r}^{r}(\tau^2+t^2)Y_{l+\tau,v+t}}{(2r-1)(2r+3)(2r+1)^2}.$$
$$(1.153)$$

For example, if $s = r = 3$ in (1.153), one can evaluate the two-dimensional third-order polynomial filter

$$\left\{\alpha_{\tau,t}^{(3,3)}\right\}_{\tau,t=-3}^{3} =$$

$$\begin{pmatrix} -0.048 & -0.014 & 0.007 & 0.014 & 0.007 & -0.014 & -0.048 \\ -0.014 & 0.020 & 0.041 & 0.048 & 0.041 & 0.020 & -0.014 \\ 0.007 & 0.041 & 0.061 & 0.068 & 0.061 & 0.041 & 0.007 \\ 0.014 & 0.048 & 0.068 & 0.075 & 0.068 & 0.048 & 0.014 \\ 0.007 & 0.041 & 0.061 & 0.068 & 0.061 & 0.041 & 0.007 \\ -0.014 & 0.020 & 0.041 & 0.048 & 0.041 & 0.020 & -0.014 \\ -0.048 & -0.014 & 0.007 & 0.014 & 0.007 & -0.014 & -0.048 \end{pmatrix}. \quad (1.154)$$

The two-dimensional filters can be superimposed (multiple smoothing) in exactly the same way as the one-dimensional filter. Such a possibility is especially important for the first-order ($q = 1$) filters because it enables us to obtain two-dimensional filters with positive weights. [If $q = 3$ and greater, the filter weight matrices have some negative values; see above example (1.154) and Fig. 1.17.] Multiple smoothing by the first-order two-dimensional filters can be used, for example, for solving an important class of problems: the spectrum estimation of the two-dimensional random field. Filtering with even a few negative weights is sometimes not appropriate for this purpose because they can give unacceptable negative estimates of the spectral density.

As an example of building a filter with positive weights, consider the procedure of two-tuples smoothing by the simplest two-dimensional first order ($q = 1$) filter of size $(2r+1) \times (2s+1) = 2 \times 2$. The resulting matrix of weights is

$$\{\alpha_{\tau,t}\}_{\tau,t=-1}^{1} = \frac{1}{16} \begin{pmatrix} 1 & 2 & 1 \\ 2 & 4 & 2 \\ 1 & 2 & 1 \end{pmatrix}. \quad (1.155)$$

This filter, the two-dimensional generalization of the Tukey filter, will be used for estimating the two-dimensional spectra of the homogeneous random fields.

1.11.2 Differentiating Filters

Now, consider the numerical differentiation of a random field (1.140). The estimate $\bar{Y}'_{lv}$ of the mean of the derivative $\partial Y/\partial \tau$, where τ is the variation along axis l, is given by

$$\bar{Y}'_{lv} = \sum_{\tau=-s}^{s} \sum_{t=-r}^{r} \gamma_{\tau,t}^{s,r} Y_{l+\tau,v+t}, \quad (1.156)$$

1.11 Two-Dimensional Filters

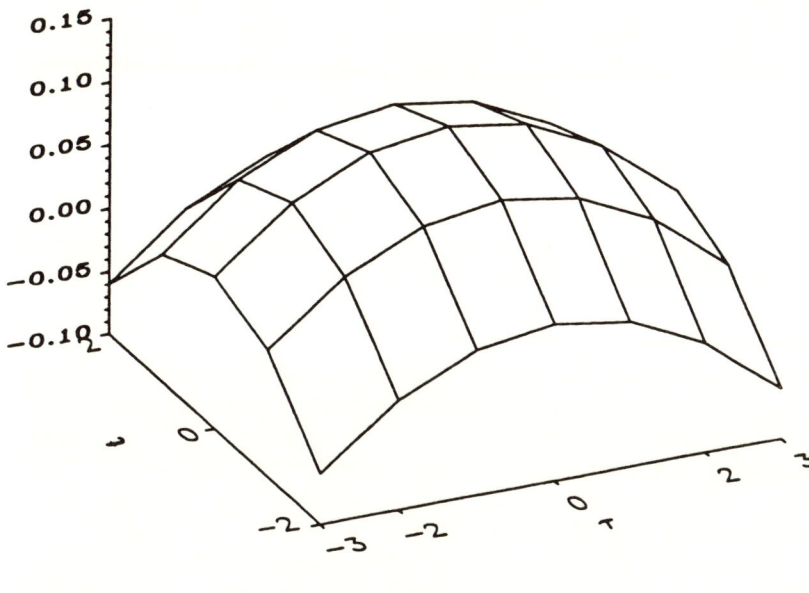

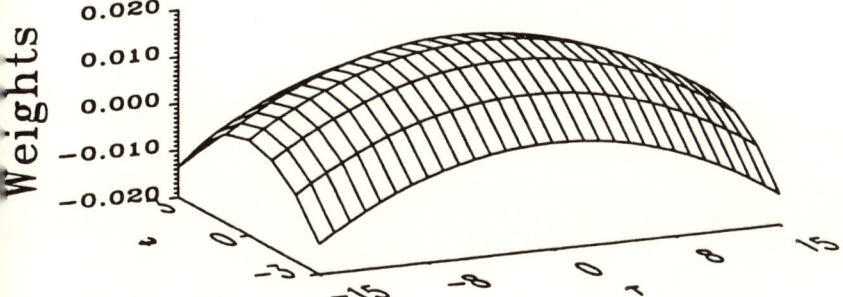

Figure 1.17: Weights of the third-order ($q = 3$) two-dimensional smoothing filter for $s = 3$, $r = 2$ (top) and $s = 15$, $r = 3$ (bottom).

where for $q = 1$ the weights are

$$\gamma_{\tau,t}^{s,r} = \frac{3\tau}{s(s+1)(2s+1)(2r+1)}. \qquad (1.157)$$

For $q = 3$ the weights are

$$\gamma_{\tau,t}^{s,r} = \tau(e_0 + e_1\tau^2 + e_2 t^2), \qquad (1.158)$$

where

$$e_0 = \\ \frac{5[15(3s^4 + 6s^3 - 3s + 1)(3r^2 + 3r - 1) - 7r(r+1)(3s^2 + 3s - 1)^2]}{s(s^2 - 1)(4s^2 - 1)(s+2)(2s+3)(4r^2 - 1)(2r+3)}, \\ \qquad (1.159)$$

$$e_1 = \frac{-35(3s^2 + 3s - 1)}{s(s^2 - 1)(4s^2 - 1)(s+2)(2s+3)(2r+1)}, \qquad (1.160)$$

$$e_2 = \frac{-45}{s(s+1)(2s+1)(4r^2 - 1)(2r+3)}. \qquad (1.161)$$

Figure 1.18 shows the shape of the window of the third-order ($q = 3$) differentiating filter (1.158) when s and r are fixed. This filter is a two-dimensional odd function with zero values for $\tau = 0$.

The variance σ_n^2 of estimate (1.156) is

$$\sigma_n^2 = \begin{cases} \dfrac{3\sigma^2}{s(s+1)(2s+1)(2r+1)} & \text{if} \quad q = 1, \\ \\ \sigma^2 e_0 & \text{if} \quad q = 3. \end{cases} \qquad (1.162)$$

And, finally, for estimating the mean of the second-order derivative $\partial^2 Y/\partial \tau^2$, if $q = 3$, one can obtain

$$\bar{Y}_{lv}'' = 2 \sum_{\tau=-s}^{s} \sum_{t=-r}^{r} \delta_{\tau,t}^{s,r} Y_{l+\tau,v+t}, \qquad (1.163)$$

where

$$\delta_{\tau,t}^{s,r} = g_0 + g_1 \tau^2, \qquad (1.164)$$

1.11 Two-Dimensional Filters

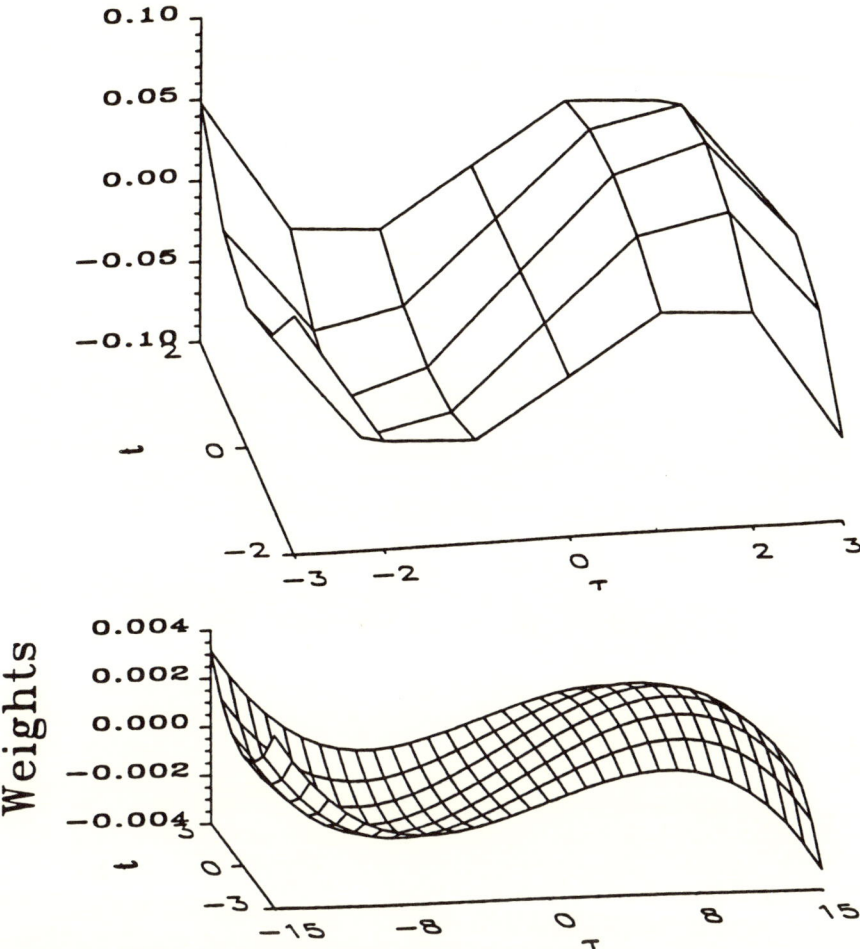

Figure 1.18: Weights of the third-order ($q = 3$) two-dimensional differentiating filter for $s = 3$, $r = 2$ (top) and $s = 15$, $r = 3$ (bottom).

$$g_0 = \frac{-15}{(4s^2 - 1)(2s + 3)(2r + 1)}, \tag{1.165}$$

$$g_1 = \frac{45}{s(s + 1)(4s^2 - 1)(2s + 3)(2r + 1)}. \tag{1.166}$$

As an example, Figure 1.19 presents the windows of this filter for the fixed widths s and r.

This filter is an even two-dimensional function with equal weights corresponding to the fixed values of τ.

The variance σ_n^2 of estimate (1.163) is

$$\sigma_n^2 = 4\sigma^2 g_1. \tag{1.167}$$

The estimators considered in this section were obtained with the aid of cumbersome algebraic arrangements, which have been omitted because of their purely formal character. The correlation and harmonic analysis of the smoothing and differentiating filters of random fields also demands complicated algebraic transformations, which can be done analogously to those provided for the one-dimensional schemes considered in Sections 1.6, 1.7, and 1.10.

1.12 Multidimensional Filters

The development and analysis of different filters is one of the interesting aspects of computational statistics with a wide area of applications. The scope of problems to be solved has not been exhausted by the analyses that have been carried out in the previous sections. For instance, we did not consider many specific filters (for example, a Kalman filter) that are widely used in many applications, or questions connected with studying the bias of estimates, or many other problems of filtration. It is necessary to develop a theory of filters with positive weights for estimating spectral density. Some of the questions of filtering of correlated observations will be considered briefly in Chapter 2. Further studies of different problems in nature, science, and engineering will inevitably lead to the creation of filters with varying dimensionality for smoothing and differentiating corresponding random fields.

The general scheme of smoothing a μ-dimensional random field $Y_{t_1, t_2, \ldots, t_\mu}$, which is given on the equally spaced system of grid points, can be noted as follows:

$$\bar{Y}_{t_1, t_2, \ldots, t_\mu} = \sum_{j_1=-r_1}^{r_1} \sum_{j_2=-r_2}^{r_2} \cdots \sum_{j_\mu=-r_\mu}^{r_\mu} \alpha_{j_1, j_2, \ldots, j_\mu} Y_{t_1+j_1, t_2+j_2, \ldots, t_\mu+j_\mu}. \tag{1.168}$$

1.12 Multidimensional Filters

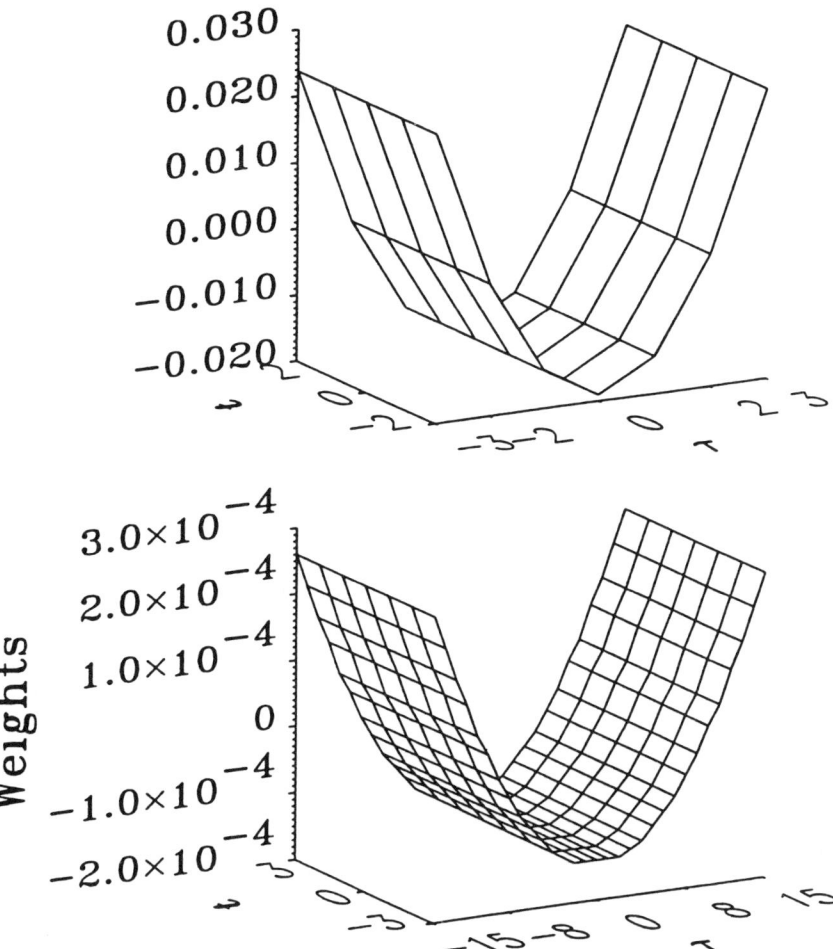

Figure 1.19: Weights of the third-order ($q = 3$) two-dimensional differentiating (for the second derivative) filter for $s = 3$, $r = 2$ (top) and $s = 15$, $r = 3$ (bottom).

One approach to determining the weights in (1.168) is to build the least squares μ-dimensional polynomial and consequently obtain the expressions for the coefficients (β_0 or any other β_i, when the derivatives are estimated) in the form of a linear combination of observations.

2

Averaging and Simple Models

Simple linear procedures for processing correlated observations are considered and interpreted in this chapter. Primarily, they present different schemes for averaging data. These procedures are important because climatology has historically dealt with spatial and temporal averaging of statistically dependent meteorological observations. The accuracy of such averaging is determined by the volume of data and by its correlation structure. The examples presented in this chapter illustrate the level of accuracy that can be achieved within the framework of some assumptions about such correlation structure.

2.1 Correlated Observations

Let us consider the principal relationships of the least squares method for the statistically dependent observations (see Rao, 1973). The basic assumptions are as follows.
We are given a vector of observations

$$\mathbf{Y} = \{Y_i\}_{i=0}^{k}, \tag{2.1}$$

with a covariance matrix

$$\mathbf{M} = \{M_{ij}\}_{i,j=0}^{k}, \tag{2.2}$$

and a system of conditions

$$E(Y_i) = x_{i0}\beta_0 + x_{i1}\beta_1 + \ldots + x_{im}\beta_m, \tag{2.3}$$

where x_{ij} ($i = 0, 1, \ldots, k$; $j = 0, 1, \ldots, m$) are fixed values; β_j ($j = 0, 1, \ldots, m$) are unknown parameters.

Estimators of these parameters can be found by minimizing (over the unknown parameters β_j) the quadratic form

$$(\mathbf{Y} - \mathbf{X}\boldsymbol{\beta})^\mathrm{T} \mathbf{M}^{-1} (\mathbf{Y} - \mathbf{X}\boldsymbol{\beta}), \tag{2.4}$$

where

$$\mathbf{X} = \{\mathbf{X}_{ij}\}_{i=0, j=0}^{k, m}, \quad \boldsymbol{\beta} = \{\beta_j\}_{j=0}^{m}.$$

The normal equation system in this case is

$$\mathbf{C}\boldsymbol{\beta} = \mathbf{X}^\mathrm{T} \mathbf{M}^{-1} \mathbf{Y}, \tag{2.5}$$

where

$$\mathbf{C} = \mathbf{X}^\mathrm{T} \mathbf{M}^{-1} \mathbf{X}.$$

Let $\widehat{\boldsymbol{\beta}}$ be the solution of the system (2.5). Then $\mathbf{C}^{-1}$ is the covariance matrix of vector $\widehat{\boldsymbol{\beta}}$.

If the covariance matrix of the observations (2.1) can be presented as $\mathbf{M} = \sigma^2 \mathbf{B}$, where $\mathbf{B}$ is the correlation matrix and σ is the variance, then the estimator of σ is

$$\widehat{\sigma} = \frac{(\mathbf{Y} - \mathbf{X}\widehat{\boldsymbol{\beta}})^\mathrm{T} \mathbf{B}^{-1} (\mathbf{Y} - \mathbf{X}\widehat{\boldsymbol{\beta}})}{k - m}. \tag{2.6}$$

For covariance matrix $\mathbf{M}_{\widehat{Y}}$ of point estimates $\widehat{\mathbf{Y}} = \mathbf{X}\widehat{\boldsymbol{\beta}}$, one can obtain the following formula:

$$\mathbf{M}_{\widehat{Y}} = \mathbf{X}\mathbf{C}^{-1}\mathbf{X}^\mathrm{T}. \tag{2.7}$$

The diagonal elements of this matrix

$$\sigma_i^2 = \{\mathbf{X}\mathbf{C}^{-1}\mathbf{X}^\mathrm{T}\}_{ii} \tag{2.8}$$

are the variances of the point estimates.

As was shown in Section 1.1, for the independent equally accurate observations the normalized variances of the least squares parameter estimates are determined by the type of the approximation function and by the number of points. In the case of statistically dependent data, such variances are additionally dependent on the covariances; therefore, each type of covariance matrix requires special consideration.

2.1 Correlated Observations

Example 1

Let us consider the graphs of the normalized variances (2.8) of the point estimates for the polynomial approximation of the correlated observations, given on an equally spaced symmetrical grid (identical to those considered in Sections 1.2 and 1.3). Let

$$E(Y_i) = f(t_i) = \beta_0 + \beta_1 t_i + \ldots + \beta_m t_i^m$$

$(i = 0, 1, \ldots, k = 2r;\ t_{i+1} - t_i = 1;\ t = -r, -(r-1), \ldots, r-1, r)$

and

$$\mathbf{M} = \{M_{ij}\}_{i,j=0}^{k} = \{e^{-\alpha|t_i - t_j|}\}_{i,j=0}^{k}.$$

For $\alpha = 0.1$ and $k = 30$, the graph of the normalized variances (2.8) of point estimates for $m = 0, 1, \ldots, 7$ are given in Figure 2.1. The location of the curves on this graph is higher than the location of the corresponding curves for independent observations in Figure 1.2 because correlated observations contain less information about an expected value than independent observations; and, consequently, the variances of the estimates are greater. Figure 2.1 shows the variance curve shapes for different m and for the given level of the correlation α. If the number of observations k increases, the point estimate variance curves, while retaining their shape for each m, shift down to the horizontal axis.

If $t = 0$, then $f(0) = \beta_0$, and the variance, which corresponds to the central point of the interval, is equal to the variance of $\sigma^2_{\beta_0}$. Figure 2.2 presents the dependence of $\sigma^2_{\beta_0}$ on the level of correlation α of the observations. The interval and the number of points are the same as in Figure 2.1. The polynomial degree m varies from 0 to 7. It is apparent that the point estimate variances of statistically dependent observations are greater than the corresponding variances of independent observations for identical m and k (compare Figure 2.2 with Figure 1.3). If k is fixed, the minimum values for each curve are equal to $\sigma^2_{\beta_0}$ computed for the independent observations.

Example 2

Let us draw the variances of the point estimates for an example with the third-degree polynomial approximation of the statistically independent unequally accurate observations given on grid points $t = -15, -14, \ldots, 15$ ($k = 30$). Suppose that the observation variances are 1 for all points except point $t = -5$. Consider two cases: In the first, the variance in point $t = -5$ is 100; in the second, this vari-

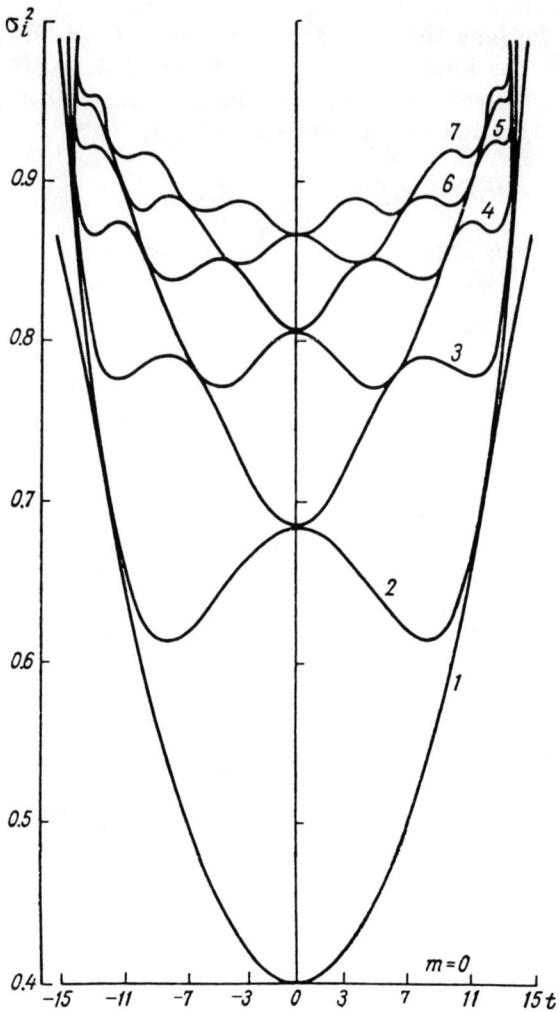

Figure 2.1: Normalized variances of the point estimates for the correlated observations for different polynomial degrees (m).

2.2 The Mean and the Linear Trend

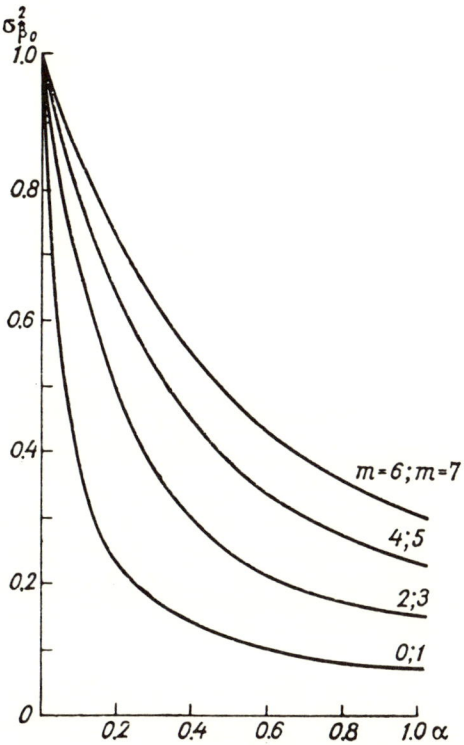

Figure 2.2: Normalized variances of the β_0 estimate for different polynomial degree (m) and correlation level (α) of the observations.

ance is 0.01. (The variances of the observations are located along the main diagonal of the covariance matrix **M**; all other elements of matrix **M** are zero.) Figure 2.3 shows the curves of the point estimate variances for these examples, as well as the curve of the variances corresponding to the independent equally accurate observations with the same number of points $k = 30$ and the same polynomial degree $m = 3$. We can see that the low accuracy of at least one of the observations ($\sigma^2_{-5} = 100$) increases the variances of the point estimates in the vicinity of the bad point (compared with the equally accurate observations). The high accuracy of at least one observation ($\sigma^2_{-5} = 0.01$) leads to an analogous decrease of the point estimate variances, which is illustrated by curve (3) in Figure 2.3.

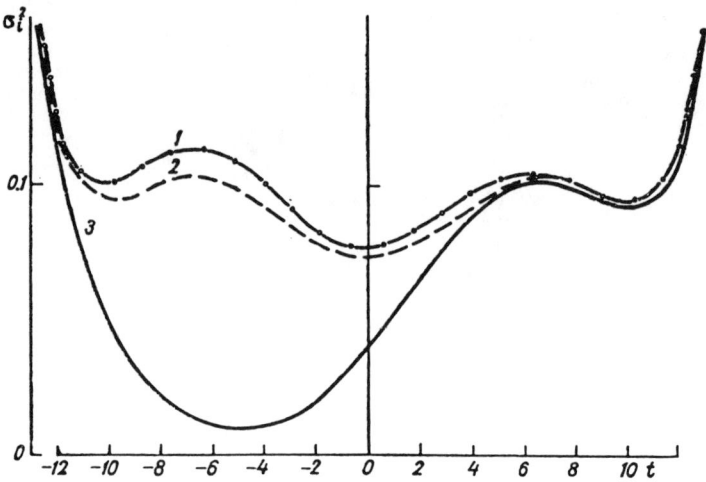

Figure 2.3: Normalized variances of the point estimates of the unequally accurate observations. (1) Variance of the observation of point $t = -5$ is 100. (2) Equal accuracy observations. (3) Variance of the observation of point $t = -5$ is 0.01.

2.2 The Mean and the Linear Trend

The most widespread application of the least squares method considered in Section 2.1 is averaging correlated data. The matrix expressions in this case are simplified, and the formulas can be presented in scalar form. A knowledge of the statistical structure of data provides the foundation for designing and developing the optimal averaging procedure. The accuracy of averages will be demonstrated by simple examples.

Let us introduce the following notation:

$$\mathbf{M}^{-1} = \{\mu_{ij}\}_{i,j=0}^{k}. \tag{2.9}$$

If one estimates the mean, matrix $\mathbf{X}$ is a vector of size $k+1$ with all elements equal to one. Vector $\boldsymbol{\beta}$ is scalar β_0. Therefore, (2.5) yields

$$\left(\sum_{i=0}^{k}\sum_{j=0}^{k}\mu_{ij}\right)\beta_0 = Y_0\sum_{i=0}^{k}\mu_{i0} + Y_1\sum_{i=0}^{k}\mu_{i1} + \ldots + Y_k\sum_{i=0}^{k}\mu_{ik}. \tag{2.10}$$

2.2 The Mean and the Linear Trend

The estimator for the mean is

$$\widehat{Y} = \widehat{\beta}_0 = \frac{\sum\limits_{q=0}^{k} Y_q \sum\limits_{i=0}^{k} \mu_{iq}}{\sum\limits_{i=0}^{k}\sum\limits_{j=0}^{k} \mu_{ij}} = \sum_{q=0}^{k} p_q Y_q, \qquad (2.11)$$

where

$$p_q = \frac{\sum\limits_{i=0}^{k} \mu_{iq}}{\sum\limits_{i=0}^{k}\sum\limits_{j=0}^{k} \mu_{ij}}. \qquad (2.12)$$

Variance $\sigma^2_{\widehat{Y}}$ of the estimator $\widehat{Y}$ is obtained by taking into account the fact that $\mathbf{C}^{-1}$ is the covariance matrix of vector $\widehat{\beta}$:

$$\sigma^2_{\widehat{Y}} = \sigma^2_{\widehat{\beta}_0} = \mathbf{C}^{-1} = \frac{1}{\sum\limits_{i=0}^{k}\sum\limits_{j=0}^{k} \mu_{ij}}. \qquad (2.13)$$

In the particular case of independent unequally accurate observations ($M_{ij} = 0$, if $i \neq j$), (2.11) and (2.13) yield

$$\widehat{Y} = \frac{\sum\limits_{i=0}^{k} \frac{1}{M_{jj}} Y_j}{\sum\limits_{j=0}^{k} \frac{1}{M_{jj}}} \qquad (2.14)$$

and

$$\sigma^2_{\widehat{Y}} = \frac{1}{\sum\limits_{j=0}^{k} \frac{1}{M_{jj}}}. \qquad (2.15)$$

Let us determine the optimal estimator for the mean of the random variables with the covariance matrix

$$\mathbf{M} = \sigma^2 \mathbf{B}, \qquad (2.16)$$

where **B** is the correlation matrix.

Example 1
First consider the observations with exponential correlations

$$\mathbf{B} = \{B_{ij}\}_{i,j=0}^{k} = \{e^{-\alpha|i-j|}\}_{i,j=0}^{k}. \tag{2.17}$$

This idealized structure is given as an illustrative model to obtain a comparative representation of the possible accuracy of the optimal and nonoptimal mean estimators in the case of a large sample.

It is easy to verify that

$$\mathbf{M}^{-1} = \frac{1}{\sigma^2}\mathbf{B}^{-1} = \frac{1}{\sigma^2(1-\gamma^2)}$$

$$\times \begin{pmatrix} 1 & -\gamma & 0 & 0 & \cdots & 0 & 0 & 0 \\ -\gamma & 1+\gamma^2 & -\gamma & 0 & \cdots & 0 & 0 & 0 \\ 0 & -\gamma & 1+\gamma^2 & -\gamma & \cdots & 0 & 0 & 0 \\ \vdots & \vdots & \vdots & \vdots & & \vdots & \vdots & \vdots \\ 0 & 0 & 0 & 0 & \cdots & -\gamma & 1+\gamma^2 & -\gamma \\ 0 & 0 & 0 & 0 & \cdots & 0 & -\gamma & 1 \end{pmatrix}, \tag{2.18}$$

where $\gamma = e^{-\alpha}$ is the correlation coefficient for the lag $i - j = 1$.

Adding all the elements of this matrix and substituting the results in (2.13) yields

$$\sigma_{\widehat{Y}}^2 = \frac{\sigma^2}{1 + k(1-\gamma)/(1+\gamma)} = \frac{\sigma^2}{1 + k\tanh(\alpha/2)}. \tag{2.19}$$

From (2.11) and (2.18) we get the mean estimator

$$\widehat{Y} = \frac{Y_0 + Y_k + (1-\gamma)\sum_{i=1}^{k-1} Y_i}{(1+\gamma) + k(1-\gamma)}. \tag{2.20}$$

According to (2.19), the accuracy of the mean estimator depends upon the number of observations (k) and on the level of correlations (α).

The normalized standard deviation

$$s1 = \frac{\sigma_{\widehat{Y}}}{\sigma} = \frac{1}{\sqrt{1 + k(1-\gamma)/(1+\gamma)}}$$

2.2 The Mean and the Linear Trend

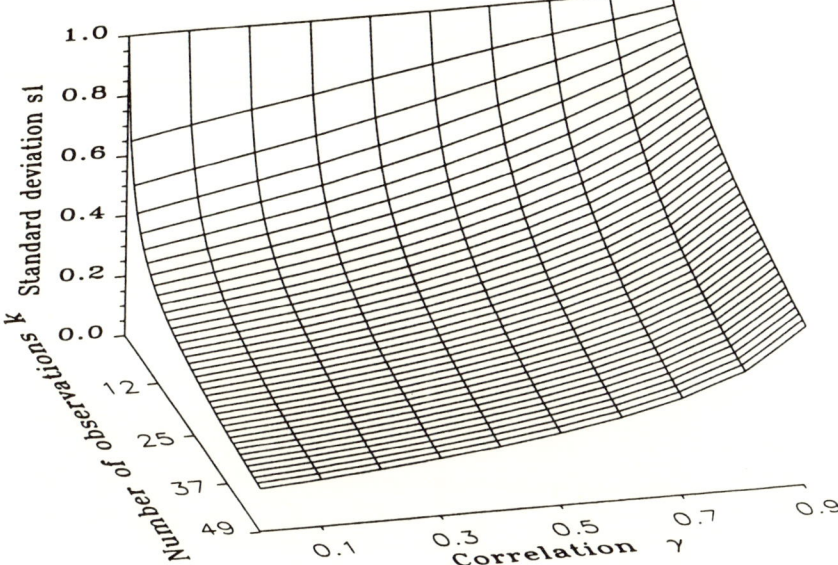

Figure 2.4: Normalized standard deviations for the optimal estimator of the mean of the observations with correlations (2.17).

of the estimator (2.20) for different correlations γ and number of points (k) is presented in Figure 2.4.

If k increases, the accuracy also increases without limit, in spite of the fact that the observations are statistically dependent.

Example 2

As a second example, let us study the accuracy of the optimal mean estimator when the elements of matrix $\mathbf{B}$ in (2.16) are determined in the following way:

$$B_{ij} = \begin{cases} 1 & \text{if } i = j, \\ \rho & \text{if } i \neq j, \end{cases} \quad (0 < \rho < 1). \tag{2.21}$$

It is possible to show that the elements of the inverse matrix

$$\mathbf{B}^{-1} = \{\mu_{ij}\}_{i,j=0}^{k} \tag{2.22}$$

are equal to

$$\mu_{ij} = \begin{cases} \dfrac{1+(k-1)\rho}{(1+k\rho)(1-\rho)} & \text{if } i = j, \\[2mm] \dfrac{-\rho}{(1+k\rho)(1-\rho)} & \text{if } i \neq j. \end{cases} \quad (2.23)$$

Substituting the μ_{ij} in (2.13) for these values yields

$$\sigma_{\widehat{Y}}^2 = \frac{\sigma^2(\rho + 1/k)}{1 + 1/k} > \sigma^2 \rho. \quad (2.24)$$

For sufficiently large k,

$$\sigma_{\widehat{Y}}^2 \approx \sigma^2 \rho, \quad (2.25)$$

and the variance of the sample mean actually does not depend on the number of observations. Therefore, if the correlations of the observations are constant, then, beginning with some value of k, increasing the number of observations in order to increase accuracy of the estimation is senseless.

The optimal estimator of the mean for this example is the same as for the independent observations

$$\widehat{Y} = \frac{1}{k+1} \sum_{i=0}^{k} Y_i. \quad (2.26)$$

The normalized standard deviation

$$s2 = \frac{\sigma_{\widehat{Y}}}{\sigma} = \sqrt{\frac{\rho + 1/k}{1 + 1/k}}$$

of the estimator (2.26) is illustrated in Figure 2.5. From (2.24) it follows that, even if ρ is small (approximately 0.1 to 0.2), the accuracy of the sample mean is very far from the accuracy which, it seems, could be obtained if one has several dozen (or several thousand) observations.

In Table 2.1 the mean and the linear trend parameter estimators and corresponding variances are given for two situations: independent and statistically dependent observations with the correlations of the above considered type (2.17).

Example 3
Now let us consider a typical climatological example.
We are given 31 daily temperature observations (in $°C$)

2.2 The Mean and the Linear Trend

Table 2.1: The mean and the linear trend parameter estimators.

Par.	Independent observations	Observations with correlation function $e^{-\alpha\|r\|}$ ($e^{-\alpha}=\gamma$)
		Mean $E(Y_i) = $ const.
β_0	$\dfrac{1}{2r+1}\sum_{i=0}^{2r} Y_i$	$\left[Y_0 + Y_k + (1-\gamma)\sum_{i=1}^{k-1} Y_i\right] / [1+\gamma+k(1-\gamma)]$
σ^2	$\dfrac{1}{2r-1}\sum_{i=0}^{2r}[Y_i - \beta_0 - \beta_1(i-r)]^2$	$\dfrac{(y_0 - \gamma y_1)y_0 + (y_k - \gamma y_{k-1})y_k + \sum_{i=1}^{k-1} y_i[(1+\gamma^2)y_i - \gamma(y_{i-1}+y_{i+1})]}{(k-1)(1-\gamma^2)}$
		where $y_i := Y_i - \hat{Y}$
$\sigma^2_{\hat{Y}}$	$\sigma^2/(2r+1)$	$\sigma^2(1+\gamma)/[1+\gamma+k(1-\gamma)]$

(Continued)

Table 2.1 (Cont.)

Linear trend $E(Y_i) = \beta_0 + \beta_1 t$

β_0	$\dfrac{1}{2r+1}\sum_{i=0}^{2r} Y_i$	$\left[Y_0 + Y_k + (1-\gamma)\sum_{i=1}^{k-1} Y_i\right] / [1 + \gamma + k(1-\gamma)]$
β_1	$\dfrac{3}{r(f+1)(2r+1)}\sum_{i=0}^{2r}(i-r)Y_i$	$\dfrac{(Y_k - Y_0)[r + \gamma/(1-\gamma)] + (1-\gamma)\sum_{i=1}^{k-1} Y_i(i-r)}{r[2(r + \gamma/(1-\gamma)) + (1-\gamma)(r-1)(2r-1)/3]}$
σ^2	$\dfrac{1}{2r-1}\sum_{i=0}^{2r}[Y_i - \beta_0 - \beta_1(i-r)]^2$	$\dfrac{(y_0 - \gamma y_1)y_0 + (y_k - \gamma y_{k-1})y_k + \sum_{i=1}^{k-1} y_i[(1+\gamma^2)y_i - \gamma(y_{i-1} + y_{i+1})]}{(k-1)(1-\gamma^2)}$
		where $y_i = Y_i - \beta_0 - \beta_1(i-r)$
$\sigma^2_{\beta_0}$	$\sigma^2/(2r+1)$	$\sigma^2(1+\gamma)/[1 + \gamma + k(1-\gamma)]$
σ_{β_1}	$3\sigma^2/r(r+1)(2r+1)$	$\sigma^2(1+\gamma)/\{r[2(r + \gamma/(1-\gamma)) + (1-\gamma)(r-1)(2r-1)/3]\}$
σ_i^2	$\sigma^2_{\beta_0} + \sigma^2_{\beta_1}(i-r)^2$	$\sigma^2_{\beta_0} + \sigma^2_{\beta_1}(i-r)^2$
	$(i = 0, 1, \ldots, k)$	$(i = 0, 1, \ldots, k)$

2.2 The Mean and the Linear Trend

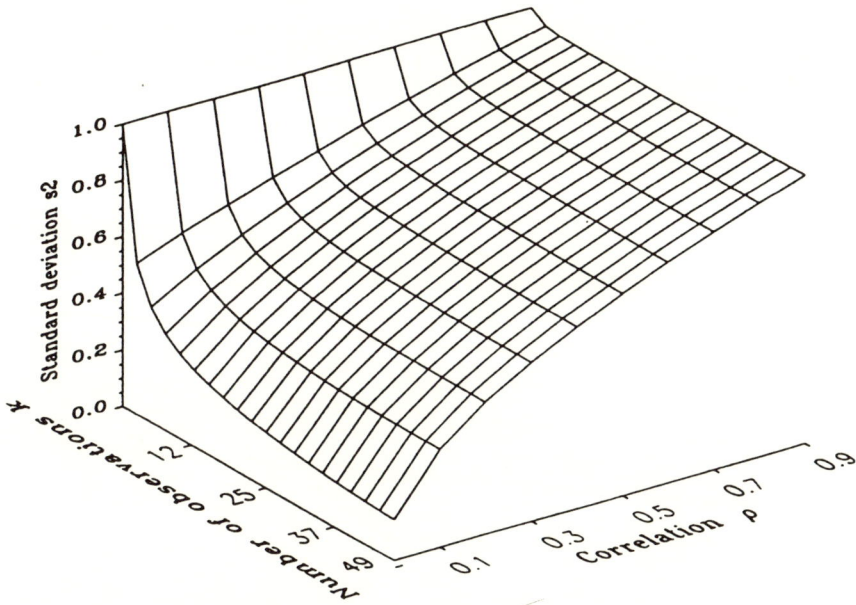

Figure 2.5: Normalized standard deviations for the optimal estimator of the mean of the observations with correlations (2.21).

7.4	6.7	3.7	4.7	3.4	1.4	2.6	2.5	3.2	5.8	
7.2	6.2	8.9	7.2	5.5	4.4	4.9	4.7	4.4	3.5	
4.2	5.0	5.2	5.5	3.6	3.4	6.5	6.4	8.8	6.1	4.3

for August 1972 in Hveravellir, Iceland (data are taken from Tong, 1990).

Let us estimate the mean, the standard deviation σ, and the standard deviation $\sigma_{\widehat{Y}}$ of the mean estimate of temperature for August 1972 in Hveravellir. Assuming that the sample presents independent and equally accurate observations of daily temperatures, we can apply corresponding formulas from Table 2.1.

The estimates are
$\widehat{Y} = (7.4 + 6.7 + \cdots + 4.3)/31 = 5.07;$
$\widehat{\sigma} = \{[(7.4 - 5.07)^2 + \cdots + (4.3 - 5.07)^2]/30\}^{1/2} = 1.8;$
$\sigma_{\widehat{Y}} = 1.8/(31)^{1/2} = 0.32.$

Now we want to estimate the climatic mean of the August temperature in Hveravellir. Assuming that the above data presents a sample of a stationary process with autocorrelation function $\gamma^{-\tau}$, where $\tau = i - j$, and i (and j) is the day number, we can apply the corresponding formulas for correlated observations from Table 2.1. For instance, setting $\gamma = 0.7$, we get

$\hat{Y} = (7.4+4.3+(1-0.7)(6.7+\cdots+6.1)/(1+0.7+30(1-0.7)) = 5.18;$
$\hat{\sigma} \approx 2.01;$
$\sigma_{\hat{Y}} = \{(2.01)^2(1 + 0.7)/[1 + 0.7 + 30(1 - 0.7)]\}^{1/2} \approx 0.80.$

The following table gives analogously derived estimates for different values of autocorrelation γ (the sample autocorrelation coefficient of this data for $\tau = 1$ day is 0.606).

γ	0.00	0.10	0.20	0.40	0.50	0.60	0.70	0.80	0.90	0.95
$\hat{Y}$	5.07	5.08	5.09	5.11	5.12	5.14	5.18	5.23	5.36	5.50
$\sigma_{\hat{Y}}$	0.32	0.34	0.36	0.43	0.50	0.61	0.80	1.17	2.15	3.70

The numerical results show that statistically dependent observations carry less information about the mean value than independent observations and their standard deviation $\sigma_{\hat{Y}}$ is greater than for independent observations.

Example 4

Consider estimates of the linear trend parameter β_1 obtained through the seasonal and annual mean surface air temperature time series of the Northern Hemisphere for the period from 1891 to 1975 (data from the archives of the World Data Center, Obninsk, Russia). The observations were spatially averaged over various $15°$ latitude bands (see Table 2.2). It was assumed that the correlation of the time series terms is exponential (2.17) with the first autocorrelation $\gamma = 0.3$ (see Chapter 6). Each time series was approximated by the straight line $\beta_0 + t\beta_1$. As the results in Table 2.2 show there are only six statistically significant (t-statistic is greater than 2) estimates for the seasonal time series and two for the annual data. Temperature rate is more noticeable for the winter and summer seasons and for the northern latitudes.

These time series were also used to find the least squares straight lines in the assumption of independence. The estimate values in

Table 2.2: Estimates of β_1 (C°/year) and corresponding t-statistics for the zonal temperature time series of different seasons ($\rho_1 = 0.3$).

Lat. band	Winter	Spring	Summer	Fall	Annual
	\multicolumn{5}{c}{β_1 estimates}				
90–75°	0.026	0.001	0.011	−0.002	0.009
75–60°	0.015	0.004	0.006	0.001	0.007
60–45°	0.004	0.005	0.002	0.002	0.004
45–30°	0.004	0.004	0.003	0.002	0.003
30–15°	0.003	−0.001	−0.003	0.001	0.000
	\multicolumn{5}{c}{t-statistics}				
90–75°	2.5	0.2	3.2	−0.3	1.8
75–60°	2.5	0.9	2.5	0.3	2.2
60–45°	0.8	1.8	1.4	0.2	1.8
45–30°	1.3	2.8	1.9	1.4	3.0
30–15°	2.4	−0.3	−1.6	0.5	0.2

this case were approximately the same as in Table 2.2 but the standard deviations were greater that changed the statistical significance markedly: overall, 16 values of t−statistics were greater than 2.

More general results for the monthly mean surface air temperatures of different geographical and political regions of the world are given in Section 4.8 (Table 4.2) and Chapters 6 and 7. Many of these estimates are statistically significant.

But no statistically significant estimates were found among linear trend parameter values of climatic times series of some Russian stations which were carefully maintained and controlled by several generations of Russian climatologists.

2.3 Nonoptimal Estimation

When the covariance matrix is unknown, the calculations are often performed by applying the same method used for independent observations (see Section 2.1). Such procedure is not statistically optimal (in other words, it does not provide the minimal variance); but, in many situations, if the volume of data is large, it is close to the optimal estimation.

Let us derive the formulas for the covariance matrices of the parameters and point estimates for the nonoptimal estimation. Assume that observations (2.1) with an unknown covariance matrix (2.2) are used for estimation of vector β under conditions (2.3). As an estimator of β, one takes the solution of the system

$$\mathbf{X}^T\mathbf{X}\overline{\beta} = \mathbf{X}^T\mathbf{Y}, \qquad (2.27)$$

which was obtained in Section 1.1.

The covariance matrix of $\overline{\beta}$ (as a linear function of observations $\mathbf{Y}$) is

$$\mathbf{M}_{\overline{\beta}} = \left(\mathbf{X}^T\mathbf{X}\right)^{-1}\mathbf{X}^T\mathbf{M}\mathbf{X}\left(\mathbf{X}^T\mathbf{X}\right)^{-1}. \qquad (2.28)$$

Covariance matrix $\mathbf{M}_{\overline{Y}}$ of point estimates $\overline{Y} = \mathbf{X}\overline{\beta}$ is

$$\mathbf{M}_{\overline{Y}} = \mathbf{X}\mathbf{M}_{\overline{\beta}}\mathbf{X}^T = \mathbf{X}\left(\mathbf{X}^T\mathbf{X}\right)^{-1}\mathbf{X}^T\mathbf{M}\mathbf{X}\left(\mathbf{X}^T\mathbf{X}\right)^{-1}\mathbf{X}^T. \qquad (2.29)$$

The diagonal elements of this matrix

$$\sigma_i^2 = \{\mathbf{M}_{\overline{Y}}\}_{ii} \quad (i = 0, 1, \ldots, k) \qquad (2.30)$$

determine the variances of the nonoptimal point estimates.

This scheme is often applied to estimating the mean of the correlated observations with unknown matrix $\mathbf{M}$. The estimator of the mean

$$\overline{Y} = \frac{1}{k+1}\sum_{i=0}^{k}Y_i \qquad (2.31)$$

in this case has a variance

$$\sigma_{\overline{Y}}^2 = \frac{1}{(k+1)^2}\sum_{i=0}^{k}\sum_{j=0}^{k}M_{ij}. \qquad (2.32)$$

We will use this scheme for averaging observations with the particular type covariances M_{ij}, which depend only on $|j - i|$ (i.e., $M_{ij} = M_{|j-i|}$). Then matrix $\mathbf{M}$ is presented as

$$\mathbf{M} = \begin{pmatrix} M_0 & M_1 & M_2 & \ldots & M_{k-1} & M_k \\ M_1 & M_0 & M_1 & \ldots & M_{k-2} & M_{k-1} \\ \vdots & \vdots & \vdots & & \vdots & \vdots \\ M_k & M_{k-1} & M_{k-2} & \ldots & M_1 & M_0 \end{pmatrix}. \qquad (2.33)$$

2.3 Nonoptimal Estimation

Such a matrix is called the Toeplitz matrix.

The variance of the mean of the observations with a covariance matrix of (2.33) type is derived from (2.32):

$$\sigma_{\bar{Y}}^2 = \frac{1}{(k+1)^2} \sum_{i=0}^{k} \sum_{j=0}^{k} M_{|j-i|}$$

$$= \frac{1}{k+1} \left[M_0 + 2 \sum_{i=1}^{k} \left(1 - \frac{i}{k+1}\right) M_i \right]. \qquad (2.34)$$

Example 1

Let us compare the variances of the sample means obtained by the various methods considered in Sections 1.1, 2.1, and 2.3. If the mean of the random variables with the unknown correlation matrix (2.17) is estimated by (2.31), then (2.34) yields

$$\sigma_{\bar{Y}}^2 = \frac{\sigma^2}{k+1} \left\{ 1 + \frac{2\gamma}{1-\gamma} \left[1 - \frac{1-\gamma^{k+1}}{(k+1)(1-\gamma)} \right] \right\}. \qquad (2.35)$$

For sufficiently large k, (2.35) gives

$$\sigma_{\bar{Y}}^2 \approx \frac{\sigma^2}{(k+1)\tanh(\alpha/2)}. \qquad (2.36)$$

The comparison of the variances (2.19) and (2.36) illustrates the possibility of obtaining a nonoptimal mean estimator with a variance that (for a large number of observations) is only slightly distinguished from the variance of the optimal estimator. Moreover, it is possible to show that for this example, the numerical values of variances (2.19) and (2.35) do not differ markedly even for a small amount of data. These examples also reveal the importance of the independence (for estimation of the expected value) of the observations, because in this case, the variances of the estimates are minimal. The distinction of the variance of the nonoptimal scheme from the optimal one are affected by the covariance values and the number of observations.

Formula (2.31) is used for estimating the mean of the time series (statistically dependent observations with an unknown covariance matrix) of the stationary random process.

As for the example (2.21), the estimator (2.31) (and its variance) is the same as that for optimal and nonoptimal approaches. To some extent, the example (2.21) is the idealized illustration of the practical meteorological situation when the spatial or temporal span of the collected data is relatively small; indeed, the correlations between the observations can be assumed to be approximately constant. For instance, rain data obtained by averaging radar measurements within a small elemental data box is consistent with the above assumptions.

Example 2

Let us draw the graphs of the variances of the nonoptimal point estimates for the polynomial approximation of the observations with exponential correlations:

$$E(Y_i) = f(t_i) = \beta_0 + \beta_1 t_i + \ldots + \beta_m t_i^m,$$

$(i = 0, 1, \ldots, k = 2r; \; t_{i+1} - t_i = 1; \; t = -r, \; -(r-1), \ldots, r-1, \; r),$

and

$$\mathbf{M} = \{M_{ij}\}_{i,j=0}^{k} = \{e^{-\alpha|t_i - t_j|}\}_{i,j=0}^{k}.$$

Figure 2.6 shows the shape of the variance curves (2.30) of the nonoptimal point estimates, when the polynomial degrees m and the level of correlation α are fixed but the number k varies. Figure 2.7 presents the graph of the normalized variances (2.30) for the fixed k and α ($k = 30, \alpha = 0.1$) and for the different polynomial degree m. Figures 2.6 and 2.7 show that the variances of the nonoptimal point estimates (for the points close to the interval ends) can be greater than the variances of the original observations.

A comparison of Figures 1.2, 2.1, and 2.7 offers a representation of the shape and location of the variance curves of the independent and correlated observations when different approaches are used.

Example 3

The examples in Figure 2.8 are provided so that the accuracy of the considered methodologies may be compared more clearly. This figure presents the graphs of the normalized variances of the point estimates, when $k = 14$, $\alpha = 0.3$, and for polynomial degrees $m = 1$ and 3. Curve 1 was obtained for independent observations (1.15); curve 2 corresponds to the optimal scheme (2.8); and curve 3 is the result of application of formula (2.30). The graphs show that for these types of correlations the variances (2.30) of the nonoptimal point estimates only slightly exceed corresponding minimal variances (2.8).

2.4 Spatial Averaging

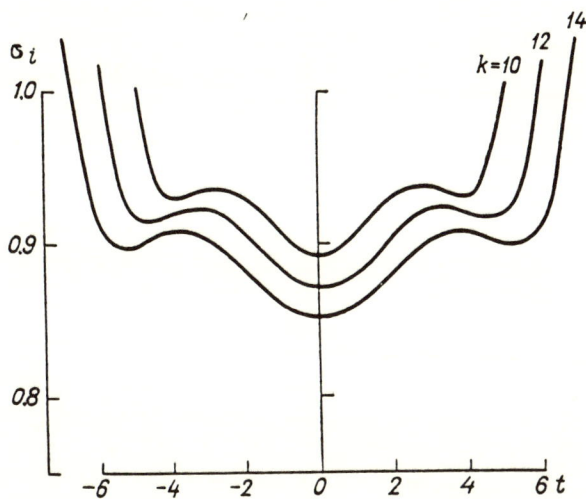

Figure 2.6: Normalized variances (2.30) of the nonoptimal point estimates when the polynomial degree m and α are fixed ($m = 3$, $\alpha = 0.1$) and the number of points k varies.

2.4 Spatial Averaging

Observations of a three-dimensional field (depending on two spatial and one temporal arguments) can be considered as multiple time series. Spatial averages of this field for each moment of time present a new univariate time series with a statistical structure different from the statistical structure of the original time series. Averaging is a linear operation; therefore, having the information about spatial-temporal correlations, one can evaluate the temporal correlations of the time series of averages.

Consider $k+1$ time series $y_0(t), y_2(t), \ldots, y_k(t)$ at different points on the earth's surface. Let

$$M_{ij}(\tau) = E[y_i(t)y_j(t+\tau)] \qquad (2.37)$$

be the spatial-temporal covariance of the observations $y_i(t)$ and $y_j(t+\tau)$ at points i and j; τ is the time interval between these two points.

For each moment of time t, we will define an area mean

$$\bar{Y}(t) = \sum_{i=0}^{k} w_i y_i(t), \qquad (2.38)$$

where

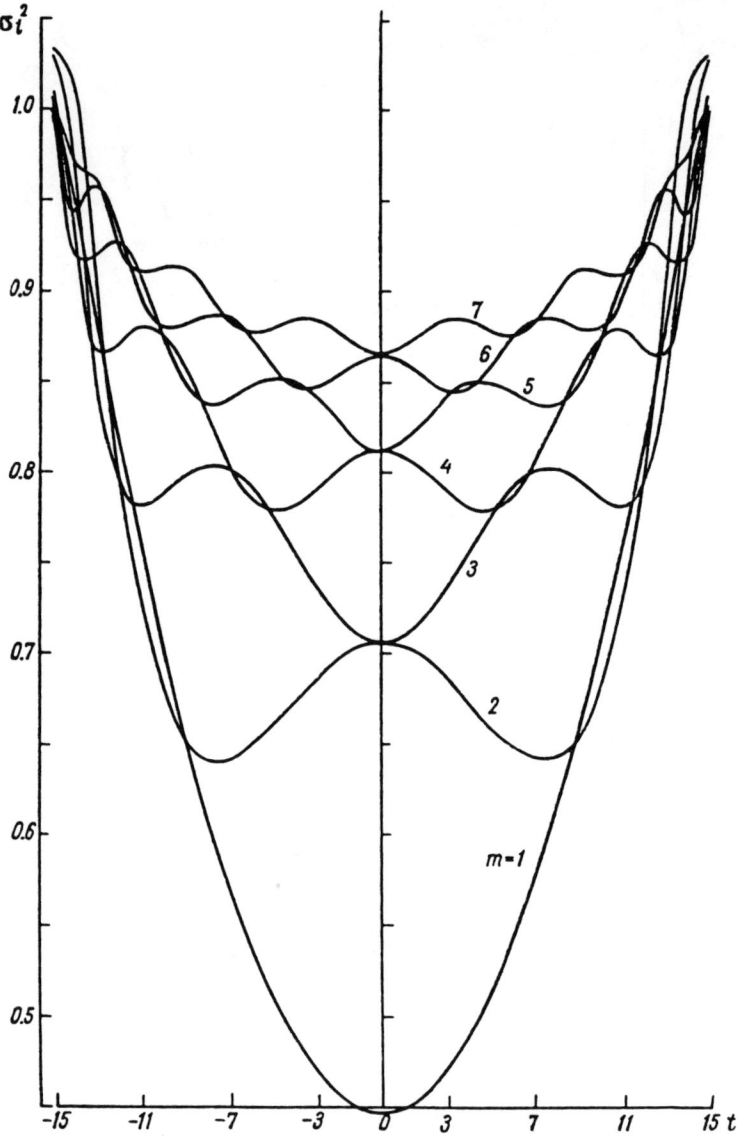

Figure 2.7: Normalized variances (2.30) of the nonoptimal point estimates when the number of points k and α are fixed ($k = 30; \alpha = 0.1$) and the polynomial degree m varies.

2.4 Spatial Averaging

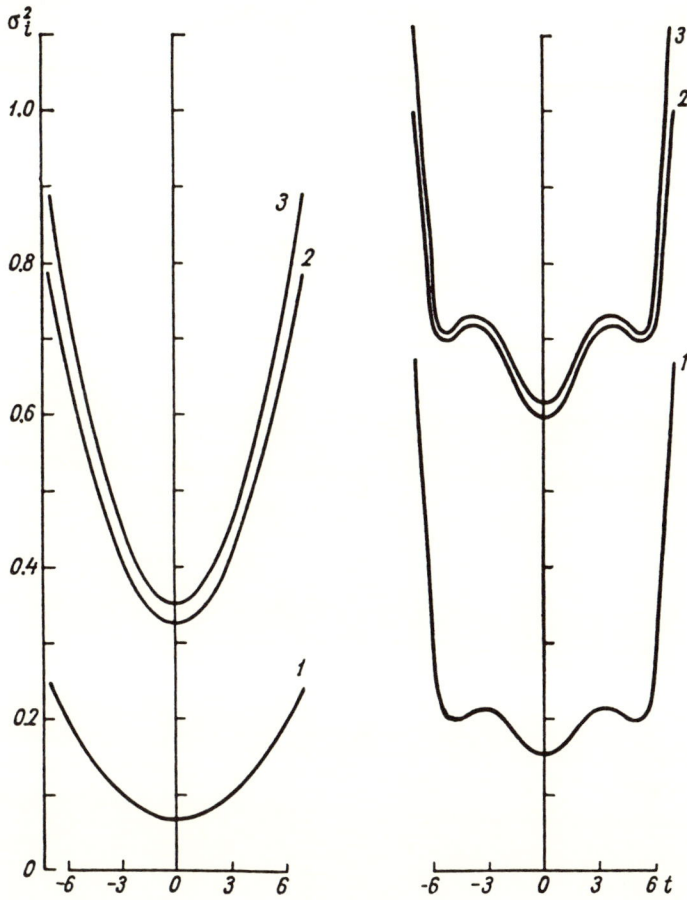

Figure 2.8: Normalized variances of the optimal polynomial point estimates for independent (1) and correlated (2) observations, and corresponding variances for the nonoptimal estimates (3); $m = 1$ (left), $m = 3$ (right).

$$w_i \ (i = 0, 1, \ldots, k; \ \sum_{i=0}^{k} w_i = 1)$$

are the weights of averaging.

Estimates $\bar{Y}(t)$ ($t = 0, 1, 2, \ldots$) present a new process, the autocovariance function ($K(\tau)$) of which is determined by spatial-temporal covariances (2.37) and by weights w_i:

$$K(\tau) = \mathbf{E}\left[\bar{Y}(t)\bar{Y}(t+\tau)\right]$$

$$= \mathbf{E}\left[\sum_{i=0}^{k} w_i y_i(t) \sum_{j=0}^{k} w_j y_j(t+\tau)\right] = \sum_{i=0}^{k}\sum_{j=0}^{k} w_i w_j M_{ij}(\tau). \quad (2.39)$$

The corresponding autocorrelation function $\rho_{cp}(\tau)$ is

$$\rho_{cp}(\tau) = \frac{\sum_{i=0}^{k}\sum_{j=0}^{k} w_i w_j M_{ij}(\tau)}{\sum_{i=0}^{k}\sum_{j=0}^{k} w_i w_j M_{ij}(0)}. \quad (2.40)$$

If

$$w_i = \frac{1}{k+1}, \quad (2.41)$$

we have

$$\rho_{cp}(\tau) = \frac{\sum_{i=0}^{k}\sum_{j=0}^{k} M_{ij}(\tau)}{\sum_{i=0}^{k}\sum_{j=0}^{k} M_{ij}(0)}. \quad (2.42)$$

Let us assume that the time series $y_i(t)$ is the sum of the signal $S_i(t)$ and the white noise $N_i(t)$,

$$y_i(t) = S_i(t) + N_i(t), \quad (2.43)$$

and that noise components $N_i(t)$ are independent in space and have identical variance σ_N^2. Let us also suppose that the statistical dependence of $y_i(t)$ is completely determined by the signal component $S_i(t)$, i.e.,

$$M_{ii}(0) = M_{ii} = \sigma_S^2 + \sigma_N^2; \quad M_{ij}(0) = \sigma_S^2 \rho_{ij}, \quad (2.44)$$

2.4 Spatial Averaging

$$M_{ij}(\tau) = \sigma_S^2 \rho_{ij} \rho(\tau) \quad (\tau > 0), \tag{2.45}$$

where σ_S^2 is the variance of process $S_i(t)$; ρ_{ij} is the spatial correlation of values $S_i(t)$ and $S_j(t)$; $\rho_{ij} = 1$ if $i = j$; and $\rho(\tau)$ is the autocorrelation function which is identical for all processes $S_i(t)$. Such spatial-temporal statistical dependence is called *separable* (Christakos, 1992). It is not something very particular because it is expressed in general form in (2.44)–(2.45) without assumptions about the types of the spatial and temporal correlations, ρ_{ij} and $\rho(\tau)$.

Spatial averaging of $y_i(t)$ leads to the equality

$$\bar{Y}(t) = \bar{S}(t) + \bar{N}(t). \tag{2.46}$$

Taking the variance of both sides of (2.46) yields

$$\sigma_{\bar{Y}}^2 = \sigma_{\bar{S}}^2 + \sigma_{\bar{N}}^2 = \frac{\sigma_S^2}{(k+1)^2} \sum_{i=0}^{k} \sum_{j=0}^{k} \rho_{ij} + \frac{\sigma_N^2}{k+1}. \tag{2.47}$$

Furthermore, we replace $\sigma_{\bar{S}}^2$ as follows:

$$\sigma_{\bar{S}}^2 = \frac{\sigma_S^2}{(k+1)^2} \sum_{i=0}^{k} \sum_{j=0}^{k} \rho_{ij} = \frac{\sigma_S^2}{n+1}, \tag{2.48}$$

where $n+1$ ($n < k$) is the number of statistically independent variables, the mean of which has the same variance, $\sigma_{\bar{S}}^2$, as the mean of the given system of correlated values $S_i(t)$. (The definition of the system of statistically equivalent independent variables can be found, for example, in Bayley and Hammersley, 1946, or in Polyak, 1975.)

We have

$$\sigma_{\bar{Y}}^2 = \sigma_{\bar{S}}^2 + \sigma_{\bar{N}}^2 = \frac{\sigma_S^2}{n+1} + \frac{\sigma_N^2}{k+1}. \tag{2.49}$$

The estimation of the autocorrelation function of separate time series $y_i(t)$ of the signal-plus-white-noise-type process actually leads to the estimation of the function

$$\rho_H(\tau) = \begin{cases} 1 & \text{if } \tau = 0 \\ \alpha \rho(\tau) & \text{if } \tau \neq 0, \end{cases} \tag{2.50}$$

where

$$\alpha = \frac{\sigma_S^2}{\sigma_S^2 + \sigma_N^2} \tag{2.51}$$

is the signal ratio (Parzen, 1966; see also Section 4.8). $\rho_H(\tau)$ have lower absolute values than $\rho(\tau)$ because $\alpha \leq 1$. The distinction between $\rho(\tau)$ and $\rho_H(\tau)$ becomes more noticeable with increasing σ_N^2. Of course, if there is no additional white noise, $\rho_H(\tau)$ coincides with $\rho(\tau)$.

The estimation of the autocorrelation function $\rho(\tau)$ of the means $\bar{Y}(t)$ leads to the estimation of the function ($\tau > 0$)

$$\rho_{cp}(\tau) = \frac{\sigma_{\bar{S}}^2 \rho(\tau)}{\sigma_{\bar{S}}^2 + \sigma_{\bar{N}}^2} = \frac{\frac{1}{n+1}\sigma_S^2 \rho(\tau)}{\frac{\sigma_S^2}{n+1} + \frac{\sigma_N^2}{k+1}} = \bar{\alpha}\rho(\tau), \qquad (2.52)$$

where

$$\bar{\alpha} = \frac{\sigma_{\bar{S}}^2}{\sigma_{\bar{S}}^2 + \sigma_{\bar{N}}^2} = \frac{\sigma_S^2}{\sigma_S^2 + \frac{n+1}{k+1}\sigma_N^2}. \qquad (2.53)$$

The autocorrelation function $\rho_{cp}(\tau)$ of $\bar{Y}(t)$ is closer to $\rho(\tau)$ than $\rho_H(\tau)$. For positive correlations, we have

$$\rho_{cp}(\tau) > \rho_H(\tau) \qquad (2.54)$$

because

$$\bar{\alpha} > \alpha. \qquad (2.55)$$

In any case, the result reveals that what is really estimated for the time series of the signal-plus-white-noise type is the autocorrelation function of the signal multiplied by α (or $\bar{\alpha}$). The reason for the relatively large autocorrelations of the means $\bar{Y}(t)$ is the spatial-temporal dependence in the signal. If $S_i(t)$ were spatially independent ($\rho_{ij} = 0$), averaging would not improve the estimate of $\rho(\tau)$, because in this case $n = k$ and $\alpha = \bar{\alpha}$.

The higher the spatial statistical dependence of the values $S_i(t)$, the smaller is $n+1$, the closer $\bar{\alpha}$ is to 1, and the closer the estimate $\rho_{cp}(\tau)$ is to $\rho(\tau)$. Non-zero spatial-temporal correlations (2.45) of the original time series are grounds for improving the estimates of the autocorrelation function of a signal (with the aid of the time series of means). Notice that a decrease of the value of $\sigma_{\bar{N}}^2$ with an increase of $k+1$ occurs faster than a decrease in the value of $\sigma_{\bar{S}}^2$.

The principal result of this section, equation (2.53), can be ex-

2.4 Spatial Averaging

pressed as follows:

$$\bar{\alpha} = \frac{\sigma_S^2}{\sigma_S^2 + \frac{n+1}{k+1}\sigma_N^2} = \frac{1}{1 + \frac{n+1}{k+1}\frac{\sigma_N^2}{\sigma_S^2}} = \frac{1}{1 + \frac{n+1}{k+1}\frac{1-\alpha}{\alpha}}, \quad (2.56)$$

where n is the number of statistically independent time series, equivalent to the given system of time series $y_i(t)$ in the sense that the variances of their spatial means are equal. The ratio $(n+1)/(k+1)$ characterizes the spatial statistical dependence of the $y_i(t)$. The closer $(n+1)/(k+1)$ is to zero, the higher this correlation is. If $(n+1)/(k+1) = 1$, the $y_i(t)$ are statistically independent in space, and their averaging does not lead to an increase in the signal ratio of the spatial means. If $(n+1)/(k+1) < 1$, then $\bar{\alpha} > \alpha$ and the signal ratio of means is larger than that of the $y_i(t)$. Thus, spatial averaging of time series with the spatial-temporal statistical structure (2.44)–(2.45) gives new time series with the same signal but with a smaller noise component.

Figure 2.9 illustrates the dependence (2.56) of $\bar{\alpha}$ on the parameter $(n+1)/(k+1)$ and the signal ratio α. The graph shows the signal ratio associated with the spatial averaging. Even with very low α (about 0.1), the series of means enables us to reliably identify a signal against a background noise with the appropriate values of ratio $(n+1)/(k+1)$, that is, with the fairly high spatial-temporal statistical dependence of the observations.

Let us consider another example of the spatial-temporal statistical dependence of a signal. Let

$$M_{ij}(\tau) = \begin{cases} \sigma_S^2 + \sigma_N^2 & \text{if } i = j, \ \tau = 0, \\ \sigma_S^2 \rho(\tau) & \text{if } i = j, \ \tau \neq 0, \\ \sigma_S^2 \rho_{ij} & \text{if } i \neq j, \ \tau = 0, \\ 0 & \text{if } i \neq j, \ \tau \neq 0, \end{cases} \quad (2.57)$$

that is, spatial-temporal covariances of the signal are equal to zero. Then from (2.42) we have

$$\rho_{cp}(\tau) = \frac{\frac{1}{k+1}\sigma_S^2 \rho(\tau)}{\frac{1}{n+1}\sigma_S^2 + \frac{1}{k+1}\sigma_N^2} = \frac{\sigma_S^2 \rho(\tau)}{\frac{k+1}{n+1}\sigma_S^2 + \sigma_N^2}. \quad (2.58)$$

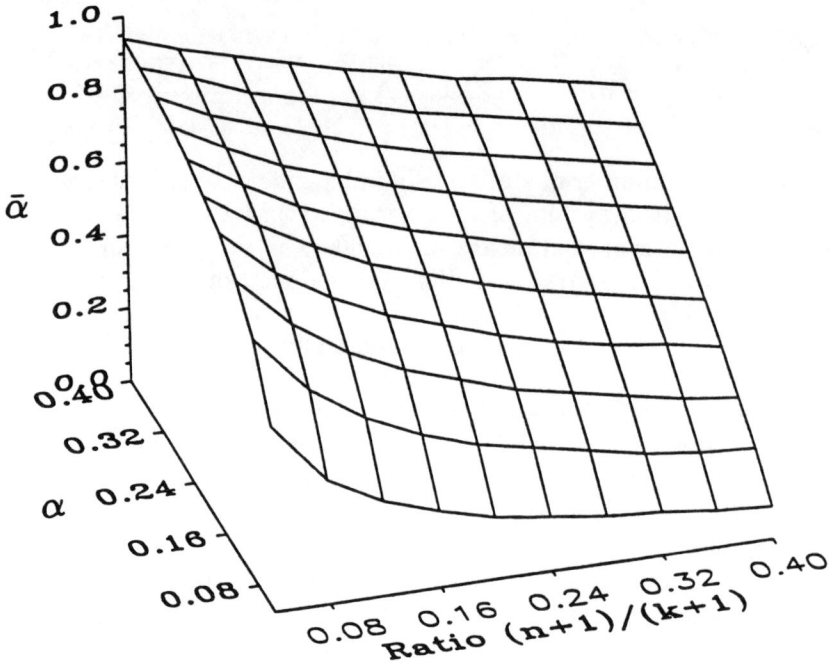

Figure 2.9: Dependence of signal ratio $\bar{\alpha}$ of the spatial means on the signal ratio α of the original time series and the ratio $(n+1)/(k+1)$.

In this case, the signal ratio

$$\hat{\alpha} = \frac{\sigma_S^2}{\frac{k+1}{n+1}\sigma_S^2 + \sigma_N^2} < \alpha < \bar{\alpha} \qquad (2.59)$$

of averages is smaller than the corresponding ratio of the original time series.

These results emphasize that the non-zero spatial-temporal correlations are the basis for detecting a signal by spatial averaging of the signal-plus-white-noise-type time series.

2.5 Smoothing of Correlated Observations

The principal requirements for design and development of a digital filter are conditioned by the geometry of the observed function as well as by the statistical structure of the observations. Obviously, the application of the filters considered in Chapter 1 for smoothing the correlated observations is not theoretically correct because those filters were obtained under the assumption of independence and equal accuracy. For correlated observations, the considered schemes do not supply the minimum of the point-estimator variance. If observations are statistically dependent, the change in the correlation structure of the time series, as a result of filtration, has a more complicated character. In this case, the covariance matrix of smoothing value vector $\bar{\mathbf{Y}}$ (1.69) is $\mathbf{FMF^T}$, where $\mathbf{M} = \{M_{ij}\}_{i,j=0}^{N}$ is the covariance matrix of observations $\mathbf{Y}$. But for certain types of statistical structures of observations (and when the width of the filter is large), there are grounds to suppose that the variances of the point estimates, obtained by applying the Chapter 1 filtering schemes, differ insufficiently from the minimal possible values [as it was for the estimates of the mean (see Section 2.3)]. This fact justified the practical application of the digital filters considered in Chapter 1 in situations when observations are not independent or do not have equal accuracy. The losses of accuracy as a result of the application of the considered filters for smoothing correlated observations will depend on the type of covariance matrix $\mathbf{M}$. The simplicity of the filters developed in Chapter 1 and the lack of information about matrix $\mathbf{M}$ do not ordinarily allow an option for the experimenter. But the reasoning about the estimator accuracy may not make any sense when the correlated data are smoothed by the digital filter derived for the independent observations.

To construct a minimal variance filter, it is necessary to know the covariances of the observations. For example, if the covariances of the observations are $k_\tau = \sigma^2 e^{-\alpha|\tau|}$, then instead of averaging with equal weights, as was done in the case of regressive filters with $m = 1$, (2.20) yields

$$\bar{Y}_t = \frac{\left[Y_{t-r} + Y_{t+r} + (1-\gamma) \sum_{j=-(r-1)}^{r-1} Y_{t+j} \right]}{1 + \gamma + k(1-\gamma)}, \qquad (2.60)$$

where $\gamma = e^{-\alpha}$ is the correlation coefficient. To construct the regressive-type filter when the statistical structure of the data is known, one

88   2 Averaging and Simple Models

must solve the system of corresponding normal equations obtained by using the covariance matrix of the observations.

The statistical structure of the point estimates, obtained by the moving average procedure, is determined by the covariances of the original data. The formal approach to developing a procedure for evaluation of the statistical structure of the point estimates for the observations with stationary correlations follows.

Consider observations

$$Y_0, Y_1, \ldots, Y_N \qquad (2.61)$$

of the stationary random processes $Y(t)$ with covariance function $M(\tau)$.

The covariances $\mathbf{R}_\tau$ of the point estimates

$$\bar{Y}_t = \sum_{j=-r}^{r} \alpha_j Y_{t+j} \quad (t = r, r+1, \ldots, N-r), \qquad (2.62)$$

(where α_j are the weights of the numerical filter, $\alpha_j = \alpha_{-j}$) are

$$R_\tau = E(\bar{Y}_t \bar{Y}_{t+\tau}) = \sum_{v=-r}^{r} \sum_{j=-r}^{r} \alpha_j \alpha_v E(Y_{t+j} Y_{t+\tau+v}). \qquad (2.63)$$

The $\bar{Y}_t$ is a stationary random process with covariance function $R(\tau)$, which is determined for points $\tau = 0, 1, \ldots, N - 2r$ as

$$R_\tau = \sum_{v=-r}^{r} \sum_{j=-r}^{r} \alpha_j \alpha_v M_{j-v-\tau}. \qquad (2.64)$$

Correlation function K_τ of the estimates (2.62) is

$$K_\tau = \frac{R_\tau}{R_0} = \frac{\sum_{v=-r}^{r} \sum_{j=-r}^{r} \alpha_j \alpha_v M_{j-v-\tau}}{\sum_{v=-r}^{r} \sum_{j=-r}^{r} \alpha_j \alpha_v M_{j-v}}. \qquad (2.65)$$

When smoothing is simple averaging

$$\bar{Y}_t = \frac{1}{2r+1} \sum_{j=-r}^{r} Y_{t+j}, \qquad (2.66)$$

we have

2.5 Smoothing of Correlated Observations

$$R_\tau = \frac{1}{(2r+1)^2} \sum_{v=-r}^{r} \sum_{j=-r}^{r} M_{j-v-\tau}$$

$$= \frac{1}{2r+1} \left[M_\tau + \sum_{i=1}^{2r} \left(1 - \frac{i}{2r+1}\right)(M_{\tau+i} + M_{\tau-i}) \right], \quad (2.67)$$

and

$$K_\tau = \frac{R_\tau}{R_0} = \frac{M_\tau + \sum_{i=1}^{2r}\left(1 - \frac{i}{2r+1}\right)(M_{\tau+i} + M_{\tau-i})}{M_0 + 2\sum_{i=0}^{2r}\left(1 - \frac{i}{2r+1}\right) M_i}. \quad (2.68)$$

The expression

$$R_0 = \frac{1}{2r+1}\left[M_0 + 2\sum_{i=0}^{2r}\left(1 - \frac{i}{2r+1}\right) M_i \right] \quad (2.69)$$

determines the variance of the point estimates.

If estimates (2.66) are computed not for all the points but only those with step equal to $2r$, then the autocorrelation function of the new time series (consisting of the consecutive means of the observations on the nonoverlapping subintervals with $2r+1$ terms) is

$$\rho_v = \frac{M_{(2r+1)v} + \sum_{i=1}^{2r}\left(1 - \frac{i}{2r+1}\right)\left[M_{(2r+1)v+i} + M_{(2r+1)v-i}\right]}{M_0 + 2\sum_{i=1}^{2r}\left(1 - \frac{i}{2r+1}\right) M_i}$$
(2.70)

For example, the autocorrelation function of the monthly (or annual) mean time series (as linear transformation) of the meteorological observations for the individual stations is determined by the autocorrelation function of daily time series. By assuming that the latter are stationary, we can apply the above methodology to calculate the correlation function of the series of the means (annual or monthly).

Let us assume that M_τ is the autocorrelation function of the time series of daily temperature anomalies measured by a point gauge. Averaging these anomalies over the consecutive nonoverlapping subintervals with, for example, $2r+1=31$ points on each, one obtains the monthly mean time series (the unit of v is one month). If $2r+1 = 365$, the result will be the time series of annual means (the unit of v is one year). It is known (Lepekhina and Fedorchenko, 1972) that the approximate autocorrelation function of daily temperature anomalies (for middle latitude stations' data) is $\mathbf{M}_\tau = \rho^\tau$; where $\rho \approx 0.8$ for $\tau = 1$ day. By substituting this value in (2.70) for $\mathbf{M}_\tau$ and assuming that $2r + 1 = 31$ (or $2r + 1 = 365$), we obtain the autocorrelation function of the monthly (or annual) mean time series of temperature anomalies.

To perform a general study of the effect of the correlation of the daily data on the correlation structure of the monthly and annual mean time series, a broader analysis can be carried out by varying ρ from, for example, 0.35 to 0.95 with the temporal step of 0.05. Substituting $\mathbf{M}_\tau = \rho^\tau, [\rho = \rho(1) = 0.95]$ for $\mathbf{M}_\tau$ in (2.70), we obtain (for $v = 1, 2,$ and 3) the first three autocorrelations of the monthly means: 0.40, 0.08, and 0.02; as well as the first three autocorrelations of the annual means: 0.03, 0.00, 0.00. In these examples, the first autocorrelation of the time series of the means significantly exceeds any other autocorrelation. If $\rho < 0.95$, this result becomes even more evident in the sense that the values of all other autocorrelations (for lags equal two and greater) have three or more zeroes after the decimal point. Therefore, it makes sense to consider only the dependence of the first autocorrelation of the time series of the means on the value of $\rho(1)$ of the original observations. This dependence is presented in Figure 2.10.

Assuming, for example, that $\rho(1) = 0.8$ for daily data yields the value of $\rho(1) \approx 0.07$ for the monthly data and $\rho(1) \approx 0.006$ for the annual data. Therefore, statistical dependence of the monthly mean temperature time series, conditioned by the synoptic fluctuations, is significantly smaller than the autocorrelation actually observed. (Such autocorrelation of the historical monthly mean temperature records is equal to approximately 0.1 to 0.3 (see Section 6.4). Analogously, the first autocorrelation (estimated in this way) for the annual means is 15 to 20 times smaller than the autocorrelations actually observed for the historical records.

The possible cause of the discrepancies obtained is that autocorrelations in the historical records were conditioned by not only a synoptic, but also a low-frequency fluctuation. If these fluctuations are not the result of measurement errors, then it is a natural temperature trend, which can be interpreted as a confirmation of the nonstationary character of climate change.

2.6 Filters of Finite Differences

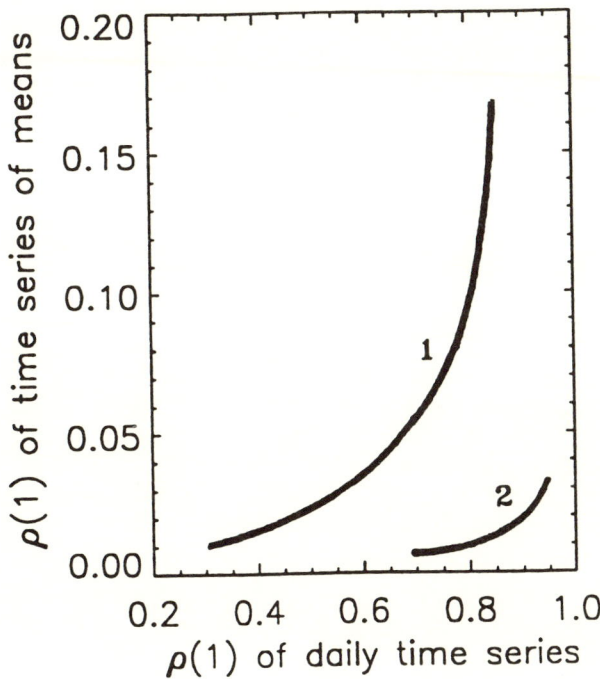

Figure 2.10: Dependence of the first autocorrelations of monthly (1) and annual (2) mean time series on the autocorrelation $\rho(1)$ of the time series of daily fluctuations.

2.6 Filters of Finite Differences

The filters considered above enable us to smooth data or to remove high-frequency components of the fluctuations. But in many problems one needs to analyze the high-frequency components and filter out the long period variations (the low-frequency part or the fluctuations). Such removal can be done with the aid of a wide range of linear filters, the simplest class of which, the finite differences filters, is considered in this section. Such filters are used, for example, in the theory of random processes with stationary increments and in its application to fitting the nonstationary stochastic models (Box and Jenkins, 1976) to the time series with trend.

The procedure for computing finite differences, as for any linear transformation, changes the statistical and harmonic structure of data. For different types of the correlation structure of time series, such changes will be different. Here we will study such transformations for the simplest types of autocorrelations of observations.

The order v finite differences $\nabla^v y_t$ of the time series $y_0, y_1, \ldots, y_k$ are determined in the following way:

$$\nabla y_t = y_t - y_{t-1}, \tag{2.71}$$

$$\nabla^2 y_t = \nabla y_t - \nabla y_{t-1} = y_t - 2y_{t-1} + y_{t-2}. \tag{2.72}$$

$$\nabla^v y_t = \nabla^{v-1} y_t - \nabla^{v-1} y_{t-1} = y_t - \binom{v}{1} y_{t-1} + \binom{v}{2} y_{t-2} - \ldots$$
$$+ (-1)^v y_{t-v}. \tag{2.73}$$

Let us compare the harmonic structure of the time series y_t and $\nabla^v y_t$. If $e^{i\omega t}$ is any of the harmonics of the Fourier transform of the time series y_t, then the frequency characteristic $I^2(\omega)$ of the transformation of (2.73) is

$$I^2(\omega) = 2^{2v} \sin^{2v}\left(\frac{\omega}{2}\right). \tag{2.74}$$

Figure 2.11 illustrates this dependence. It shows that difference filters suppress low-frequency power and that, the greater the finite differences order v, the wider the low-frequency interval of such suppression.

The correlation structure of the time series of finite differences $\nabla^v y_t$ depends on the statistical and harmonic structure of the original time series y_t.

If y_t is a stationary time series with an autocovariance function $M(\tau)$, then the time series of the first differences ∇y_t is also stationary with the autocovariance function,

$$M_1(\tau) = \mathbf{E}(\nabla y_t \nabla y_{t+\tau}) = 2M(\tau) - M(\tau+1) - M(\tau-1). \tag{2.75}$$

From this, we get the expression for the variance

$$\sigma_1^2 = 2[M(0) - M(1)] \tag{2.76}$$

and for the autocorrelation function

$$K_1(\tau) = \frac{M(\tau) - 0.5[M(\tau+1) + M(\tau-1)]}{M(0) - M(1)} \tag{2.77}$$

2.6 Filters of Finite Differences

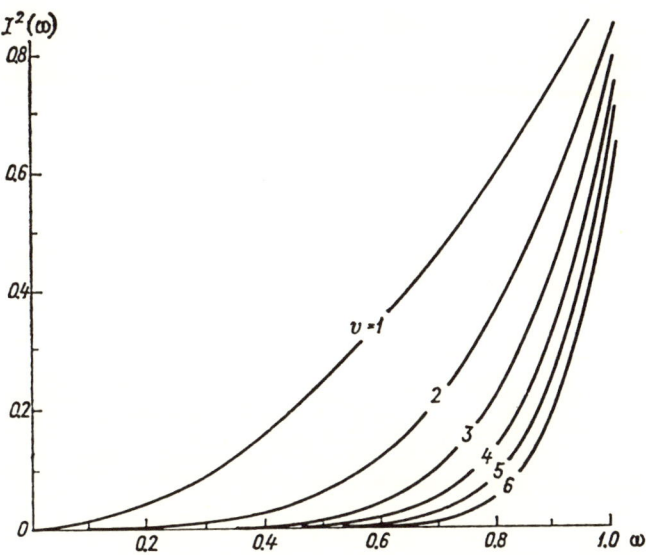

Figure 2.11: Frequency characteristics of the different order v finite differences.

of the time series of ∇y_t. By using analogous reasoning for the differences $\nabla^v y_t$ of order v, we get the recursion relations

$$M_v(\tau) = 2M_{v-1}(\tau) - M_{v-1}(\tau+1) - M_{v-1}(\tau-1), \qquad (2.78)$$

$$\sigma_v^2 = 2\left[\sigma_{v-1}^2 - M_{v-1}(1)\right], \qquad (2.79)$$

$$K_v(\tau) = \frac{M_{v-1}(\tau) - 0.5[M_{v-1}(\tau+1) + M_{v-1}(\tau-1)]}{M_{v-1}(0) - M_{v-1}(1)}. \qquad (2.80)$$

These formulas make it possible to study the correlation structure of the finite differences time series.

Example 1

Let y_t be the sequence of independent random variables with

$$M(\tau) = \begin{cases} 1 & \text{if } \tau = 0, \\ 0 & \text{if } \tau \neq 0. \end{cases} \qquad (2.81)$$

Table 2.3: Variances and autocorrelation functions $K_v(\tau)$ of the white noise process finite differences.

Time series	σ_v^2	$K_v(\tau)$ for τ			
		1	2	3	4
y_t	1	0	0	0	0
∇y_t	2	-1/2	0	0	0
$\nabla^2 y_t$	6	-2/3	1/6	0	0
$\nabla^3 y_t$	20	-3/4	3/10	-1/20	0
$\nabla^4 y_t$	70	-4/5	2/5	-4/35	1/70

Using (2.78)–(2.80) in sequence, we find the variances σ_v^2 as well as the autocorrelation functions $K_v(\tau)$ of the time series of finite differences (for instance, up to the fourth order). Functions $K_v(\tau)$ are given in Figure 2.12 and in Table 2.3. The results show that application of the finite difference filters to the white-noise time series forms a random process with a negative autocorrelation for the first lag as the autocorrelations approach zero, subsequently changing their sign.

The variance of the finite difference time series is increased (with increasing v) in accordance with the formula

$$\sigma_v^2 = \sum_{j=0}^{v} \binom{v}{j}^2 = \binom{2v}{v}, \tag{2.82}$$

which can be obtained by taking variances of the right and left parts of (2.73).

Example 2

Let y_t be the normalized Markov random process with autocorrelation function

$$M(\tau) = \rho^{|\tau|}, \quad |\rho| \leq 1.$$

2.6 Filters of Finite Differences

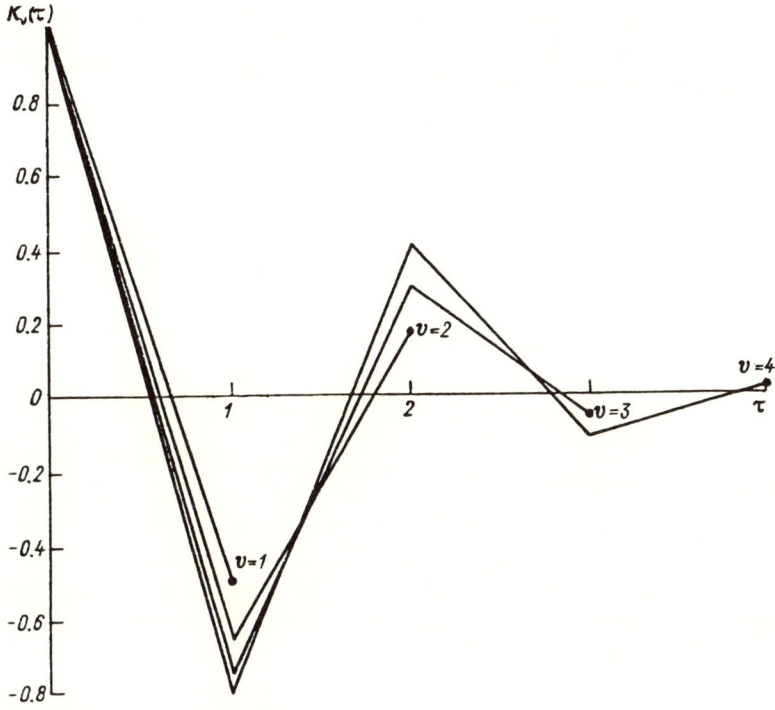

Figure 2.12: Autocorrelation functions of the different order v finite differences of the white noise.

In the case of the first differences, we get [from (2.75) to (2.77)]

$$M_1(\tau) = -\rho^{\tau-1}(1-\rho)^2 \quad (\tau > 0),$$

$$\sigma_1^2 = 2(1-\rho),$$

$$K_1(\tau) = 0.5\rho^{\tau-1}(\rho - 1) \quad (\tau > 0). \tag{2.83}$$

If $v = 2$, we have

$$\begin{aligned}
M_2(\tau) &= \rho^{\tau-2}(1-\rho)^4 \quad (\tau > 0), \\
\sigma_2^2 &= 2(1-\rho)(3-\rho), \\
K_2(\tau) &= 0.5\frac{(1-\rho)^3}{3-\rho}\rho^{\tau-2}.
\end{aligned} \tag{2.84}$$

Table 2.4: Variances and autocorrelation functions $K_1(\tau)$ of the first finite differences of the Markov process.

		\multicolumn{4}{c}{$K_1(\tau)$ for τ}			
ρ	σ_1^2	1	2	3	4
0.25	1.5	-0.37	-0.01	0.00	0.00
0.50	1.0	-0.25	-0.12	-0.06	-0.03
0.75	0.50	-0.12	-0.09	-0.07	-0.05

Given different values of ρ ($\rho = 0.75, 0.50, 0.25$) and using formulas (2.83) and (2.84), one can calculate the autocorrelation functions of the first and second differences of the Markov process. The results of the calculations are presented in Tables 2.4 and 2.5 and in Figure 2.13.

Figure 2.13 illustrates the dependence of the statistical structure of the first-order finite differences of the Markov process on the ρ value.

Let us consider a more general case for a time series, which is the sum of the deterministic and random components:

$$y_t = P(t) + \delta_t, \qquad (2.85)$$

where $P(t)$ is the polynomial of degree v, and where δ_t is the stationary random process. The finite differences are presented by the formula

$$\nabla^v y_t = \nabla^v P(t) + \nabla^v \delta_t. \qquad (2.86)$$

Assuming that $\mathbf{E}(\delta_t) = 0$, we have

$$\mathbf{E}(\nabla^v y_t) = \nabla^v P(t) = \beta_v, \qquad (2.87)$$

where β_v is the coefficient of t^v of the polynomial $P(t)$.

Each term of the finite difference time series (2.86) consists of two summands: the constant β_v (mean value) and the random $\nabla^v \delta_t$ component. In other words, the procedure for computing the finite differences filters out the polynomial trend.

2.6 Filters of Finite Differences

Table 2.5: Variances and autocorrelation functions $K_2(\tau)$ of the second finite differences of the Markov process.

		$K_1(\tau)$ for τ			
ρ	σ_2^2	1	2	3	4
0.25	4.125	0.31	0.08	0.02	0.01
0.50	2.500	0.05	0.02	0.01	0.00
0.75	1.125	0.005	0.003	0.002	0.001

Stochastic modeling of the finite differences of the observations (as well as its spectral and correlation analysis) brings into focus some questions. What properties of the results are imposed by the procedure of finding the finite differences, and what properties correspond to the nature of the original data? As a rule, it is not known beforehand whether or not the time series of the finite differences is the sample of a stationary process. The formal procedure of filtering out a trend by finding the finite differences of the adjacent terms could be nonoptimal for the consequent fitting of a stochastic model.

It should be noted that in some problems the estimate of the mean of the finite difference time series $\nabla^v y_t$ could be important. For example, accepting

$$\bar{\mu} = \frac{1}{k+1-v} \sum_{t=v}^{k} \nabla^v y_t \qquad (2.88)$$

as the estimate of a mean value can lead to very large errors. For instance, when $v = 1$

$$\bar{\mu} = \frac{1}{k} \sum_{t=1}^{k} \nabla y_t = \frac{1}{k}(y_k - y_0). \qquad (2.89)$$

In other words, the estimate of the mean of the first differences depends only on the first and last terms, and ignores all other observations.

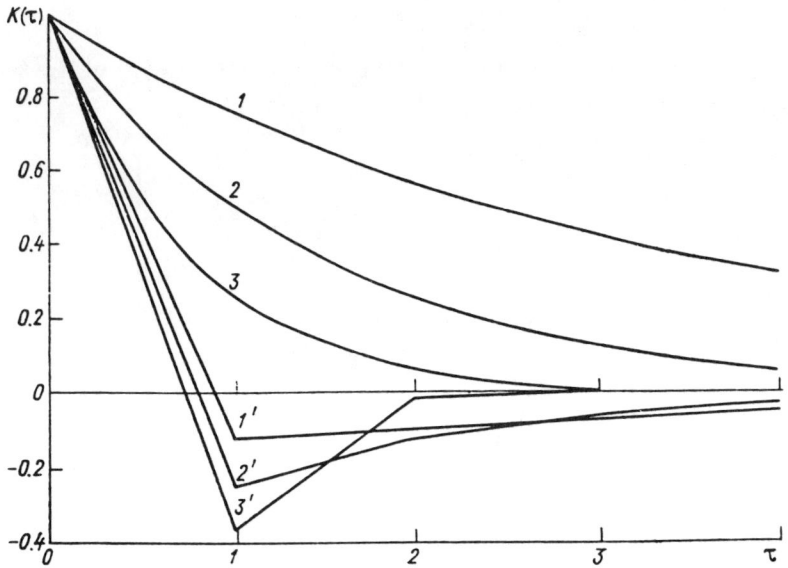

Figure 2.13: Autocorrelation functions (1–3) of the time series and the corresponding autocorrelation functions $K_1(\tau)$ ($1'-3'$) of their first finite differences.

Curve...	1	2	3
$M(\tau)$...	0.75^τ	0.5^τ	0.25^τ

Generally, the mean of $\nabla^v y_t$ determines the asymptotic behavior of the nonstationary model forecast (see Section 4.9); consequently, the problem of choosing its appropriate estimate cannot be underestimated. If $v = 1$, the slope of the straight line determines the asymptote to which the forecast approaches when there is an increase in the lead time or when there is a decrease in the autocorrelations.

The optimal scheme of estimating β_v by the least squares method depends on the unknown autocorrelation structure of random values δ_t. Therefore, taking $\widehat{\beta}_v$ as an estimate of the mean of the corresponding finite differences does not reduce the difficulties. As a preliminary estimate of such mean, it is possible to take the estimate of β_v, when the assumption about independence of δ_t is true. The optimal estimation of the mean can be accomplished by the least squares method along with other parameters of the model.

2.7 Regression and Instrumental Variable

The linear regression and the instrumental variable methods (Kendall and Stuart, 1963) are the next step (after averaging and the linear trend) toward more complicated statistical modeling of data. In this section we will analyze the normalized standard deviations of the point estimates of two simple examples of these methods; this analysis is preliminary to the autoregressive-moving average models studied in Chapter 4. Autoregressive and moving average schemes (see Chapter 4) are the results of application of linear regression and instrumental variable methods to the analysis of time series.

Our purpose here is to numerically interpret and compare the corresponding accuracy characteristics of the point estimates of these methods.

2.7.1 Linear Regression

Let us begin with linear regression, considering three random variables $x, y,$ and z with the unit variances and correlation matrix

$$\mathbf{M} = \begin{pmatrix} 1 & \rho_{xy} & \rho_{xz} \\ \rho_{xy} & 1 & \rho_{yz} \\ \rho_{xz} & \rho_{yz} & 1 \end{pmatrix}. \tag{2.90}$$

The linear regression equation y on x and z is

$$y = ax + bz + e, \tag{2.91}$$

where a and b are parameters that must be estimated and e is the random error, independent of x and z. The least squares method provides the estimators for a and b:

$$a = \frac{\rho_{xy} - \rho_{yz}\rho_{xz}}{1 - \rho_{xz}^2} \tag{2.92}$$

$$b = \frac{\rho_{yz} - \rho_{xz}\rho_{xy}}{1 - \rho_{xz}^2}. \tag{2.93}$$

The error variance ε_e^2 is

$$\varepsilon_e^2 = 1 - a\rho_{xy} - b\rho_{yz} = 1 - \frac{\rho_{xy}^2 + \rho_{yz}^2 - 2\rho_{xy}\rho_{xz}\rho_{yz}}{1 - \rho_{xy}^2}. \tag{2.94}$$

The domain of the permissible correlations of the linear regression model is determined by the equality

$$0 \leq 1 - \frac{\rho_{xy}^2 + \rho_{yz}^2 - 2\rho_{xy}\rho_{xz}\rho_{yz}}{1 - \rho_{xy}^2} \leq 1. \qquad (2.95)$$

By fixing the value of one of these three correlations ρ_{xy}, ρ_{xz}, or ρ_{yz} (for example ρ_{xz}), it is possible to compute the field of standard deviations ε_e by (2.94) and to outline the domains (2.95) on the plane of two other correlations. If ρ_{xz} is fixed (for example, $\rho_{xz} = 0.7$), these domains have an elliptical form, which is presented in Figure 2.14.

Isolines of ε_e in this figure characterize the accuracy of the linear regression model in different points of the domain of permissible correlations. More general results, which demonstrate the dependence of ε_e on all three correlations, are presented by the sequence of illustrations in Figure 2.15a. As in the case of the polynomial approximations, the standard deviations ε_e can be considered for evaluating the linear regression accuracy when one has numerical values of correlations ρ_{xy}, ρ_{xz}, and ρ_{yz}.

If absolute values of the correlations are small (for example, less than 0.3), Figure 2.15a shows that the linear regression model gives estimates of y that are close to the expected value (because the value of ε_e is close to one). Setting $\rho_{xy} = \rho_{yz} = \rho$ in (2.94), we have

$$\varepsilon_e^2 = 1 - \frac{2\rho^2}{1 + \rho_{xz}}. \qquad (2.96)$$

Isolines of this standard deviation ε_e (see Figure 2.16) illustrate that it is possible to obtain estimates with $\varepsilon_e < 0.8$, when ρ is sufficiently large ($|\rho| > 0.4$). For small ρ, the dependence of ε_e on the correlations of x and z is very low.

2.7.2 Instrumental Variable Method

In the case of the instrumental variable method (see Kendall and Stuart, 1963) for the variables x, y, and z with the same correlation matrix (2.90), the following model must be built:

$$y = ax + b + e, \quad x = m + \delta, \qquad (2.97)$$

where m is the expected value of random variable x; e and δ are independent random errors; and a and b are the parameters, which must

2.7 Regression and Instrumental Variable

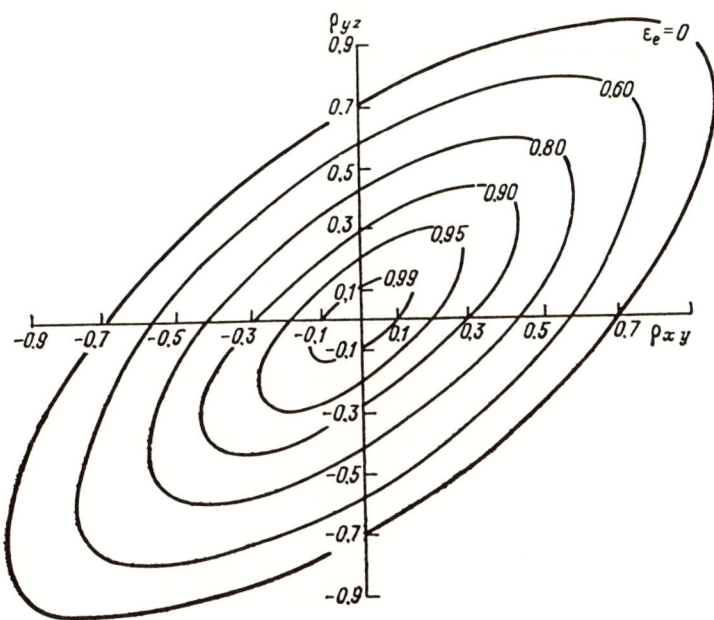

Figure 2.14: Domain of the permissible correlations and isolines of the normalized standard deviations ε_e of the two-variate linear regression model ($\rho_{xz} = 0.7$).

be estimated. The third variable z is considered an instrumental variable.

The estimator of parameter a is given by the formula

$$a = \rho_{yz}/\rho_{xz}. \qquad (2.98)$$

Assuming that $b = \beta e_1$, where β is a unknown parameter, and e_1 is independent of e random variable, one obtains $b + e = \beta e_1 + e$; that is, y is presented as a sum of ax and of a linear combination of two independent random variables:

$$y = ax + \beta e_1 + e. \qquad (2.99)$$

(Notice that this scheme, applied to the time series modeling, produces the first-order autoregressive-moving average model ARMA(1,1), see Chapter 4).

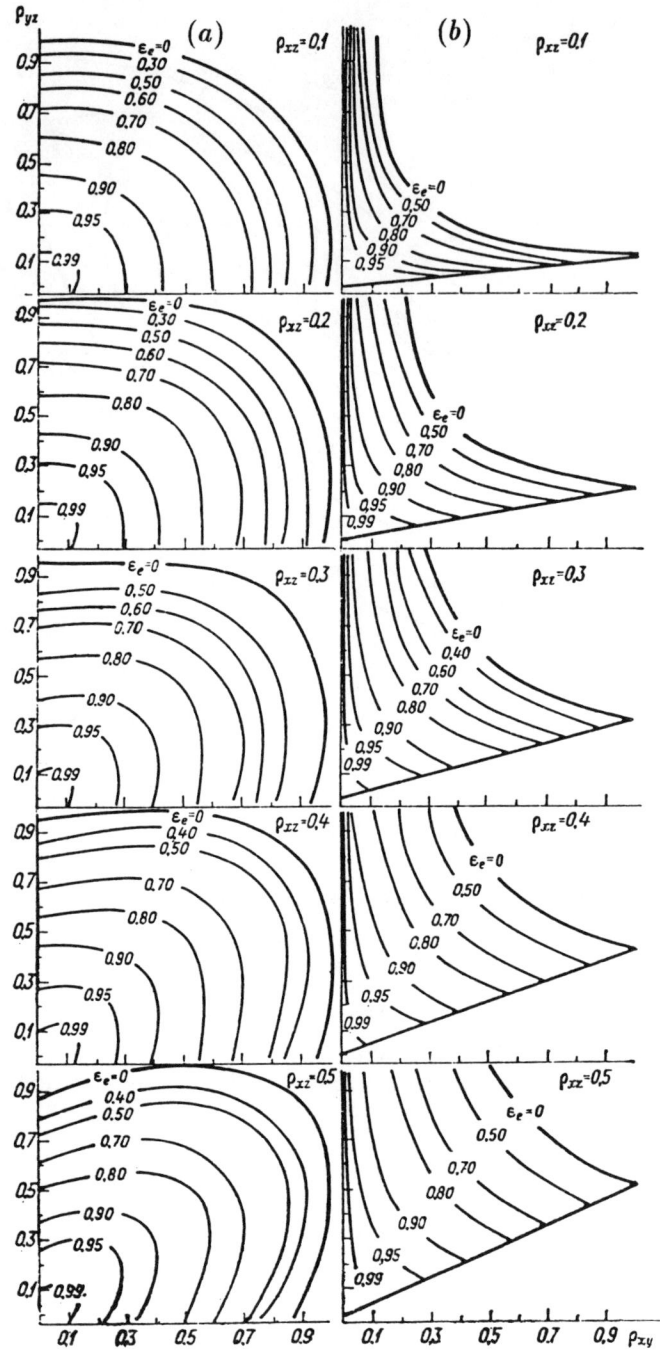

2.7 Regression and Instrumental Variable

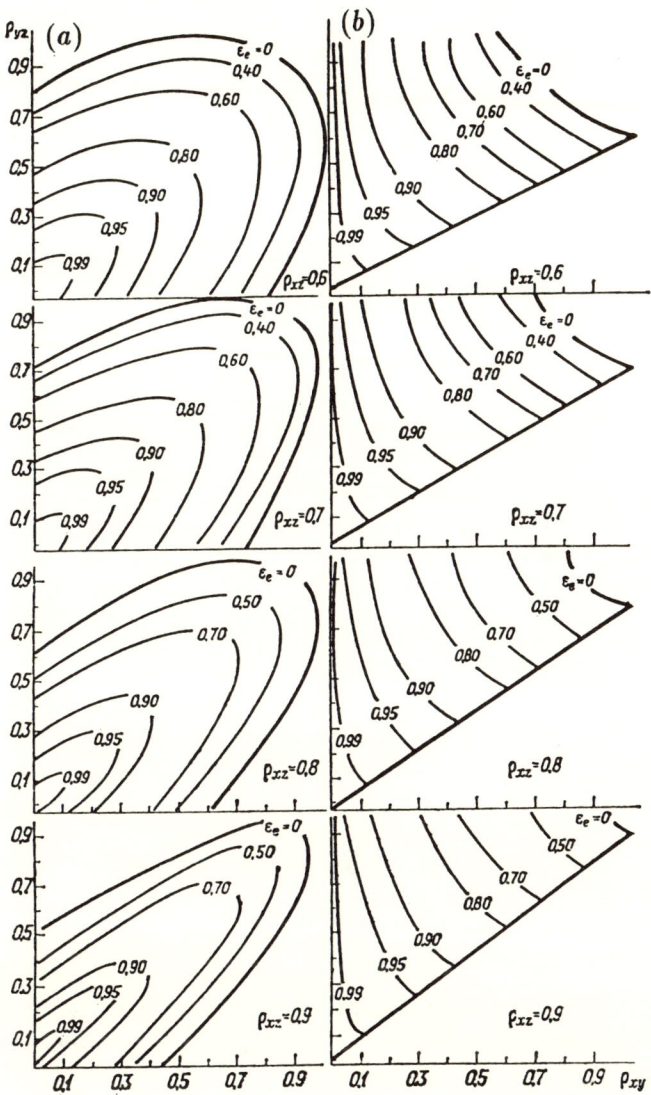

Figure 2.15: Domains of the permissible correlations and isolines of the normalized standard deviations ε_e of the two-variate linear regression (a) and of the method of instrumental variable (b) for different values of correlations ρ_{xy}, ρ_{xz}, ρ_{yz}.

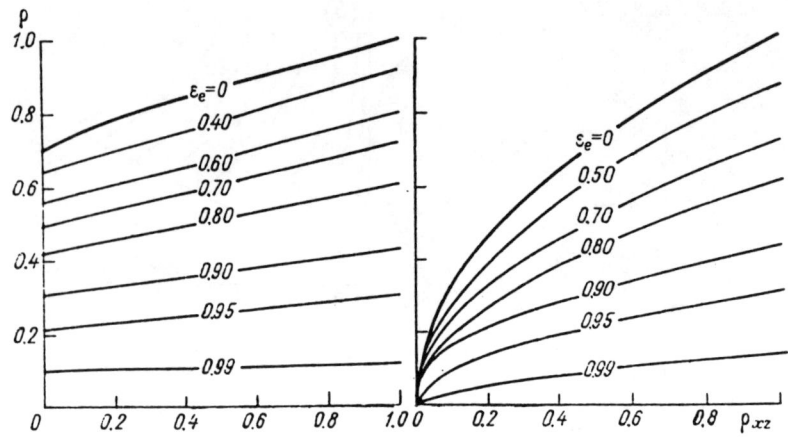

Figure 2.16: Domains of the permissible autocorrelations and isolines of the normalized standard deviation ε_e of the two-variate linear regression (left) and of the instrumental variable method (right) if $\rho_{xy} = \rho_{yz} = \rho$.

According to the instrumental variable theory (Kendall and Stuart, 1963), the following formulas for the variances of random variables e and δ are correct:

$$\varepsilon_e^2 = 1 - \frac{\rho_{yz}\rho_{xy}}{\rho_{xz}}, \quad \varepsilon_\delta^2 = 1 - \frac{\rho_{xz}\rho_{xy}}{\rho_{yz}}. \qquad (2.100)$$

The domain of the permissible correlations of the model (2.97) is determined by the inequalities

$$0 \leq \varepsilon_e^2 \leq 1, \quad 0 \leq \varepsilon_\delta^2 \leq 1. \qquad (2.101)$$

The right set of inequalities in (2.101) occurs because, in the inverse case, instead of needing to get estimates (2.97), one must get the mean values of x and y.

So we have

$$0 \leq 1 - \frac{\rho_{xz}\rho_{xy}}{\rho_{yz}} \leq 1, \quad 0 \leq 1 - \frac{\rho_{yz}\rho_{xy}}{\rho_{xz}} \leq 1. \qquad (2.102)$$

It follows that the equivalent system of inequalities is

$$1 - \frac{\rho_{xz}\rho_{xy}}{\rho_{yz}} \geq 0; \quad 1 - \frac{\rho_{yz}\rho_{xy}}{\rho_{xz}} \geq 0; \quad \frac{\rho_{xy}\rho_{xz}}{\rho_{yz}} \geq 0; \quad \frac{\rho_{xy}\rho_{yz}}{\rho_{xz}} \geq 0. \qquad (2.103)$$

2.7 Regression and Instrumental Variable

These inequalities determine the domain of permissible correlations ρ_{xy}, ρ_{xz}, and ρ_{yz} for the model (2.97). By fixing the value of any of these correlations, it is possible to outline the boundary of this domain on the plane of two other correlations. For example, assuming that $\rho_{xz} = 0.7$, inequalities (2.103) give

$$\begin{aligned}
\rho_{yz} - 0.7\rho_{xy} &\geq 0, \\
0.7 - \rho_{yz}\rho_{xy} &\geq 0, \\
\rho_{xy}\rho_{yz} &\geq 0.
\end{aligned} \qquad (2.104)$$

This part of the domain of permissible correlations [together with the isolines of the standard deviations ε_e computed by (2.100)] is illustrated in Figure 2.17. Isolines characterize the accuracy of the point estimates y in the domain of the permissible correlations.

A comparison between Figures 2.14 and 2.17 reveals that the domain of the permissible correlations for the linear regression is significantly larger than the corresponding domain of the model (2.97). Actually, the model (2.97) is determined in the domain, the total size of which does not exceed the unit quadrant of the first quarter of the coordinate system.

In spite of the clear distinction in the form of the domains of the permissible correlations for the two considered models and the character of the variation of corresponding isolines of ε_e, it is possible to make the following remarks: For the coinciding parts of the domains and small values of the correlations (if ρ_{xy} and ρ_{yz} are less than 0.4), values ε_e of both models are approximately the same, that is, in this part of the domain, the accuracies of these approximations are virtually indistinguishable. The instrumental variable model is slightly more accurate on the periphery of its domain of permissible correlations (when ρ_{xy} and ρ_{yz} are greater than 0.7).

Figure 2.15b presents the variations of the sizes of the domain of the permissible correlations for the model (2.97) and isolines ε_e as a function of all three correlations ρ_{xy}, ρ_{yz}, and ρ_{xz}.

A comparison of the accuracies of these two models, presented in Figure 2.15, confirms the inferences obtained by the analysis of Figures 2.14 and 2.17. Moreover, if the values of correlations ρ_{xy}, ρ_{xz}, and ρ_{yz} are known, one can use Figure 2.15 to establish the possibility of building corresponding models and to determine their normalized standard deviations ε_e.

If point $(\rho_{xy}, \rho_{xz}, \rho_{yz})$ is outside the domain of the postulated model, then the corresponding formally computed point estimates will have a negative variance, as happened in an example in Kendall and Stuart (1963).

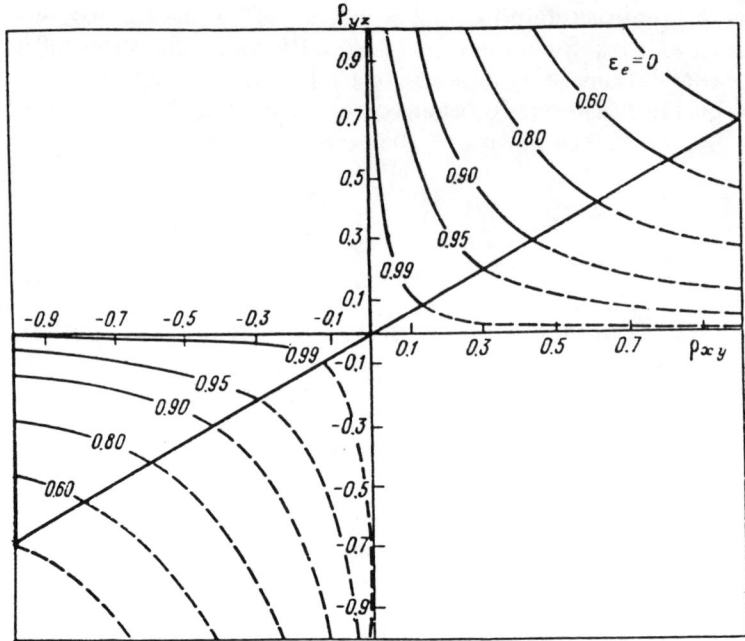

Figure 2.17: Domain of the permissible correlations and isolines of the normalized standard deviations ε_e of the instrumental variable method ($\rho_{xz} = 0.7$).

Let us assume that $\rho_{xy} = \rho_{yz} = \rho$ in (2.94), (2.95), (2.100), and (2.103). Then, the domains of the permissible correlations of both models are the two-dimensional domains presented in Figure 2.16. This figure shows especially clearly the distinction between the accuracies of the methods considered in the sizes of the domain of the permissible correlations as well as in the shapes of isolines ε_e.

As we conclude this comparative analysis, let us notice that linear regression is used more often in applications than the scheme of the instrumental variable because its domain of permissible correlations is greater. The slight advantage in accuracy (on the periphery of the domain of permissible correlations) of the instrumental variable method proves important for specific problems. The solutions of these problems must be accompanied by testing to be sure that the correlations are inside the domain of permissible correlations. Figures 2.14 through 2.17 can be used for this purpose.

Let us also notice the analogy of the above consideration with the analysis of the variances of the point estimates, which was carried out in Chapter 1 for the polynomial approximation, and with the comparison (which will be given in Chapter 4) of the theoretical accuracy of different ARMA models.

In general, theoretical analysis of normalized variances of the point estimates can be done for any linear statistical method prior to actual application. Such powerful methodology helps to get representation about the entire class of corresponding models, to provide an optimum way of obtaining the necessary estimates, or to draw a conclusion without computations.

Moreover, it is possible to state the approximate requirements for the accuracy of the model considered in terms of the normalized variance values. For instance, the model can be considered reasonable only if $\varepsilon_e < 0.9$. Such condition leads to an additional limitation on the correlations (ρ_{xy}, ρ_{xz}, ρ_{yz}) even if they are inside the corresponding domain of their permissible values: Their absolute values must be greater than 0.3–0.4. If the correlations are smaller than these values, building a linear regression equation (as well as the instrumental variable equation) does not make sense.

2.8 Nonlinear Processes

The statistical methods based on the operations with the covariance structure of the random variables makes it possible to evaluate a level of corresponding linear statistical dependence. The relationships of the elements within the climatic system are nonlinear. This can be seen, for instance, from the hydrodynamic equations of the atmospheric circulation (because the latter are nonlinear). The covariance analysis of the nonlinear objects can be inaccurate (or only a very rough approximation), not always reflecting the real character of the dependence between variables.

A systematic presentation of the nonlinear theory of random processes is given, for example, by Raibman (1983). The principal concept of this theory (corresponding to the concept of the covariance function) is the dispersion function, which is determined by the second-order conditional moments. With the aid of the concepts, defined by Raibman (1983), it is possible to carry out more in-depth investigations, as compared with the analysis of the covariance structure.

In the simplest case of the univariate process $z(t)$, the core of the nonlinear theory is the definition of the autodispersion function

$$\theta(t,\tau) = \mathbf{E}\{[\mathbf{E}\{z(t)|z(\tau)\} - \mathbf{E}\{z(t)\}]^2\}, \qquad (2.105)$$

where $\mathbf{E}\{z(t)|z(\tau)\}$ is the conditional mathematical expectation of random process $z(t)$ given $z(\tau)$.

The standardized autodispersion function of $z(t)$ is

$$\eta(t,\tau) = \frac{[\theta(t,\tau)]^{1/2}}{\sigma(t)}, \qquad (2.106)$$

where $\sigma^2(t)$ is the variance of process $z(t)$.

If the values of $z(t)$ and $z(\tau)$ are connected in a deterministic way, $z(t) = f[z(\tau)]$, then the autodispersion function $\theta(t,\tau)$ equals the variance of the random process $z(t)$:

$$\theta(t,\tau) = \sigma_z^2(t), \qquad (2.107)$$

and the standardized autodispersion function $\eta(t,\tau)$ equals one. The values of $z(t)$ and $z(\tau)$ are called nondispersive for some values of t and τ, if $\theta(t,\tau) = 0$. By definition, the autodispersion function is non-negative and satisfies the following inequalities:

$$\theta(t,\tau) \leq \sigma_z^2(t), \qquad (2.108)$$

$$\eta(t,\tau) \leq 1. \qquad (2.109)$$

The standardized autodispersion function is greater than or equal to the absolute value of the corresponding autocorrelation function $\rho(t,\tau)$, that is,

$$\eta(t,\tau) \geq |\rho(t,\tau)|. \qquad (2.110)$$

If $z(t)$ and $z(\tau)$ are connected by the linear dependence, then

$$\eta(t,\tau) = |\rho(t,\tau)|. \qquad (2.111)$$

The indication of the existence of the nonlinear dependence between $z(t)$ and $z(\tau)$ can be based upon the strict inequality in (2.110).

In order to understand the possibility of using the standardized autodispersion function $\eta(t,\tau)$ in practice as a measure of the nonlinear statistical dependence of different values of a random process, the following fact, proved by Raibman (1983), can be important: For the Gaussian processes, the equality

$$\theta(t,\tau) = \rho^2(t,\tau)\sigma^2(t) \qquad (2.112)$$

is true. It follows that

$$\eta(t,\tau) = |\rho(t,\tau)|. \qquad (2.113)$$

2.8 Nonlinear Processes

Therefore, for the Gaussian processes, the measurement and analysis of the nonlinear statistical dependence lead to the estimation of the nonstationary autocorrelation function $\rho(t,\tau)$.

The sample distributions of various climatic time series have been estimated in many papers. In most cases, the normality of these distributions is beyond question. Theoretically such normality is also ensured by the corresponding Central Limit Theorem because annual and monthly means are the results of averaging many separate daily observations.

Talking about stationarity, one must notice that climatic time series do not always satisfy this requirement (for example, it is possible to mention the time series of the meteorological elements with a statistically significant linear trend; see Chapters 6 and 7). Therefore, the problem of considering the nonlinearity does not seem absolutely hopeless, especially if we take into account the constantly increasing volume of the climatic observations. In any case, the methods of the nonlinear statistics, it seems, offer a future direction for studies, especially in the case of analyzing the point gauge observations.

3

Random Processes and Fields

In this chapter, the nonparametric methods of estimating the spectra and correlation functions of stationary processes and homogeneous fields are considered. It is assumed that the principal concepts and definitions of the corresponding theory are known (see Anderson, 1971; Box and Jenkins, 1976; Jenkins and Watts, 1968; Kendall and Stuart, 1967; Loeve, 1960; Parzen, 1966; Yaglom, 1986); therefore, only questions connected with the construction of numerical algorithms are studied.

The basic results ranged from univariate process to multidimensional field are presented in Tables 3.1 and 3.2. These formulas make it possible to compare and trace the formal character of developing estimation procedures when the dimensionality is increasing. The schemes in these tables, as well as the formulas in the previous chapters, can be used for software development without any rearrangement. In part, this approach presents the application of the methods of Chapters 1 and 2 in evaluating random function characteristics. Of course, the final identification of the algorithm parameters (for example, the spectral window widths) can be made only through trial and error and by taking into account the character of the problem under study, that is, the physical properties of the processes and fields observed. The last section of this chapter presents results of the application of these methods to the analysis of some climatological fields.

3.1 Stationary Process

Here the basic results of the univariate spectral analysis are briefly discussed in order to develop algorithms for a multidimensional case by analogous reasoning. The complete description of the estimation procedures of the spectral and correlation analysis for univariate stationary process can be found, for example, in Jenkins and Watts, 1968.

It is known (Yaglom, 1986) that the autocovariance function $M(\tau)$ and the spectral density $S(\omega)$ of stationary random process $Y(t)$ are interconnected by the Fourier transforms

$$S(\omega) = \frac{1}{2\pi} \int_{-\infty}^{\infty} M(\tau) e^{-i\omega\tau} d\tau \qquad (3.1)$$

$$M(\tau) = \int_{-\infty}^{\infty} S(\omega) e^{i\omega\tau} d\omega, \qquad (3.2)$$

where $M(\tau)$ is an even function of τ, and $S(\omega)$ is a non-negative even function of ω.

The variance of stationary process $Y(t)$ is $\sigma^2 = M(0)$.

Together with $M(\tau)$ and $S(\omega)$, we will consider the autocorrelation function

$$\widetilde{M}(\tau) = M(\tau)/\sigma^2$$

and the normalized (standardized) spectral density

$$\widetilde{S}(\omega) = S(\omega)/\sigma^2,$$

which are interconnected by equations analogous to (3.1) and (3.2).

The main problem of applied spectral and autocorrelation analysis of a random stationary process $Y(t)$ consists of estimating the mean, the spectrum $S(\omega)$, and autocovariance function $M(\tau)$. For the ergodic stationary process, such a problem can be solved with the aid of a single sample (time series) of $Y_i = Y(t_i)$ given on the finite grid $t_0, t_1, \ldots, t_{k-1}$; $\Delta t = t_{i+1} - t_i = $ constant, if the number of observations is sufficiently large.

The estimation of the mean was discussed in detail in Chapter 2. Here, let us suppose that the mean is known and that Y_i are the anomalies, that is,

$$E(Y_i) = 0. \qquad (3.3)$$

A numerical methodology for estimating spectral density consists of the application of different filtering procedures to the smoothing of the periodogram (see Jenkins and Watts, 1968).

The periodogram of observations $Y_0, Y_1, \ldots, Y_{k-1}$ is determined by the following formula:

$$H(\omega_p) = H_p = \frac{k\Delta t}{2\pi}|A_p|^2 = \frac{\Delta t}{2\pi}\left(M_0 + 2\sum_{\tau=1}^{k-1} M_\tau \cos\omega_p\tau\right), \quad (3.4)$$

where

$$\omega_p = \frac{2\pi p}{k} \quad (p = 1, 2, \ldots, (k-1)), \quad H_0 = 0,$$

$$A_p = \frac{1}{k}\sum_{t=0}^{k-1} Y_t e^{-i\omega_p t}, \quad (3.5)$$

$$M_\tau = \begin{cases} \dfrac{1}{k}\displaystyle\sum_{i=0}^{k-1-|\tau|} Y_i Y_{i+|\tau|} & \text{if } |\tau| \le k-1, \\ 0 & \text{if } |\tau| \ge k. \end{cases} \quad (3.6)$$

A graph of a periodogram (amplitudes of the Fourier expansion of the observations) as a function of frequency consists of separate, random-height lines, each of which presents the part of the variance conditioned by the fluctuation with the corresponding frequency.

Formula (3.6) presents the estimator of autocovariance function $M(\tau)$. Furthermore, we assume that $M(\tau)$ converges to zero if τ goes to infinity. We also assume that k is sufficiently large, so that for

$$\tau > n = \text{INT}(k/2),$$

the values of M_τ are close to zero. Then, from (3.4), we can obtain

$$H_p \approx \frac{\Delta t}{2\pi}\left(M_0 + 2\sum_{\tau=1}^{n} M_\tau \cos\omega_p\tau\right), \quad (3.7)$$

$$M_\tau \approx \frac{2\pi}{k\Delta t}\left(H_0 + 2\sum_{p=1}^{n} H_p \cos\omega_p\tau\right). \quad (3.8)$$

It has been shown (Jenkins and Watts, 1968) that for sufficiently large k, the covariance $\mathbf{C}_{pq}^H$ of statistics H_p and H_q is

$$\mathbf{C}_{pq}^H \approx \begin{cases} S^2(\omega_p) & \text{if } p = q, \\ 0 & \text{if } p \ne q. \end{cases} \quad (3.9)$$

For white noise process, this equality is precise.

3.1 Stationary Process

If $2r+1$ is the width of the filter and values H_p $(p = 1, 2, \ldots)$ are approximately independent and have approximately equal accuracy on each $\frac{2\pi(2r+1)}{k}$ length frequency subinterval, then the smoothing procedure, described in Chapter 1, can be applied for estimating the spectral density. Smoothing is performed with the aid of the spectral (filter) window $\{\alpha_j\}_{j=-r}^{r}$ by the formula

$$\bar{H}_p = \sum_{j=-r}^{r} \alpha_j H_{p+j}. \tag{3.10}$$

Estimates $\bar{H}_p$ are statistically dependent. The study of their covariance and correlation functions can be carried out by the approach developed in Chapter 1.

It is easy to show that

$$\bar{H}_p = \frac{\Delta t}{2\pi} \sum_{\tau=-n}^{n} \bar{M}_\tau e^{-i\omega_p \tau}, \tag{3.11}$$

where

$$\bar{M}_\tau = M_\tau w_\tau, \tag{3.12}$$

$$w_\tau = \sum_{j=-r}^{r} \alpha_j e^{i\omega_p \tau}, \tag{3.13}$$

and w_τ is the correlation window corresponding to the chosen spectral window α_j. Any specific weights of a spectral window α_j in (3.10) have a corresponding correlation window.

To build an optimal digital filter, one must know the correlation structure of the values of periodogram. Such a structure is known only approximately (3.9), and in practice we apply different empirical approaches to find the corresponding type of window and its parameters.

Therefore, the core of spectral estimation methodology is harmonic analysis of the time series with consecutive smoothing of the amplitudes of the Fourier expansion. The representation of the sample variance as a sum of the amplitudes of the harmonics with different frequencies [Parseval equation, see (3.8) for $\tau=0$] makes it possible to compare their power. Because the number of observations (k) is finite and the step is equal to Δt, we can analyze fluctuations with periods ranging from $2\Delta t$ (where Δt is the time step) to $k\Delta t$. A periodogram of normalized time series gives the variance proportion for each frequency that forms a random variation of the time series. Every point of such a representation characterizes the power of the corresponding fluctuations. For different frequencies the value of power is a random value. Periodogram is a random sample that has the same relationship to the spectral density as a separate observation of a random variable to its expected value. The principal theoretical characteristic (spectral density) presents the mean spread (over all possible samples) of the variance over the frequency axis.

It is interesting to notice that, long before the stationary random processes theory was created, the need to smooth (or to average over different subintervals of the frequency axis) periodogram values was discovered by Albert Einstein (Einstein, 1986). Indeed, Einstein anticipated the development of the modern methodology of the spectral estimations.

The principal aspects of the smoothing (by a digital filter) were discussed in Chapter 1, where the necessity of the trial and error method for obtaining the approximately unbiased estimates with small variance was shown. Smoothing in the frequency space has some features that contrast with the application of the filters for independent observations in the time domain. First, different points of a periodogram can have different variances, which can cause significant errors (bias) in the spectral estimates. For example, as will be shown, for the annual mean surface air temperature time series, the power of fluctuations in the low-frequency part of the periodogram (with the periods in several dozens of years) is larger than the corresponding power in the higher frequency part. The spectrum of the air temperature time series has a low-frequency maximum (in the vicinity of zero), and, according to (3.9), corresponding values of the periodogram have greater variances than in other parts of the spectrum. In the process of smoothing, when the left part of the spectral window is applied to points with large variances and the right part of this window covers the points with small variances, it is inevitable that the point estimates will have bias. This fact can be illustrated by a graph with superimposed curves of the periodogram and corresponding spectral estimates (analogous to those given in Chapter 6). A general approach for improving the estimates is to increase the number of observations and, as a result, the number of points of the periodogram, which allows us to vary the width of the filter.

The selection of the spectral window is affected by many factors, one of which is the necessity of obtaining non-negative spectral estimates. The values of the periodogram are always greater than or equal to zero, so negative spectral estimates can be obtained only by using an inappropriate filter, which has some negative weights. Many different spectral windows for univariate time series that guarantee the positiveness of the spectral estimates have been developed (Jenkins and Watts, 1968). As becomes clear from the filtering schemes considered in Chapter 1, only regression filters with $m = 1$ have positive weights. Therefore, the superposition of these filters [as was done by Tukey, (1.105)] is one of the approaches in constructing smoothing procedures for spectrum estimation.

Along with the harmonic and statistical analysis of the spectral estimation procedure, the purely computational aspects of the chosen numerical scheme are also important. There were two principal ways to develop a computational procedure: the Blackman-Tukey method and the Cooley-Tukey method.

The Blackman-Tukey method. An algorithm suggested by Blackman and Tukey (1958) is the estimation using (3.6) of the first ν values of M_τ; $(\tau = 0, 1, \ldots, \nu; \nu << n)$ of the covariance function. After this, the computation of $\bar{H}_p$ is performed by (3.11), where the appropriate correlation window w_τ ($w_\tau = 0$ if $\tau \geq \nu$) is applied.

The Cooley-Tukey method. The sequence of computation by this scheme supposes the following: immediately finding the periodogram by the Fourier transform (3.5) of the time series; computing and smoothing (by using a numerical filter) the periodogram (3.4); and estimating the covariance function by another Fourier transform (3.8). The widespread application of this methodology became possible after the publication of the paper by Cooley and Tukey (1965), where an algorithm of the fast Fourier transform (FFT) was obtained.

For large k, the second scheme is thousands of times more economical than the first one; for this reason, it has been widely used in the last decades. Additionally, with the aid of FFT, it is possible to analyze two time series simultaneously and to significantly increase the effectiveness of the second method.

However, the Blackman-Tukey approach (with some of the well-developed spectral windows) is not only an effective numerical scheme, it is also a rational estimation methodology, which (for some problems) can be better than the other approaches. Therefore, the comprehensive coexistence of both schemes proves inevitable when one creates a package of statistical computer programs.

The three Fourier transformation method. Our practice showed that the combination of the above methodologies (Fourier transform of the time series → periodogram → Fourier transform of the periodogram → correlation function → truncating and weighting of the correlation function → its Fourier transform → spectrum) with three

Fourier transformations utilizes the advantages of both approaches. The efficiency of this scheme has become especially apparent in the case of multidimensional spectral analysis when direct smoothing of the points close to the boundary of the multidimensional frequency domain presents real algorithmic and programming difficulties.

These approaches do not exhaust all of the smoothing schemes which may be applied for estimating the spectrum of the stationary random process. Selecting a spectral window or developing a new one is dictated by many factors (the statistical properties of data, a physical feature of the problem, the sample size, and so on).

Along with the covariance function and spectral density, it is interesting to consider the cumulative spectral function. Letting $\tau = 0$ in (3.2), we have the equality

$$\sigma^2 = \int_{-\infty}^{\infty} S(\omega)d\omega = 2\int_0^{\infty} S(\omega)d\omega, \qquad (3.14)$$

which summarizes the spectral values of all the frequencies.

When the spectral density is analyzed, it is often necessary to estimate a part of the variance corresponding to a finite frequency subinterval $[0, \omega]$. The function

$$I(\omega) = \int_{-\omega}^{\omega} S(\omega')d\omega' = 2\int_0^{\omega} S(\omega')d\omega', \qquad (3.15)$$

which gives that part of σ^2, is called the cumulative spectral function.

It is a convenient way to consider the normalized (standardized) cumulative spectral function

$$\tilde{I}(\omega) = 2\int_0^{\omega} \tilde{S}(\omega')d\omega'. \qquad (3.16)$$

$\tilde{I}(\omega)$ is a positive, nondecreasing function, and

$$\tilde{I}(0) = 0, \qquad (3.17)$$

$$\tilde{I}(\infty) = 2\int_0^{\infty} \tilde{S}(\omega')d\omega' = 1. \qquad (3.18)$$

The estimator of the normalized cumulative function (3.16) is

$$i(\omega_p) = \frac{4\pi}{k\Delta t} \sum_{q=1}^{p} H_q/M_0 \quad (p = 1, 2, \ldots, n). \qquad (3.19)$$

On the graph of the sample cumulative spectral function, the jumps of discontinuity, equal to the periodogram value, can be seen.

For testing the hypothesis that the time series is the sample of a white noise process, the well-known Kolmogorov-Smirnov criterion (van der Waerden, 1960) of constructing the confidence interval for the sample distribution function can be applied. Actually, the standardized cumulative spectral density of the discrete white noise is equal to a constant

$$\tilde{S}(\omega) = \frac{\Delta t}{2\pi} \quad (-\pi/\Delta t \leq \omega \leq \pi \Delta t) \tag{3.20}$$

and the corresponding $\tilde{I}(\omega)$ is

$$\tilde{I}(\omega) = \frac{\Delta t}{\pi}\omega \quad (0 \leq \omega \leq \pi/\Delta t). \tag{3.21}$$

This function presents a linear segment from point (0,0) to $(\pi/\Delta t, 1)$. By choosing the appropriate scale along the ω axis, the straight line segment bisecting the coordinate angle can be obtained. If estimates (3.19), plotted in such a coordinate system, are far from the theoretical line of the white noise, then it is plausible that the observations are not the sample of the white noise. Therefore, the white noise model is a useful standard with which the real processes of nature and of technique can be compared.

According to the Kolmogorov-Smirnov criterion, if $k > 200$, the confidence intervals for the standardized spectral density of the white noise are determined by the strips $\pm \delta/\sqrt{k/2 - 1}$, where values of δ, equal to 1.63, 1.36, and 1.02, correspond to significance levels of 0.01, 0.05, 0.25. If $k < 200$, the confidence intervals can be determined with the aid of a table, which can be found, for example, in Jenkins and Watts (1968).

3.2 Cross-Statistical Analysis

In the case of the two time series, the mutual statistical characteristics must be estimated together with the spectral densities and autocovariance functions of each of the processes.

Let us assume that observations $Y_i = Y(t_i)$ and $Z_i = Z(t_i)$ of stationary processes $Y(t)$ and $Z(t)$ are given in grid $t_0, t_1, \ldots, t_{k-1}$; $\Delta t = t_{i+1} - t_i = $ const.

Cross-covariance function $M^{YZ}(\tau)$ and cross-spectral density $S^{YZ}(\omega)$ of these processes are interconnected by the Fourier transforms

$$S^{YZ}(\omega) = \frac{1}{2\pi} \int_{-\infty}^{\infty} M^{YZ}(\tau) e^{-i\omega\tau} d\tau \qquad (3.22)$$

and

$$M^{YZ}(\tau) = \int_{-\infty}^{\infty} S^{YZ}(\omega) e^{i\omega\tau} d\omega. \qquad (3.23)$$

The cross-covariance function can be presented as the sum of its even and odd parts

$$M^{YZ}(\tau) = M_1(\tau) + M_2(\tau), \qquad (3.24)$$

where

$$M_1(\tau) = \frac{1}{2}\left[M^{YZ}(\tau) + M^{YZ}(-\tau)\right],$$

$$M_2(\tau) = \frac{1}{2}\left[M^{YZ}(\tau) - M^{YZ}(-\tau)\right]. \qquad (3.25)$$

Then (3.22) gives

$$S^{YZ}(\omega) = \Lambda(\omega) - i\Psi(\omega), \qquad (3.26)$$

where

$$\Lambda(\omega) = \frac{1}{2\pi}\int_{-\infty}^{\infty} M_1(\tau)e^{-i\omega\tau}d\tau = \frac{1}{2\pi}\int_{-\infty}^{\infty} M_1(\tau)\cos\omega\tau\, d\tau \qquad (3.27)$$

is called the cospectrum and

$$\Psi(\omega) = \frac{1}{2\pi}\int_{-\infty}^{\infty} M_2(\tau)\sin\omega\tau\, d\tau \qquad (3.28)$$

is called the quadrature spectrum of the processes $Y(t)$ and $Z(t)$.

With the aid of $\Lambda(\omega)$ and $\Phi(\omega)$, the phase spectrum is determined as

$$\Theta(\omega) = \arctan\left[-\frac{\Psi(\omega)}{\Lambda(\omega)}\right]. \qquad (3.29)$$

$\Theta(\omega)$ shows the lag of the harmonic with frequency ω of the process $Z(t)$ with respect to the corresponding harmonic of $Y(t)$.

The next characteristic of the bivariate random process is called the coherency function; it is determined by the formula

$$\Gamma(\omega) = \frac{|S^{YZ}(\omega)|^2}{S^Y(\omega)S^Z(\omega)} = \frac{\Lambda^2(\omega) + \Psi^2(\omega)}{S^Y(\omega)S^Z(\omega)}, \qquad (3.30)$$

3.2 Cross-Statistical Analysis

where $S^Y(\omega)$ and $S^Z(\omega)$ are the spectral densities of $Y(t)$ and $Z(t)$. $\Gamma(\omega)$ is a measure (the squared correlation coefficient) of the statistical dependence of the amplitudes of the harmonics with frequency ω of the Fourier expansions of processes $Y(t)$ and $Z(t)$.

The cross-spectral density estimation is performed by applying an appropriate digital filter (spectral window) to the smoothing of the cross-periodogram, which is determined as follows:

$$G(\omega_p) = G_p = \frac{k\Delta t}{2\pi} A_p^* B_p = \frac{\Delta t}{2\pi} \sum_{\tau=-(k-1)}^{k-1} V_\tau e^{-i\omega_p \tau}, \quad (3.31)$$

where $\omega_p = 2\pi p/k$ $(p = 0, 1, \ldots, k-1)$,
A_p^* is a complex conjugate of A_p,

$$A_p = \frac{1}{k} \sum_{t=0}^{k-1} Y_t e^{-i\omega_p t}, \quad (3.32)$$

$$B_p = \frac{1}{k} \sum_{t=0}^{k-1} Z_t e^{-\omega_p t}, \quad (3.33)$$

$$V_\tau = \begin{cases} \dfrac{1}{k} \displaystyle\sum_{i=0}^{k-1-\tau} Y_i Z_{i+\tau} & \text{if } 0 \leq \tau \leq k-1, \\ \dfrac{1}{k} \displaystyle\sum_{i=0}^{k-1+\tau} Z_i Y_{i-\tau} & \text{if } 0 > \tau > -k+1, \\ 0 & \text{if } |\tau| \geq k. \end{cases} \quad (3.34)$$

The estimates of the cospectrum $\Lambda(\omega)$ and quadrature spectrum $\Phi(\omega)$ are obtained by smoothing the sample cospectrum

$$G_p' = \frac{k\Delta t}{2\pi}(A_p' B_p' + A_p'' B_p'') = \frac{\Delta t}{2\pi}\left[V_0 + \sum_{\tau=1}^{k-1}(V_\tau + V_{-\tau})\cos\omega_p\tau\right] \quad (3.35)$$

(where $'$ and $''$ designate real and imaginary parts respectively of corresponding complex numbers) and the sample quadrature spectrum

$$G_p'' = \frac{k\Delta t}{2\pi}\left(A_p' B_p'' - A_p'' B_p'\right) = \frac{\Delta t}{2\pi}\sum_{\tau=1}^{k-1}(V_\tau - V_{-\tau})\sin\omega_p\tau. \quad (3.36)$$

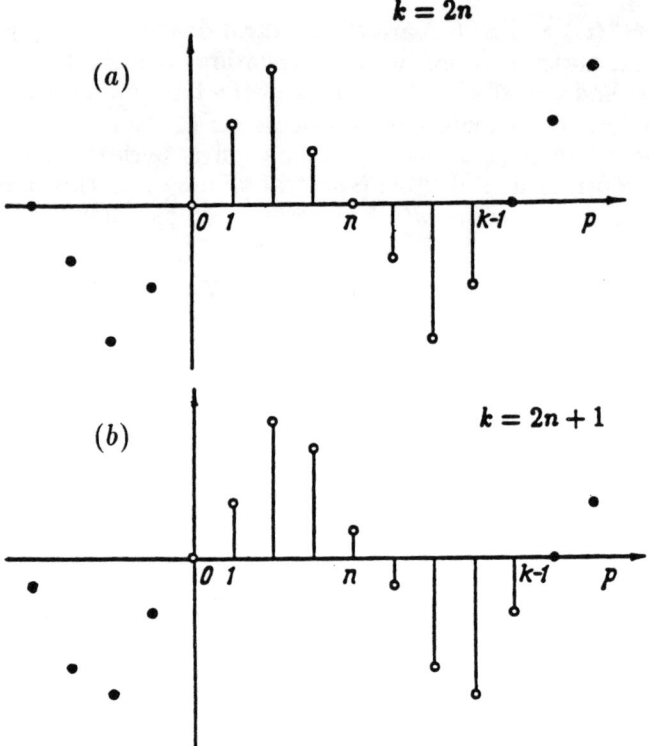

Figure 3.1: Location of the discrete points of the even periodic function. (a) k is even, (b) k is odd.

If k is large, the formulas for the variances of (3.35) and (3.36) are (see Jenkins Watts, 1968)

$$\sigma^2(\omega_p) \approx \begin{cases} \dfrac{1}{2}\{S^Y(\omega_p)S^Z(\omega_p) + \Lambda^2(\omega_p) - \Psi^2(\omega_p)\} \text{ for } G'_p, \\ \\ \dfrac{1}{2}\{S^Y(\omega_p)S^Z(\omega_p) - \Lambda^2(\omega_p) + \Psi^2(\omega_p)\} \text{ for } G'''_p. \end{cases} \quad (3.37)$$

Quantities G'_p and G'''_p are considered preliminary estimates, which must be smoothed by applying the appropriate digital filters. It is possible to show that smoothing values $\bar{G}'_p$ and $\bar{G}'''_p$, considered as the estimators of the cospectrum and quadrature spectrum, are asymp-

3.2 Cross-Statistical Analysis

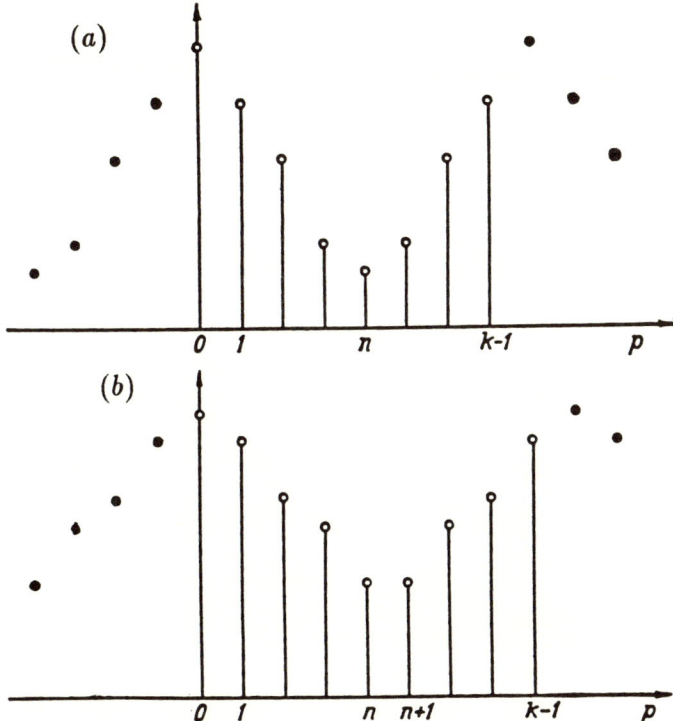

Figure 3.2: Location of the discrete points of the even periodic function. (a) k is even, (b) k is odd.

totically unbiased.

When smoothing the values of G'_p and G''_p, which are close to the ends of the interval, their even or odd character must be taken into account and corresponding periodical extensions must be performed. The illustrations in Figures 3.1 and 3.2 can be useful for such extension in developing computer routines.

The estimators of phase spectrum $\Theta(\omega)$ and coherency function $\Gamma(\omega)$ are

$$\bar{\Theta}(\omega_p) = \arctan\left(-\frac{\bar{G}''_p}{\bar{G}'_p}\right) \qquad (3.38)$$

and

$$\bar{\Gamma}(\omega_p) = \frac{(\bar{G}'_p)^2 + (\bar{G}''_p)^2}{\bar{H}_p^Y \bar{H}_p^Z}, \qquad (3.39)$$

where $\bar{H}_p^Y$, $\bar{H}_p^Z$, $\bar{G}'_p$, and $\bar{G}''_p$ are the estimates of spectral densities, cospectrum, and quadrature spectrum [of processes $Y(t)$ and $Z(t)$], which were obtained by smoothing the corresponding periodograms.

3.3 Nonstationary Processes

The principal statistical characteristics of the nonstationary process $Y(t)$ are the mean $C(t)$ and the covariance function $M(t,\theta)$, which are determined by formulas

$$C(t) = E[Y(t)] \qquad (3.40)$$

and

$$M(t,\theta) = E\{[Y(t) - C(t)][Y(\theta) - C(\theta)]\}, \quad t,\theta \in (-\infty, \infty). \quad (3.41)$$

$M(t,\theta)$ is the function of two variables, and for its estimation, an ensemble of samples is required.

If $t = \theta$,
$$M(t,t) = \sigma^2(t) \qquad (3.42)$$

is the variance of the process. In general, a nonstationary process has no spectral density. Some classes of a nonstationary random processes that do have spectral density are considered in this section.

Different random processes can have nonstationary moments of different order. For example, the time series of the monthly means of meteorological elements have a mean and a variance with an obvious nonstationary component, a seasonal cycle. Sometimes, the study of such nonstationary processes can be performed by simple transformation (removing the seasonal cycle) of the samples, which makes it possible to obtain the time series of normalized (standardized) anomalies, which can be considered as approximately stationary.

Before we consider the numerical scheme for the $M(t,\tau)$ estimation, it is necessary to mention some of the approaches that are used for analyzing some types of nonstationarities having only one sufficiently long time series.

Let us denote the observations

3.3 Nonstationary Processes

$$Y_0, Y_1, \ldots, Y_{k-1} \tag{3.43}$$

given in the temporal moments

$$t_0, t_1, \ldots, t_{k-1} \quad (\Delta t = t_{i+1} - t_i = \text{const.}) \tag{3.44}$$

as a random vector

$$\mathbf{Y} = \{Y_i\}_{i=0}^{k-1}. \tag{3.45}$$

Let the covariance matrix of $\mathbf{Y}$ be

$$\mathbf{M} = \{M_{ij}\}_{i,j=0}^{k-1}, \tag{3.46}$$

where $M_{ij} = M(t_i, \theta_j)$

1. *Estimation of the mean that is variable in time.* The least squares numerical schemes described in Chapter 1 can be used to approximate the nonstationary mean if it can be presented as a linear function

$$C(t) = \sum_{j=0}^{m} \beta_j x_j(t), \tag{3.47}$$

where β_j are unknown parameters that have to be estimated; $x_j(t)$, $(j = 0, 1, \ldots, m)$ is the chosen system of functions. For example, the simplest variation in time is the linear trend,

$$C(t) = \beta_0 + \beta_1 t. \tag{3.48}$$

Let us consider a general case (3.47). If parameters β_j are estimated by the least squares method, we must minimize (with respect to β) the quadratic form

$$(\mathbf{Y} - \mathbf{X}\beta)^{\mathrm{T}}(\mathbf{Y} - \mathbf{X}\beta), \tag{3.49}$$

where

$$\beta = \{\beta_j\}_{j=0}^{m}, \tag{3.50}$$

$$\mathbf{X} = \{X_{ij}\}_{i,j=0}^{k-1,m}, \tag{3.51}$$

$$X_{ij} = x_i(t_j). \tag{3.52}$$

In this case the estimates $\bar{\beta}$ are not optimal because matrix (3.46) is unknown and is not used in the process of estimation.

Sometimes, further spectral and correlation analyses are applied to the residuals,

$$Y'_t = Y_i - \bar{C}(t_i), \qquad (3.53)$$

where $\bar{C}(t)$ is the estimated mean.

Whatever method has been used for estimating a nonstationary mean, the second moments of the original observations and the residuals are different. The scale of this difference depends on the statistical structure of the process as well as on the methodology that is used to estimate the mean. This statement will be illustrated by the formula for covariance matrix of the residuals (3.53) of the least squares method. These residuals are the elements of vector

$$\mathbf{Y} - \mathbf{X}\bar{\beta} = \mathbf{Y}^T[\mathbf{I} - \mathbf{X}(\mathbf{X}^T\mathbf{X})^{-1}\mathbf{X}^T], \qquad (3.54)$$

the covariance matrix $\mathbf{M}'$ of which is

$$\mathbf{M}' = [\mathbf{I} - \mathbf{X}(\mathbf{X}^T\mathbf{X})^{-1}\mathbf{X}^T]\mathbf{M}[\mathbf{I} - \mathbf{X}^T(\mathbf{X}^T\mathbf{X})^{-1}\mathbf{X}]. \qquad (3.55)$$

It is clear that matrices $\mathbf{M}$ and $\mathbf{M}'$ are different. The conditions, when matrices $\mathbf{M}$ and $\mathbf{M}'$ can be asymptotically ($k \to \infty$) close, are given in Anderson (1971).

If k is not large, the correlation structures of the processes, presented by residuals $(\mathbf{Y} - \mathbf{X}\bar{\beta})$ and by $Y(t)$, are significantly different even when $Y(t)$ is the white noise or the Markov process. In spite of the fact that the derivation of the matrix expression (3.55) is simple enough, the practical aspects of the evaluation of the closeness of matrices $\mathbf{M}$ and $\mathbf{M}'$ are difficult.

Therefore, estimating the second moments of residuals $\mathbf{Y} - \mathbf{X}\bar{\beta}$ by any method, we obtain the estimates of the elements of matrix $\mathbf{M}'$ (not the $\mathbf{M}$). It is clear that scheme (3.47) should be used very carefully and only in situations when the primary interest is the estimation of the mean, not in the accurate analysis of the second moments.

If parameters β_j are estimated by some other method than the least squares method, the correlation structure of vector (3.54) differs from that of (3.55). For the linear estimation scheme, the deduction of the expression for the covariance matrix of the residuals does not present any formal difficulties.

2. *Moving average harmonic analysis.* The process with variable-in-time autocorrelation function and spectral density has a more complicated nonstationary structure. Sometimes, such variation in time is slow, and on certain subintervals (of the observational interval) the process can be considered as approximately stationary. It

3.3 Nonstationary Processes

is possible to get a qualitative picture of the variable-in-time statistical structure of the process if the sample size is sufficiently large. In this case one must carry out the standard spectral analysis of data on separate (possibly overlapping) subintervals and compare the results. Estimates of the correlation function and spectral density, which correspond to separate subintervals, are referred to as the local estimates. Their consecutive analysis can show such interesting features as the transfer of the power from one frequency band to another; the process of the formation of the spectral maxima; the temporal localization of separate observed quasi-periodic fluctuations; and the variation of the white noise level. In this case, the estimates of the spectra can be suitably presented in the form of a two-dimensional (frequency and subinterval number) function. In a sense, this algorithm is the generalization of the moving average procedure, because on each subinterval, the estimates of the mean, as well as some other statistical characteristics, are obtained.

3. *Harmonizable random processes.* The most general class of nonstationary processes that have spectral representations is referred to as the harmonizable processes. Their practical spectral analysis is possible if the corresponding covariance function has the Fourier expansion

$$M(t,\theta) = \int_{-\infty}^{\infty} \int_{-\infty}^{\infty} S(\Omega,\omega) e^{i(\Omega t + \omega \theta)} d\Omega d\omega, \qquad (3.56)$$

where

$$S(\Omega,\omega) = \frac{1}{4\pi^2} \int_{-\infty}^{\infty} \int_{-\infty}^{\infty} M(t,\theta) e^{-i(\Omega t + \omega \theta)} dt d\theta. \qquad (3.57)$$

The function $S(\Omega,\omega)$ is called the double frequency spectral density of the nonstationary random processes $Y(t)$. The estimation of $M(t,\theta)$ and $S(\Omega,\omega)$ can be performed with the aid of the ensemble of samples.

Consider n time series of the harmonizable random processes with k observations in each. These observations are given for the same moments of time, and they can be presented in the form of matrix

$$\{Y_{jq}\}_{j=1, q=0}^{n, k-1}. \qquad (3.58)$$

Let us suppose that the random process has a zero mean.

The estimator $R_{t\theta}$ of covariance function $M(t,\theta)$ is

$$R_{t\theta} = \frac{1}{n}\sum_{j=1}^{n} Y_{jt}Y_{j\theta}, \quad (t = 0, 1, \ldots, k-1; \ \theta = 0, 1, \ldots, k-1). \quad (3.59)$$

The estimator $H_{pq} = H(\Omega_p, \omega_q)$ of function $S(\Omega_p, \omega_q)$ is defined in an analogy with a definition of the double frequency spectrum (3.57):

$$H_{pq} = \frac{1}{k^2}\sum_{t=0}^{k-1}\sum_{\theta=0}^{k-1} R_{t\theta} e^{-i(\Omega_p t + \omega_q \theta)}, \quad (\Omega_p = 2\pi p/k; \ \omega_q = 2\pi q/k).$$
(3.60)

Substituting (3.59) for $R_{t\theta}$ in (3.60), we have

$$H_{pq} = \frac{1}{n}\sum_{j=1}^{n}\left(\frac{1}{k}\sum_{t=0}^{k-1} Y_{jt} e^{-i\Omega_p t}\right)\left(\frac{1}{k}\sum_{\theta=0}^{k-1} Y_{j\theta} e^{-i\omega_q \theta}\right). \quad (3.61)$$

The last three formulas demonstrate the possibility of finding the double frequency spectrum estimates by one of the following ways: the Fourier transform of weighed estimates $R_{t\theta}$, or the Fourier transform of the separate samples with consequent computation by formula (3.61).

To find the estimate of the double frequency spectrum, it is necessary to smooth the field of values H_{pq} by an appropriate two-dimensional digital filter. An estimator of the covariance function in this case is

$$R_{t\theta} = \sum_{p=0}^{k-1}\sum_{q=0}^{k-1} H_{pq} e^{i(\Omega_p t + \omega_q \theta)}. \quad (3.62)$$

We considered a formal numerical approach for estimating the double frequency spectrum of the harmonizable process. This approach does not contain any new numerical ideas. But the physical interpretation of the estimates obtained by this method is not simple, and this might be why relatively few papers discuss such results. Notice, furthermore, that the generalization of the considered numerical approach on the cross-analysis of the harmonizable processes has no difficulty in principal. However, in an effort to avoid encumbering the text and repeating very similar ideas, we do not present such a generalization. A review of some of the algorithms of the numerical analysis of nonstationary processes is provided by Anderson (1971).

3.4 Nonergodic Stationary Process

The methodology of spectral and correlation analysis for the stationary random process (Section 3.1) can be applied when the condition of ergodicity is satisfied. This condition theoretically makes it possible to find the estimates with the aid of the only sample. If this condition is not satisfied, then the estimation of the spectral density and correlation function requires an ensemble of time series. We will discuss the estimation scheme, which can be applied in this case. Let

$$\{Y_{jq}\}_{j=1,q=0}^{n,k-1} \tag{3.63}$$

be n time series (with k observations in each, given in the same time grid) of the nonergodic stationary random process with a zero mean.

The sample covariance matrix,

$$\mathbf{R} = \{R_{t\theta}\}_{t=0,\theta=0}^{k-1,k-1}, \tag{3.64}$$

of these observations is formed by the estimates

$$R_{t\theta} = \frac{1}{n} \sum_{j=1}^{n} Y_{jt} Y_{j\theta}. \tag{3.65}$$

The process is stationary, and its covariance function depends on only one argument (lag). Therefore, the elements, situated on any matrix (3.64) diagonal that is parallel to the main diagonal, present the estimates of the same correlation coefficient. (An exact equality of these estimates has not taken place because of the sample variability.)

The estimates R_τ of covariance function $M(\tau)$ can be found by averaging the estimates on the diagonals, which are parallel to the main diagonal of matrix (3.64):

$$R_\tau = \frac{1}{k-\tau} \sum_{t=0}^{k-1-\tau} R_{t,\,t+\tau} (\tau = 0, 1, \ldots, k-1). \tag{3.66}$$

Obtaining $R_{t,t+\tau}$ by (3.65) and substituting it in (3.66), we have

$$R_\tau = \frac{1}{n} \sum_{j=1}^{n} \frac{1}{k-\tau} \sum_{t=0}^{k-1-\tau} Y_{jt} Y_{jt+\tau} = \frac{1}{n} \sum_{j=1}^{n} R_\tau^j, \tag{3.67}$$

where

$$R_\tau^j = \frac{1}{k-\tau} \sum_{t=0}^{k-1-\tau} Y_{jt} Y_{jt+\tau}. \qquad (3.68)$$

Formula (3.67) is the estimator of the covariance function for an ergodic stationary processes. The estimator (3.67) is the mean of the estimates of the covariance functions, obtained for the separate time series of the ensemble by applying scheme (3.6), which was developed for the ergodic processes. The estimation of the spectrum by (3.11) yields

$$\bar{H}_p = \frac{\Delta t}{2\pi} \left[R_0 + 2 \sum_{\tau=1}^{k-1} R_\tau w_\tau \cos \omega_p \tau \right] = \frac{1}{n} \sum_{j=1}^{n} \bar{H}_p^j, \qquad (3.69)$$

where $\bar{H}_p^j$ is the estimate of the spectral density of sample j by the scheme (3.10). Therefore, the estimate $\bar{H}_p$ (3.69) is the average of the spectral estimates for the separate time series, and R_τ (3.67) is the average of the corresponding sample covariance functions.

In the situation considered, all of the samples of the ensemble had the same size. In practice, however, this is not often the case. For example, the time series of the meteorological elements of different stations can have different sizes. In order not to lose the information by reducing the sizes of time series to the common overlapping subinterval, it is possible to use the following estimation procedure.

1. Compute the periodogram for each time series.

2. Divide the frequency interval (from zero to the Nyquist frequency) to a certain number of nonoverlapping equal-width bands.

3. Average the values of each separate periodogram inside such bands.

4. Average the obtained estimates for each frequency band over the whole ensemble.

5. Find the estimates of a spectrum by smoothing the set of the results using a digital filter.

6. Estimate the covariance function by the Fourier transform of the spectrum estimates.

This estimation scheme can be easily generalized for the case of the cross-correlation analysis of the two stationary nonergodic random processes, each of which is presented by the ensemble of samples. The estimates of the cross-statistical characteristics in this case are the ensemble means of the estimates, which are found for the separate sample pairs under assumption of their ergodicity.

3.5 Time Series with Missing Data

One of the features of the geophysical time series is the possible existence of random gaps. These gaps are often caused by the nature of the phenomena being studied. For example, observations of direct solar radiation cannot be obtained when clouds cover the sun. As an added example, the discontinuity of the precipitation or cloud data is part of their physical nature.

There are numerous methodological procedures in this field, some of which can be found in Parzen's references (1984).

Consider a general approach to the problem of a discrete harmonic analysis of observations

$$Y_0, Y_1, \ldots, Y_{k-1}, \tag{3.70}$$

given on irregular grid points $t_0, t_1, \ldots, t_{k-1}$; $t_{i+1} > t_i$; $t_{i+1} - t_i \neq$ constant (so the time steps are not equal).

The estimates of the coefficients of the approximating trigonometrical polynomial

$$Y(t) = A_0 + A_1 \cos \Omega_1 t + A_2 \cos \Omega_2 t + \ldots + A_\nu \cos \Omega_\nu t$$
$$+ B_1 \sin \Omega_1 t + B_2 \sin \Omega_2 t + \ldots + B_\nu \sin \Omega_\nu t, \tag{3.71}$$

(where $2\nu + 1 < k$; $\Omega_1, \Omega_2, \ldots, \Omega_\nu$ are an irregular system of grid points of some frequency interval $[0, \Omega_\nu]$) must be found.

In this case, the harmonic analysis is connected with the identification of the proper system of points $\{\Omega_p\}_{p=0}^{\nu}$ of a frequency interval. This situation is analogous to the identification of the algebraic polynomial degree, when least squares polynomial approximation of a set of observed data is performed.

The proper analysis of the physical contents of the problem can be helpful in such identification. It is difficult to make general recommendations; nonetheless, the following qualitative reasoning can assist in some practical situations.

For the reliable estimation of coefficients A_p and B_p corresponding to any frequency

$$\Omega_p = 2\pi/T_p,$$

it is necessary that the time interval $[0, t_{k-1}]$ contain a sufficient number of nonoverlapping subintervals of T_p length with at least two observations on each subinterval.

If $\Delta t_{\min}$ is the minimal distance between observations, then interval $[0, \Omega_\nu] \subset [0, \pi/\Delta t_{\min}]$. In searching for Ω_ν, one can use the value

$$\Omega_{cp} = \pi/\Delta t_{cp} \qquad (3.72)$$

(where $\Delta t_{cp} = (t_{k-1} - t_0)/k$) as an approximate high possible value of frequency (Ω_ν) in (3.71).

The coefficients (A_p and B_p) of the expansion (3.71) can be estimated by the least squares method. The amplitudes obtained for frequencies Ω_p correspond to the nonequal frequency steps. The smoothing of such estimates can be done by approximation using an appropriate function.

This approach is very cumbersome. On irregular grid points, the trigonometrical functions have no orthogonal properties, and estimation of parameters A_p and B_p requires that one construct and solve the system of normal equations. For large ν, this problem requires a huge number of computations.

The analysis of the equally spaced time series with a small number of missing data can be carried out with the methods described, for example, by Parzen (1984).

The following qualitative reasoning characterizes the possible inaccuracies in the spectral estimates arising as a result of missing data in the time series (given on the regular grid points, if $\Delta t = t_{i+1} - t_i$ is a constant). According to the Nyquist theorem, the minimum period of the Fourier expansion is

$$T_{\min} = 2\Delta t, \qquad (3.73)$$

and it contains two observations. Therefore, gaps affect the estimates of the spectrum corresponding to the periods, for which condition (3.73) does not hold all the time. Thus, one can expect that a small number of gaps will not significantly distort the results for the low and middle frequencies. The closer the estimates are to the Nyquist frequency, the lower their accuracy is. With an increasing amount of missing data, the improperly estimated part of the spectrum is shifted to the domain of the middle frequencies.

When the interest is only in the low-frequency part of the spectrum, it is possible to intentionally reduce the frequency interval by temporally averaging the terms of the original time series. For example, instead of observations with step Δt, one can take the values with step $\lambda \Delta t$ (where λ is a positive integer); these new values are the averages of at least λ consecutive observations of the original time series. In this case, the Nyquist frequency is reduced λ times (from

$\pi/\Delta t$ to $\pi/\lambda\Delta t$), but the newly created time series contains fewer missing terms than the original one. The frequency characteristic

$$\left[\frac{\sin(\lambda\Delta t\omega/2)}{\lambda\sin(\Delta t\omega/2)}\right]^2 \qquad (3.74)$$

of the above averaging does not accurately reflect the real modification of the amplitudes because the real averaging of the original data was performed only for the existing observations.

The main idea of the statistical formulation of the problem with missing data (Parzen, 1984) is in introducing a random function $\eta(t)$, which equals zero if the observation is absent and which equals one in all other points. Thus, the sample can be presented in the form of the amplitude-modulated time series

$$Y(t) = X(t)\eta(t), \qquad (3.75)$$

where $X(t)$ is the sample with no missing data. Under some conditions (first, if the number of observations is sufficiently large), the estimates of the covariance function and the spectral density, found with the aid of $Y(t)$, will be close to the corresponding characteristics of $X(t)$.

Notice here that an additional deterministic consideration of this problem is possible. The effects imposed by function $\eta(t)$ are conditioned by the convolution theorem. For each observed sample, the function $\eta(t)$ is known, and one can perform its harmonic analysis. Therefore, the values of the Fourier expansion of $\eta(t)$ reveal the character of the distortion of the amplitudes of the original time series. If such distortions are randomly and evenly distributed along the frequency interval, one can choose the appropriate spectral window, taking into account the information obtained about the frequency characteristics of $\eta(t)$. If these distortions lead to significant misrepresentation of the information on the separate frequency subintervals, obtaining reliable results can be impossible. In any case, detailed analysis is labor-intensive.

3.6 Two-Dimensional Fields

The formal generalization of the theoretical concepts and methods of the stationary random processes to homogeneous random fields does not present any difficulties. At the same time, however, the estimation methodology (together with the natural complication of the numerical algorithms) of the two-dimensional analysis contains some peculiarities. The purely formal succession to the analogy with the estimation theory for stationary processes can lead to loss of a part of the information contained in the samples.

1. We will consider the algorithms for the cross-statistical analysis of two-dimensional homogeneous random fields presented by two samples. For the single sample, the required characteristics will be obtained as a particular case of the methodology carried out. For the two homogeneous random fields

$$Y(\theta, t) \text{ and } X(\theta, t),$$

$$\theta \in (-\infty, \infty), \quad t \in (-\infty, \infty),$$

with zero means, the cross-covariance function is

$$M(v, \tau) = E[Y(\theta, t) X(\theta + v, t + \tau)],$$
$$v \in (-\infty, \infty), \quad \tau \in (-\infty, \infty). \tag{3.76}$$

This function can be presented by Fourier integral

$$M(v, \tau) = \int_{-\infty}^{\infty} \int_{-\infty}^{\infty} S(\Omega, \omega) e^{i(\Omega v + \omega \tau)} d\Omega d\omega, \tag{3.77}$$

where

$$S(\Omega, \omega) = \frac{1}{4\pi^2} \int_{-\infty}^{\infty} \int_{-\infty}^{\infty} M(v, \tau) e^{-i(\Omega v + \omega \tau)} dv d\tau \tag{3.78}$$

is called the cross-spectral density. If $Y(\theta, t)$ is equal to $X(\theta, t)$, then (3.76) determines the covariance function, and expansion (3.77) determines the spectral density of the homogeneous random field $Y(\theta, t)$.

2. Consider two finite discrete samples

$$Y_{\theta t} \text{ and } X_{\theta t},$$
$$(\theta = 0, 1, \ldots, N - 1; \ t = 0, 1, \ldots, k - 1) \tag{3.79}$$

of random fields $Y(\theta, t)$ and $X(\theta, t)$.

3.6 Two-Dimensional Fields

The two-dimensional discrete Fourier expansions of these samples can be written as

$$Y_{\theta t} = \sum_{p=0}^{N-1} \sum_{q=0}^{k-1} A_{pq} e^{i(\Omega_p \theta + \omega_q t)}, \tag{3.80}$$

where

$$A_{pq} = \frac{1}{Nk} \sum_{\theta=0}^{N-1} \sum_{t=0}^{k-1} Y_{\theta t} e^{-i(\Omega_p \theta + \omega_q t)}, \tag{3.81}$$

$$\Omega_p = 2\pi p/N, \quad \omega_q = 2\pi q/k,$$

and

$$X_{\theta t} = \sum_{p=0}^{N-1} \sum_{q=0}^{k-1} B_{pq} e^{i(\Omega_p \theta + \omega_q t)}, \tag{3.82}$$

where

$$B_{pq} = \frac{1}{Nk} \sum_{\theta=0}^{N-1} \sum_{t=0}^{k-1} X_{\theta t} e^{-i(\Omega_p \theta + \omega_q t)}. \tag{3.83}$$

Let the step along the θ axis be equal to $\Delta\theta$, and let the step along the t axis be Δt. Then in the frequency domain, the step along the Ω axis is

$$\Delta\Omega = 2\pi/(N\Delta\theta) \tag{3.84}$$

and the step along the ω axis is

$$\Delta\omega = 2\pi/(k\Delta t). \tag{3.85}$$

The expression

$$G_{pq} = \frac{N\Delta\theta k\Delta t}{4\pi^2} A_{pq}^* B_{pq} \tag{3.86}$$

determines the cross-periodogram of $Y(\theta, t)$ and $X(\theta, t)$.

With the aid of the following definitions for the sample cross-covariance function,

$$V_{\upsilon\tau} =$$

$$\begin{cases} \dfrac{1}{Nk} \sum_{\theta=0}^{N-1-v} \sum_{t=0}^{k-1-\tau} Y_{\theta t} X_{\theta+v, t+\tau} & \text{if } 0 \le v \le N-1; \\ & \quad 0 \le \tau \le k-1, \\[2mm]
\dfrac{1}{Nk} \sum_{\theta=0}^{N-1-v} \sum_{t=0}^{k-1-|\tau|} Y_{\theta, t+|\tau|} X_{\theta+v, t} & \text{if } 0 < v \le N-1; \\ & \quad 0 > \tau \ge -(k-1), \\[2mm]
\dfrac{1}{Nk} \sum_{\theta=0}^{N-1-|v|} \sum_{t=0}^{k-1-\tau} Y_{\theta+|v|, t} X_{\theta, t+\tau} & \text{if } -(N-1) \le v < 0; \\ & \quad k-1 \ge \tau > 0, \\[2mm]
\dfrac{1}{Nk} \sum_{\theta=0}^{N-1-|v|} \sum_{t=0}^{k-1-|\tau|} Y_{\theta+|v|, t+|\tau|} X_{\theta, t} & \text{if } -(N-1) \le v \le 0; \\ & \quad 0 \ge \tau \ge -(k-1), \\[2mm]
0 & \text{if } |v| \ge N \text{ or } |\tau| \ge k. \end{cases}$$

(3.87)

(3.86) can be presented (Polyak, 1979) as

$$G_{pq} = \frac{\Delta\theta \Delta t}{4\pi^2} \sum_{v=-(N-1)}^{N-1} \sum_{\tau=-(k-1)}^{k-1} V_{v\tau} e^{-i(\Omega_p v + \omega_q \tau)}. \quad (3.88)$$

Now, let us assume that the cross-covariance function

$$M(v, \tau) \to 0, \quad \text{if} \quad v \to \infty \quad \text{or} \quad \tau \to \infty,$$

and N and k are so large that for

$$v > \nu = INT(N/2) \quad \text{or} \quad \tau > n = INT(k/2),$$

$M(v, \tau)$ is close to zero. Then

$$G_{pq} \approx \frac{\Delta\theta \Delta t}{4\pi^2} \sum_{v=-\nu}^{\nu} \sum_{q=-n}^{n} V_{v\tau} e^{-i(\Omega_p v + \omega_q \tau)}. \quad (3.89)$$

3.6 Two-Dimensional Fields

The inverse transform is

$$V_{\upsilon\tau} \approx \frac{4\pi^2}{N\Delta\theta k\Delta t} \sum_{p=0}^{N-1} \sum_{q=0}^{k-1} G_{pq} e^{i(\Omega_p \upsilon + \omega_q \tau)}. \qquad (3.90)$$

We can see that the cross-periodogram and the cross-covariance function are interconnected by the two-dimensional Fourier transforms.

By analogy with the stationary random processes, the two-dimensional cross-spectral density is estimated by smoothing the corresponding cross-periodogram with the aid of the two-dimensional digital filter (spectral window):

$$\bar{G}_{pq} = \sum_{j=-r}^{r} \sum_{l=-s}^{s} \alpha_{jl} G_{p+j,q+l}. \qquad (3.91)$$

Substituting (3.89) for $G_{p+j,q+l}$ in (3.91) we have

$$\bar{G}_{pq} = \frac{\Delta\theta\Delta t}{4\pi^2} \sum_{\gamma=-\nu}^{\nu} \sum_{\tau=-n}^{n} \bar{V}_{\upsilon\tau} e^{-i(\Omega_p \upsilon + \omega_q \tau)} \qquad (3.92)$$

where

$$\bar{V}_{\upsilon\tau} = w_{\upsilon\tau} V_{\upsilon\tau}, \qquad (3.93)$$

$$w_{\upsilon\tau} = \sum_{j=-r}^{r} \sum_{l=-s}^{s} \alpha_{jl} e^{-i(\Omega_j \upsilon + \omega_l \tau)}. \qquad (3.94)$$

Filter α_{jl} is called the two-dimensional spectral window and function $w_{\upsilon\tau}$ is a corresponding two-dimensional correlation window.

To get estimates for points G_{pq} close to the boundaries, field G_{pq} must be periodically expanded in accordance with the symmetry of the Fourier coefficients (see Figures 3.1 and 3.2) and the fact that

$$\bar{G}_{pq}^* = \bar{G}_{N-p,k-q} \; ; \; \bar{G}_{p0} = \bar{G}_{N-p,0},$$

$$(p = \nu+1, \ldots, N-1; \; q = 1, \ldots, k-1).$$

The accuracy of the estimates can be found under the assumption that, in each rectangular area equal to the size of the filter, the values of G_{pq} are approximately independent and have equal accuracy with variance σ_{pq}^2. In this case, variance $\bar{\sigma}_{pq}^2$ of estimate $\bar{G}_{pq}$ is

$$\bar{\sigma}_{pq}^2 \approx \sigma_{pq}^2 \sum_{l=-r}^{r} \sum_{j=-s}^{s} \alpha_{jl}^2. \qquad (3.95)$$

3. To obtain spectral and correlation estimators for one random field, we can assume that $Y_{\theta t} = X_{\theta t}$. Then (3.87) yields

$$M_{v\tau} =$$

$$\begin{cases} \dfrac{1}{Nk} \displaystyle\sum_{\theta=0}^{N-1-|v|} \sum_{t=0}^{k-1-|\tau|} Y_{\theta,t} Y_{\theta+|v|,t+|\tau|} & \text{if } v\tau \geq 0 \,;\, |v| \leq N-1; \\ & \qquad |\tau| \leq k-1, \\[2mm] \dfrac{1}{Nk} \displaystyle\sum_{\theta=0}^{N-1-|v|} \sum_{t=0}^{k-1-|\tau|} Y_{\theta,t+|\tau|} Y_{\theta+|v|,t} & \text{if } v\tau < 0 \,;\, |v| \leq N-1; \\ & \qquad |\tau| \leq k-1, \\[2mm] 0 & \text{if } |v| \geq N \text{ or } |\tau| \geq k. \end{cases}$$
(3.96)

The statistic $M_{v\tau}$ is called the sample covariance function of a two-dimensional random field. The corresponding periodogram is

$$H_{pq} = \frac{N\Delta\theta k\Delta t}{4\pi^2} |A_{pq}|^2, \tag{3.97}$$

and the estimation of the spectral density is performed by smoothing H_{pq} with the aid of a two-dimensional digital filter.

The main feature of the two-dimensional numerical scheme is that the cross-covariance function is determined by the four different types expressions (3.87) (one for each quadrant), instead of the two types, as in the one-dimensional case. This fact must be considered when developing methodology that estimates the covariance function as the first step and the spectral density as its Fourier transform.

The key to the practical application of the algorithms considered lies in the construction of the appropriate two-dimensional filter. The simplest approach is the superposition of the simplest filters (1.143). The high-order filter (1.144) cannot be applied because it has negative weights, which can lead to negative spectral estimates.

Solving specific physical problems connected with an estimation of the two-dimensional spectral density makes it necessary to create different classes of two-dimensional digital filters with positive weights.

Let us conclude this section with some methodological remarks.
1. When representing the graphs of the estimates of correlation and spectral fields, one must place the origin of the coordinate system in

3.6 Two-Dimensional Fields

the center of the field. The correlation functions must be placed in the lags' domain (v, τ), where

$$(-\nu \leq v \leq \nu', -n \leq \tau \leq n'),$$

and

$$\nu' = \begin{cases} \nu & \text{if } N \text{ is odd,} \\ \nu - 1 & \text{if } N \text{ is even,} \end{cases}$$

$$n' = \begin{cases} n & \text{if } k \text{ is odd,} \\ n - 1 & \text{if } k \text{ is even.} \end{cases}$$

The periodograms and spectra must be placed in the frequency domain (Ω_p, ω_q), where

$$(-\nu \leq p \leq \nu', \ -n \leq q \leq n'). \tag{3.98}$$

This presentation emphasizes the symmetry of the estimates in the case of the only field. The values of the correlation functions with the negative lags correspond to the values of the periodograms and spectra with the negative frequencies.

2. Increasing the dimensionality creates some new problems that must be solved. For example, different two-dimensional meteorological fields can have different deterministic trends (latitudinal, seasonal cycle, and so on) that must be removed beforehand in order that the sample analyzed will satisfy the condition of homogeneity. Therefore, one must begin with estimation of the trend and separation of the deterministic and random components, or filtering out a two-dimensional trend of the mean and the variance. Although the estimation algorithms can be different, their analysis must be carried out simultaneously with the estimation of the spectral and correlation functions. Different filtration methodologies can lead to different final results. After filtration, the random field must have a zero mean and a variance that is constant for all points of the field. Further calculations must be done in the standard way.

All of the two-dimensional periodograms and spectra given in this book were obtained with the aid of the normalized observations for which the sample mean equals zero and the sample variance equals one.

The methodology considered presents the standard numerical technique for estimating the correlation functions and the spectra of the two-dimensional fields. The convenient way for software development is to use the fast Fourier transform of the normalized observations, adding the zero pads to make the sizes of the field equal to the powers of two.

3. Reasoning about univariate computational methodologies given in Section 3.1 remains valid for the two-dimensional case. The combination of the Blackman-Tukey and Cooley-Tukey approaches (Fourier transform of the field → periodogram → Fourier transform of the periodogram → correlation function → truncating and weighting of the correlation function → its Fourier transform → spectrum) with three Fourier transformations is the most effective algorithm for software development. Using this algorithm, one can avoid direct smoothing, which presents algorithmic and programming difficulties for the points close to the boundary of the two-dimensional frequency domain.

Currently in climatology, the univariate spectral and correlation analysis of time series are widely used. Many results, devoted to the study of the character of the temporal fluctuations of different meteorological elements with periods ranging from seconds to millenniums, have been published. Applying multidimensional spectral analysis will make it possible, as we will see shortly, to obtain more general and interesting results.

3.7 Multidimensional Fields

The estimators for the spectral and correlation characteristics of multidimensional fields can be obtained by analogy with the one- and two-dimensional cases.

Let us consider multidimensional random field

$$Y_{t_1, t_2, \ldots, t_\mu} \quad (t_j = 0, 1, \ldots, k_j - 1; \; j = 1, 2, \ldots, \mu) \quad (3.99)$$

where μ is dimensionality; k_j is the size (number of points) along the t_j axis.

All the formulas for the spectral and correlation analysis of this field (together with the analogous formulas for the one- and two-dimensional cases) are given in Tables 3.1 and 3.2.

To smooth the multidimensional periodogram, one must create a corresponding multidimensional filter (see Sections 1.12). As a first step in developing such a filter, it is possible to use the simple moving average procedures analogous to (1.143).

Table 3.1: Discrete schemes of the Fourier analysis of time series and fields.

	Univariate time series	Two-dimensional field
Sample	$y_t(t = 0, 1, \ldots, k-1)$	$y_{\theta t}(\theta = 0, 1, \ldots, N-1; t = 0, 1, \ldots, k-1)$
Inverse Fourier transform	$a_p = \dfrac{1}{k} \sum_{t=0}^{k-1} y_t e^{-i\omega_p t}$	$a_{pq} = \dfrac{1}{Nk} \sum_{\theta=0}^{N-1} \sum_{t=0}^{k-1} y_{\theta t} e^{-i(\Omega_p \theta + \omega_q t)}$
Fourier transform	$y_t = \sum_{p=0}^{k-1} a_p e^{i\omega_p t}$	$y_{\theta t} = \sum_{p=0}^{N-1} \sum_{q=0}^{k-1} a_{pq} e^{i(\Omega_p \theta + \omega_q t)}$
	$\omega_p = 2\pi p/k$	$\Omega_p = 2\pi p/N,\ \omega_q = 2\pi q/k$
Periodogram	$h_p = \dfrac{k \Delta t}{2\pi} |a_p|^2$	$h_{pq} = \dfrac{Nk \Delta \theta \Delta t}{4\pi^2} |a_{pq}|^2$

(Continued)

Table 3.1 (Cont.)

Parseval's identity	$\dfrac{1}{k}\sum_{t=0}^{k-1} y_t^2 = \sum_{p=0}^{k-1} |a_p|^2$	$\dfrac{1}{Nk}\sum_{\theta=0}^{N-1}\sum_{t=0}^{k-1} y_{\theta t}^2 = \sum_{p=0}^{N-1}\sum_{q=0}^{k-1} |a_{pq}|^2$
Fourier expansion of periodogram	$h_p = \dfrac{\Delta t}{2\pi} \sum_{\tau=-(k-1)}^{k-1} m_\tau e^{-i\omega_p \tau}$	$h_{pq} = \dfrac{\Delta\theta \Delta t}{4\pi^2} \sum_{\nu=-(N-1)}^{N-1}\sum_{\tau=-(k-1)}^{k-1} m_{\nu\tau} e^{-i(\Omega_p \nu + \omega_q \tau)}$
Approximate Fourier expansion of periodogram	$h_p \approx \dfrac{\Delta t}{2\pi}\sum_{\tau=-n}^{n} m_\tau e^{-i\omega_p \tau}$	$h_{pq} \approx \dfrac{\Delta\theta \Delta t}{4\pi^2}\sum_{\nu=-\nu}^{\nu}\sum_{\tau=-n}^{n} m_{\nu\tau} e^{-i(\Omega_p \nu + \omega_q \tau)}$
Fourier expansion of covariance function	$m_\tau \approx \dfrac{2\pi}{k\Delta t}\sum_{p=0}^{k-1} \bar{h}_p e^{i\omega_p \tau}$	$m_{\nu\tau} \approx \dfrac{4\pi^2}{N\Delta\theta k\Delta t}\sum_{p=0}^{N-1}\sum_{q=0}^{k-1} \bar{h}_{pq} e^{i(\Omega_p \nu + \omega_q \tau)}$
Estimator of spectral density	$\bar{h}_p = \sum_{j=-r}^{r} \alpha_j h_{p+j}$	$\bar{h}_{pq} = \sum_{j=-r}^{r}\sum_{i=-s}^{s} \alpha_{ji} h_{p+j,q+l}$

	Multidimensional field
Sample	$y_{t_1 t_2 \ldots t_\mu}(t_j = 0, 1, \ldots, k_j - 1; j = 1, 2, \ldots, \mu)$
Inverse Fourier transform	$\alpha_{p_1 p_2 \ldots p_\mu} = \dfrac{1}{\prod_{j=1}^{\mu} k_j} \sum_{t_1=0}^{k_1-1} \sum_{t_2=0}^{k_2-1} \cdots \sum_{t_\mu=0}^{k_\mu-1} y_{t_1 t_2 \ldots t_\mu} \exp\left(-i \sum_{j=1}^{\mu} \omega_{p_j} t_j\right)$
Fourier transform	$y_{t_1 t_2 \ldots t_\mu} = \sum_{p_1=0}^{k_1-1} \sum_{p_2=0}^{k_2-1} \cdots \sum_{p_\mu=0}^{k_\mu-1} \alpha_{p_1 p_2 \ldots p_\mu} \exp\left(i \sum_{j=1}^{\mu} \omega_{p_j} t_j\right)$ $\omega_{p_j} = 2\pi p_j / k_j \ (j = 1, 2, \ldots, \mu)$
Periodogram	$h_{p_1 p_2 \ldots p_\mu} = \dfrac{\prod_{j=1}^{\mu}(\Delta t_j k_j)}{(2\pi)^\mu} |\alpha_{p_1 p_2 \ldots p_\mu}|^2$
Parseval's identity	$\dfrac{1}{\prod_{j=1}^{\mu} k_j} \sum_{t_1=0}^{k_1-1} \sum_{t_2=0}^{k_2-1} \cdots \sum_{t_\mu=0}^{k_\mu-1} y_{t_1 t_2 \ldots t_\mu}^2 = \sum_{p_1=0}^{k_1-1} \sum_{p_2=0}^{k_2-1} \cdots \sum_{p_\mu=0}^{k_\mu-1} |\alpha_{p_1 p_2 \ldots p_\mu}|^2$

(Continued)

Table 3.1 *(Cont.)*

Fourier expansion of periodogram	$h_{p_1 p_2 \ldots p_\mu} =$ $\dfrac{\prod\limits_{j=1}^{\mu}(\Delta t_j k_j)}{(2\pi)^\mu} \sum\limits_{\tau_1=-(k_1-1)}^{k_1-1} \sum\limits_{\tau_2=-(k_2-1)}^{k_2-1} \cdots \sum\limits_{\tau_\mu=-(k_\mu-1)}^{k_\mu-1} m_{\tau_1 \tau_2 \ldots \tau_\mu} \exp\left(-i\sum\limits_{j=1}^{\mu} \omega_{p_j}\tau_j\right)$
Approximate Fourier expansion of periodogram	$h_{p_1 p_2 \ldots p_\mu} \approx \dfrac{\prod\limits_{j=1}^{\mu}(\Delta t_j k_j)}{(2\pi)^\mu} \sum\limits_{\tau_1=-n_1}^{n_1} \sum\limits_{\tau_2=-n_2}^{n_2} \cdots \sum\limits_{\tau_\mu=-n_\mu}^{n_\mu} m_{\tau_1 \tau_2 \ldots \tau_\mu} \exp\left(-i\sum\limits_{j=1}^{\mu} \omega_{p_j}\tau_j\right)$
Fourier representation of covariance function	$m_{\tau_1 \tau_2 \ldots \tau_\mu} \approx \dfrac{(2\pi)^\mu}{\prod\limits_{j=1}^{\mu}(\Delta t_j k_j)} \sum\limits_{p_1=0}^{k_1-1} \sum\limits_{p_2=0}^{k_2-1} \cdots \sum\limits_{p_\mu=0}^{k_\mu-1} \bar{h}_{p_1 p_2 \ldots p_\mu} \exp\left(i\sum\limits_{j=1}^{\mu} \omega_{p_j}\tau_j\right)$
Estimator of spectral density	$\bar{h}_{p_1 p_2 \ldots p_\mu} = \sum\limits_{j_1=-r_1}^{r_1} \sum\limits_{j_2=-r_2}^{r_2} \cdots \sum\limits_{j_\mu=-r_\mu}^{r_\mu} \alpha_{j_1 j_2 \ldots j_\mu} h_{p_1+j_1, p_2+j_2 \ldots p_\mu+j_\mu}$

Table 3.2: Discrete schemes of the cross-Fourier analysis of time series and fields.

	Univariate time series	Two-dimensional fields
Sample	$y_t, x_t (t = 0, 1, \ldots, k-1)$	$y_{\theta t}, x_{\theta t} (\theta = 0, 1, \ldots, N-1; t = 0, 1, \ldots, k-1)$
Inverse Fourier transforms	$a_p = \dfrac{1}{k} \sum_{t=0}^{k-1} y_t e^{-i\omega_p t}$ $b_p = \dfrac{1}{k} \sum_{t=0}^{k-1} x_t e^{-i\omega_p t}$	$a_{pq} = \dfrac{1}{Nk} \sum_{\theta=0}^{N-1} \sum_{t=0}^{k-1} y_{\theta t} e^{-i(\Omega_p \theta + \omega_q t)}$ $b_{pq} = \dfrac{1}{Nk} \sum_{\theta=0}^{N-1} \sum_{t=0}^{k-1} x_{\theta t} e^{-i(\Omega_p \theta + \omega_q t)}$
Fourier transforms	$y_t = \sum_{p=0}^{k-1} a_p e^{i\omega_p t}$ $x_t = \sum_{p=0}^{k-1} b_p e^{i\omega_p t}$	$y_{\theta t} = \sum_{p=0}^{N-1} \sum_{q=0}^{k-1} a_{pq} e^{i(\Omega_p \theta + \omega_q t)}$ $x_{\theta t} = \sum_{p=0}^{N-1} \sum_{q=0}^{k-1} b_{pq} e^{i(\Omega_p \theta + \omega_q t)}$
Cross-periodogram	$g_p = \dfrac{k \Delta t}{2\pi} a_p^* b_p = g_p' + i g_p''$	$g_{pq} = \dfrac{N k \Delta \theta \Delta t}{4\pi^2} a_{pq}^* b_{pq} = g_{pq}' + i g_{pq}''$
Fourier representation of cross-periodogram	$g_p = \dfrac{\Delta t}{2\pi} \sum_{\tau=-(k-1)}^{k-1} V_\tau e^{-i\omega_p \tau}$	$g_{pq} = \dfrac{\Delta \theta \Delta t}{4\pi^2} \sum_{\upsilon=-(N-1)}^{N-1} \sum_{\tau=-(k-1)}^{k-1} V_{\upsilon \tau} e^{-i(\Omega_p \upsilon + \omega_q \tau)}$

(Continued)

Table 3.2 *(Cont.)*

Approximate Fourier representation of cross-periodogram	$g_p \approx \dfrac{\Delta t}{2\pi} \sum_{\tau=-n}^{n} V_\tau e^{-i\omega_p \tau}$	$g_{pq} \approx \dfrac{\Delta\theta \Delta t}{4\pi^2} \sum_{v=-\nu}^{\nu} \sum_{\tau=-n}^{n} V_{v\tau} e^{-i(\Omega_p v + \omega_q \tau)}$
Parseval's identity	$\dfrac{1}{k}\sum_{t=0}^{k-1} y_{t}z_{t} = \sum_{p=0}^{k-1} a_p^* b_p$	$\dfrac{1}{Nk}\sum_{\theta=0}^{N-1}\sum_{t=0}^{k-1} y_{\theta t} x_{\theta t} = \sum_{p=0}^{N-1}\sum_{q=0}^{k-1} a_{pq}^* b_{pq}$
Fourier representation of cross-covariance function	$V_\tau \approx \dfrac{2\pi}{k\Delta t} \sum_{p=0}^{k-1} \bar{g}_p e^{i\omega_p \tau}$	$V_{v\tau} \approx \dfrac{4\pi^2}{N\Delta\theta k\Delta t} \sum_{p=0}^{N-1}\sum_{q=0}^{k-1} \bar{g}_{pq} e^{i(\Omega_p v + \omega_q \tau)}$
Estimator of cross-spectral density	$\bar{g}_p = \sum_{j=-r}^{r} \alpha_j g_{p+j}$	$\bar{g}_{pq} = \sum_{j=-r}^{r}\sum_{l=-s}^{s} \alpha_{jl} g_{p+j,q+l}$
Coherency function	$\gamma_p = \dfrac{|\bar{g}_p|^2}{h_p^y h_p^x}$	$\gamma_{pq} = \dfrac{|\bar{g}_{pq}|^2}{h_{pq}^y h_{pq}^x}$
Phase difference	$\Delta\psi_p = \arctan\left(-\dfrac{\bar{g}_p''}{\bar{g}_p'}\right)$	$\Delta\psi_{pq} = \arctan\left(-\dfrac{\bar{g}_{pq}''}{\bar{g}_{pq}'}\right)$

	Multidimensional fields
Sample	$y_{t_1 t_2 \ldots t_\mu}$; $x_{t_1 t_2 \ldots t_\mu}(t_j = 0, 1, \ldots, k_j - 1; j = 1, 2, \ldots, \mu)$
Inverse Fourier transforms	$a_{p_1 p_2 \ldots p_\mu} = \dfrac{1}{\prod_{j=1}^{\mu} k_j} \sum_{t_1=0}^{k_1-1} \sum_{t_2=0}^{k_2-1} \cdots \sum_{t_\mu=0}^{k_\mu-1} y_{t_1 t_2 \ldots t_\mu} \exp\left(-i \sum_{j=1}^{\mu} \omega_{p_j} t_j\right)$ $b_{p_1 p_2 \ldots p_\mu} = \dfrac{1}{\prod_{j=1}^{\mu} k_j} \sum_{t_1=0}^{k_1-1} \sum_{t_2=0}^{k_2-1} \cdots \sum_{t_\mu=0}^{k_\mu-1} x_{t_1 t_2 \ldots t_\mu} \exp\left(-i \sum_{j=1}^{\mu} \omega_{p_j} t_j\right)$
Fourier transforms	$y_{t_1 t_2 \ldots t_\mu} = \sum_{p_1=0}^{k_1-1} \sum_{p_2=0}^{k_2-1} \cdots \sum_{p_\mu=0}^{k_\mu-1} a_{p_1 p_2 \ldots p_\mu} \exp\left(i \sum_{j=1}^{\mu} \omega_{p_j} t_j\right)$ $x_{t_1 t_2 \ldots t_\mu} = \sum_{p_1=0}^{k_1-1} \sum_{p_2=0}^{k_2-1} \cdots \sum_{p_\mu=0}^{k_\mu-1} b_{p_1 p_2 \ldots p_\mu} \exp\left(i \sum_{j=1}^{\mu} \omega_{p_j} \tau_j\right)$
Cross-periodogram	$g_{p_1 p_2 \ldots p_\mu} = \dfrac{\prod_{j=1}^{\mu}(\Delta t_j k_j)}{(2\pi)^\mu} a^*_{p_1 p_2 \ldots p_\mu} b_{p_1 p_2 \ldots p_\mu} = g'_{p_1 p_2 \ldots p_\mu} + i g''_{p_1 p_2 \ldots p_\mu}$
Fourier representation of cross-periodogram	$g_{p_1 p_2 \ldots p_\mu} = \dfrac{\prod_{j=1}^{\mu} \Delta t_j}{(2\pi)^\mu} \sum_{\tau_1=-(k_1-1)}^{k_1-1} \sum_{\tau_2=-(k_2-1)}^{k_2-1} \cdots \sum_{\tau_\mu=-(k_\mu-1)}^{k_\mu-1} V_{\tau_1 \tau_2 \ldots \tau_\mu} \exp\left(-i \sum_{j=1}^{\mu} \omega_{p_j} \tau_j\right)$

(Continued)

Table 3.2 (Cont.)

Approximate Fourier representation of cross-periodogram	$g_{p_1p_2\ldots p_\mu} \approx \dfrac{\prod\limits_{j=1}^{\mu} \Delta t_j}{(2\pi)^\mu} \sum\limits_{\tau_1=-n_1}^{n_1} \sum\limits_{\tau_2=-n_2}^{n_2} \cdots \sum\limits_{\tau_\mu=-n_\mu}^{n_\mu} V_{\tau_1\tau_2\ldots\tau_\mu} \exp\left(-i\sum\limits_{i=1}^{\mu}\omega_{p_i}\tau_j\right)$
Parseval's identity	$\dfrac{1}{\prod\limits_{j=1}^{\mu} k_j} \sum\limits_{t_1=0}^{k_1-1}\sum\limits_{t_2=0}^{k_2-1}\cdots\sum\limits_{t_\mu=0}^{k_\mu-1} y_{t_1 t_2\ldots t_\mu} x_{t_1 t_2\ldots t_\mu} = \sum\limits_{p_1=0}^{k_1-1}\sum\limits_{p_2=0}^{k_2-1}\cdots\sum\limits_{p_\mu=0}^{k_\mu-1} a^*_{p_1p_2\ldots p_\mu} b_{p_1p_2\ldots p_\mu}$
Fourier representation of cross-covariance function	$V_{\tau_1\tau_2\ldots\tau_\mu} \approx \dfrac{(2\pi)^\mu}{\prod\limits_{j=1}^{\mu}(\Delta t_j k_j)} \sum\limits_{p_1=0}^{k_1-1}\sum\limits_{p_2=0}^{k_2-1}\cdots\sum\limits_{p_\mu=0}^{k_\mu-1} \bar{g}_{p_1p_2\ldots p_\mu} \exp\left(i\sum\limits_{j=1}^{\mu}\omega_{p_j}\tau_j\right)$
Estimator of cross-spectral density	$\bar{g}_{p_1p_2\ldots p_\mu} = \sum\limits_{j_1=-r_1}^{r_1}\sum\limits_{j_2=-r_2}^{r_2}\cdots\sum\limits_{j_\mu=-r_\mu}^{r_\mu} \alpha_{j_1j_2\ldots j_\mu} g_{p_1+j_1, p_2+j_2,\ldots, p_\mu+j_\mu}$
Coherency function	$\gamma_{p_1p_2\ldots p_\mu} = \dfrac{|\bar{g}_{p_1p_2\ldots p_\mu}|^2}{\bar{h}^y_{p_1p_2\ldots p_\mu}\bar{h}^x_{p_1p_2\ldots p_\mu}}$
Phase difference	$\Delta\psi_{p_1p_2\ldots p_\mu} = \arctan\left(-\dfrac{\bar{g}''_{p_1p_2\ldots p_\mu}}{\bar{g}'_{p_1p_2\ldots p_\mu}}\right)$

Because there are standard routines for multidimensional Fourier transformations, it is not difficult to develop software for spectral and correlation analysis of multidimensional fields. The methodological problem (even for a three-dimensional case) one can meet in the pictorial representation of the results. The principle conceptual problems are in an appropriate transformation of different non-homogeneous fields into homogeneous ones.

3.8 Examples of Climatological Fields

Climatological studies can consider many different types of observed and simulated fields. The analysis of each field requires a special approach. First, a special scheme to filter out two-dimensional trends (of the mean and the variance) and to transform the original field into the form with the zero mean and unit variance is required. But even after conducting such a procedure, a meteorological field would remain nonhomogeneous because, as a rule, any of its correlations are spatially (latitude and/or altitude) and temporally (climate change) dependent. The practical way to overcome this inconvenience is by reducing the field size and increasing its resolution.

Let us enumerate some types of fields found in climatology and give examples of their spectral and correlation estimates. We denote a random field as $Y(\xi, \eta)$.

3.8.1 The 500 mb Surface Geopotential Field.

Variables ξ and η are the rectangular coordinates x and y, obtained by projecting the geographical latitude-longitude coordinates onto the plane of the standard isobaric surface. Examples of such fields are the hemisphere fields of air temperature, atmospheric pressure, and so on in the fixed moment of time. It is clear that such fields can have a latitude trend. They are also distorted because the earth's spherical shape restricts the possibility of accurate study of the part of the spatial spectrum with high wave numbers. Such fields occur frequently in climatology, and their spectral and correlation analysis has significant interest in many areas.

Let us consider observations (in the fixed moment of time) of the 500 mb level geopotential given as a 32×32 (x, y) rectangular field with the origin in the North Pole and the spatial steps equal to about 570 km along each of the axes. Such a field does not satisfy the homogeneous conditions because of its latitude trend (and, may be,

Table 3.3: The 500 mb geopotential field spatial correlation function estimates.

Lags (10^3 km)	Two-dimensional analysis	Fortus	Udin
0.0	1.00	1.00	1.00
0.6	0.83	0.78	0.90
1.2	0.60	0.47	0.67
1.7	0.33	0.17	0.47
2.3	0.10	−0.08	0.32
2.9	−0.02	−0.20	0.21
3.5	−0.03	—	—

latitude dependent other statistical characteristics). To remove this trend, the field was approximated by the truncated two-dimensional Fourier set in which only the zero and the first two-dimensional harmonics were retained. The field of the deviations of the original observations from the point estimates of this truncated Fourier set was then analyzed.

The spectral estimates (Figure 3.3, top) of this field of deviations was found by smoothing the periodogram by the simplest (1.143) two-dimensional scheme with the widths $s = r = 2$. The spectrum has a widespread maximum, centered in the origin and slightly stretched along one of the diagonals. The corresponding two-dimensional correlation function (Figure 3.3, bottom) shows that positive statistical dependence can be traced up to 2000 km, and the isolines are only slightly different from the circles. Therefore, this field is close to the isotropic one that confirms the approximate validity of the assumption about the isotropic character of such 500 mb surface geopotential fields, which has been accepted in some publications (Gandin, 1965; Gandin and Kagan, 1976). The isotropic analog (of the correlation function in Figure 3.3), obtained by averaging two-dimensional estimates equally distant from the origin, is given in Table 3.3.

For comparison, two other isotropic correlation functions (from Gandin and Kagan, 1976) are also presented in this table. These two correlation functions were obtained by M. I. Fortus and M. I. Udin

3.8 Examples of Climatological Fields

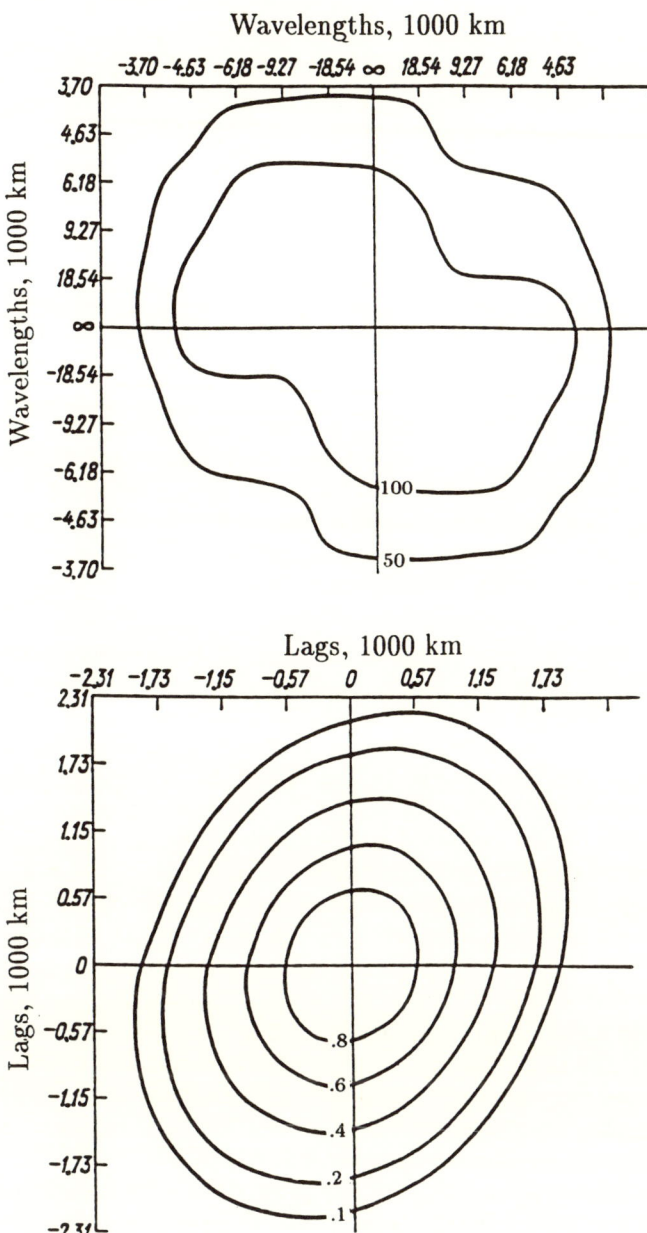

Figure 3.3: Two-dimensional spectral density (top) and correlation function (bottom) of the 500 mb geopotential surface.

from a long-term time series of observations. Comparison shows that the values of our estimates are situated between the results computed by Fortus and Udin. Therefore, in this case the two-dimensional approach was very effective because the volume of long-term observations, used by Fortus and Udin, is greater by many times than the number of observations of the above considered field.

3.8.2 Latitude-Temporal Fields

Another type of two-dimensional structure widely used in climatology is the latitude-temporal field, in which ξ is latitude and η is time. For instance, such fields can be obtained by spatially averaging data in different latitude bands (see, for example, Oort, 1983). These fields can have spatial (latitudinal) as well as temporal (seasonal cycle) trends that must be filtered out. As a rule, such field must be split along the latitude direction into (at least) three subfields corresponding to the tropical, midlatitude, and polar regions for each of which the violation of the homogeneity would not be very significant (see Chapter 7).

Let us consider three monthly mean observed fields of thickness (500–1000 mb), geopotential, and zonal component of geostrophic wind of the 500 mb surface for the midlatitudes of the Northern Hemisphere with the observational interval 1948 through 1978. The source of data is the archives of the World Data Center, Obninsk, Russia. These fields were obtained by data averaging in different 5°-width latitude bands, centered in the points 30°, 35°, ..., 80°. More elaborate analysis of these data has been provided by Polyak, 1979.

Here we will not discuss questions connected with the data accuracy conditioned by the temporal and spatial variation of the volume of observations. Such discussion can be found in Oort, 1978.

The filtering out of the trend of this data was done by estimating the monthly means and standard deviations for the time series of each latitude band and standardizing the corresponding deviations from these means. Spectral estimates, obtained by smoothing corresponding two-dimensional periodograms by the simplest scheme (1.143), are given in Figures 3.4 through 3.6 (top).

The geopotential and wind spectra have a very small maxima centered in the origin, while analogous spectral maximum of relative topography is slightly more noticeable. Therefore, the thickness fluctuations are mainly realized by means of the low-wave-number spatial and, to lesser degree, by low-frequency temporal waves. Corresponding correlation functions, presented in Figures 3.4 through

3.8 Examples of Climatological Fields

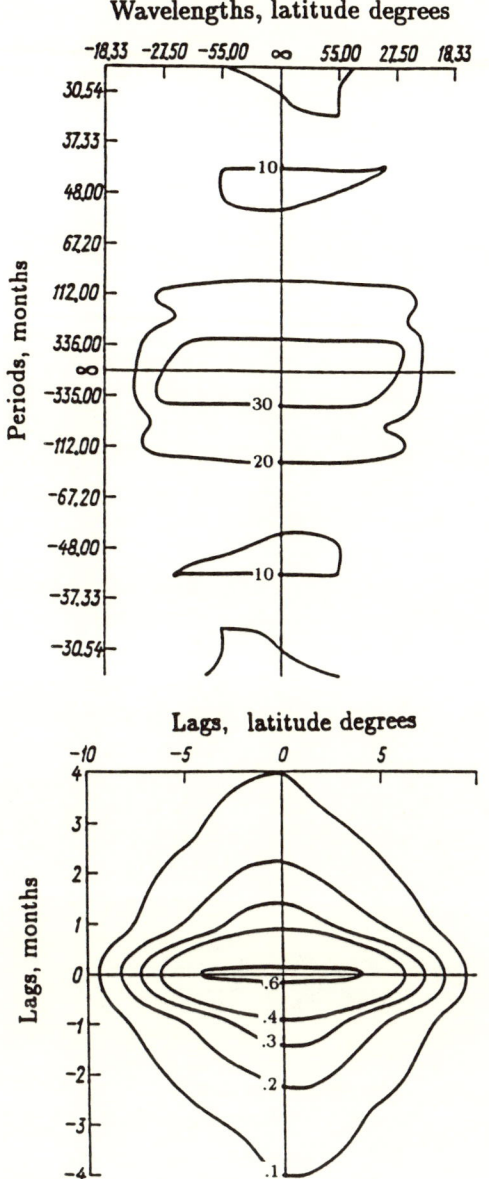

Figure 3.4: Central parts of two-dimensional spectrum (top) and correlation function (bottom) of the thickness (500 mb–1000 mb) latitude-temporal field.

152　　　　　　　　　　　　　　　　3 Random Processes and Fields

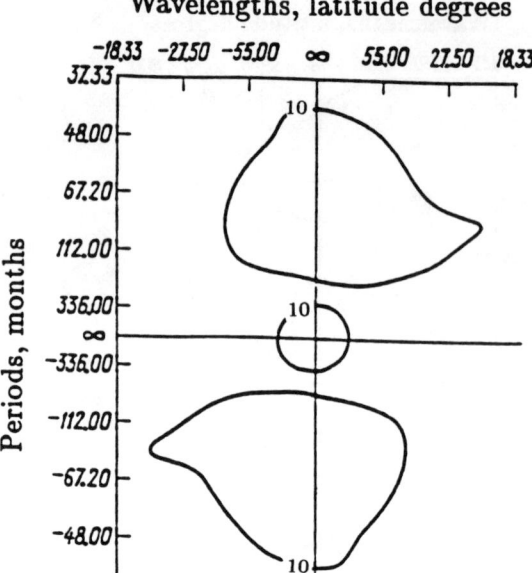

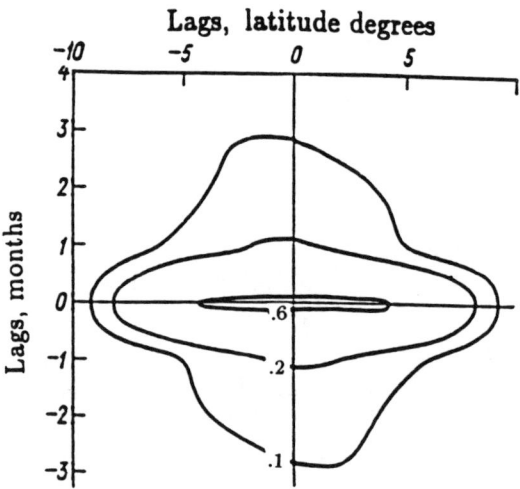

Figure 3.5: Central parts of two-dimensional spectrum (top) and correlation function (bottom) of the 500 mb surface geopotential latitude-temporal field.

3.8 Examples of Climatological Fields

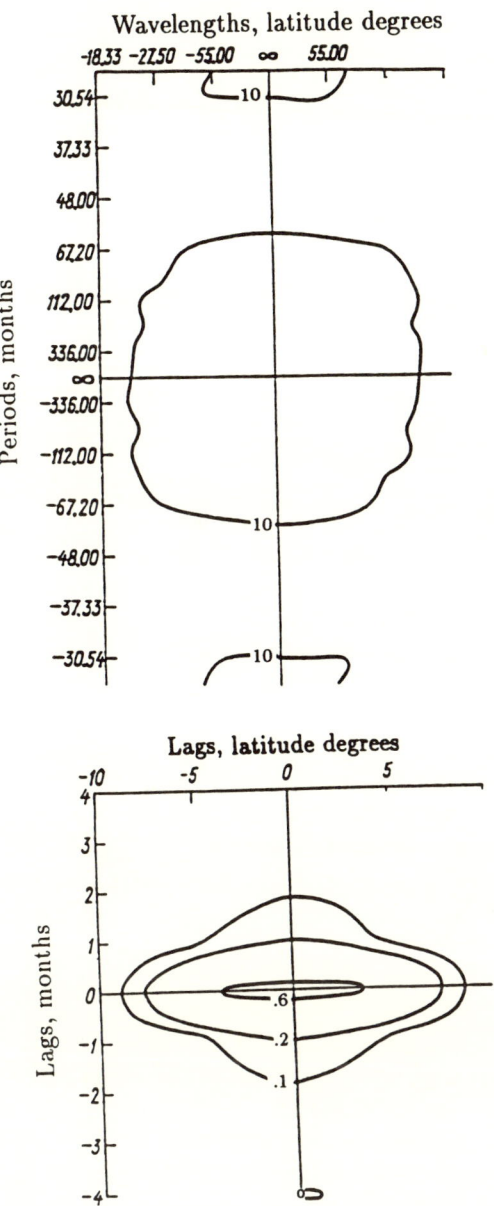

Figure 3.6: Central parts of two-dimensional spectrum (top) and correlation function (bottom) of the 500 mb surface geostrophic wind latitude-temporal field.

3.6 (bottom), demonstrate significant spatial statistical dependence of data. Spatial correlation coefficients for the 5° latitude lag are 0.53 for the relative topography and geopotential and 0.46 for the wind. Temporal statistical dependence is smaller: Correlations of the same variables for the one month lag are 0.37, 0.21, and 0.20 respectively.

3.8.3 Monthly-Annual Fields

The next type of climatological field is the monthly-annual field of the monthly mean time series when ξ is months (12 points) and η is years. Considering a monthly time series as a field enables us to split the interannual and intraannual variability for comparative study. This type of field has a temporal (seasonal cycle) trend that must be removed before the spectral analysis is done.

Time series of monthly means for each of the latitude band of any of the three fields considered above (thickness, geopotential, and zonal component of geostrophic wind of the 500 mb level of the Northern Hemisphere) can be presented and analyzed as a separate field. Because the total number of such time series for the variables considered above is too large, we will study only one of them (corresponding to the 55° latitude band) for each of the variables.

Spectra of these three fields are given in Figures 3.7 through 3.9 (top). An interesting feature of the thickness and geopotential spectra is the well-known small high-frequency maxima corresponding to the period of about two years. Additionally, the thickness spectrum has another small, wide maximum centered in the origin.

As for the geostrophic wind, the power distribution over the frequency domain is more or less even, and corresponding observations are very close to the sample of the two-dimensional white noise. The same conclusions can be drawn from consideration of the correlation functions in Figure 3.9 (bottom), because estimates of interannual (as well as intraannual) correlations are very close to zero.

Geopotential (Figure 3.8) exhibits small autocorrelation (0.2) at one-month lag and even smaller negative correlations (-0.12) for the one-year lag, which is the cause of the high-frequency maximum.

The two-dimensional correlation function of the relative topography (Figure 3.7) reveals a small temporal correlation (about 0.3) for the one-month lag.

As will be shown in Section 4.7, the most plausible conclusion from the above study is that the analyzed information has no forecasting value.

3.8 Examples of Climatological Fields

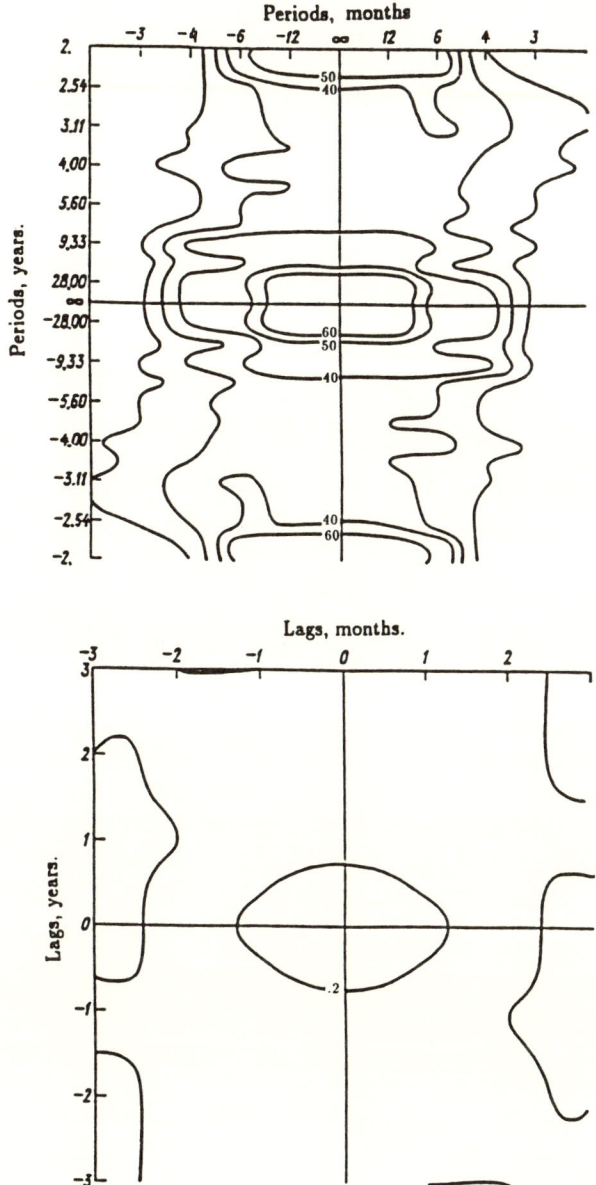

Figure 3.7: Central parts of two-dimensional spectrum (top) and correlation function (bottom) of the thickness (500 mb - 1000 mb) monthly mean time series.

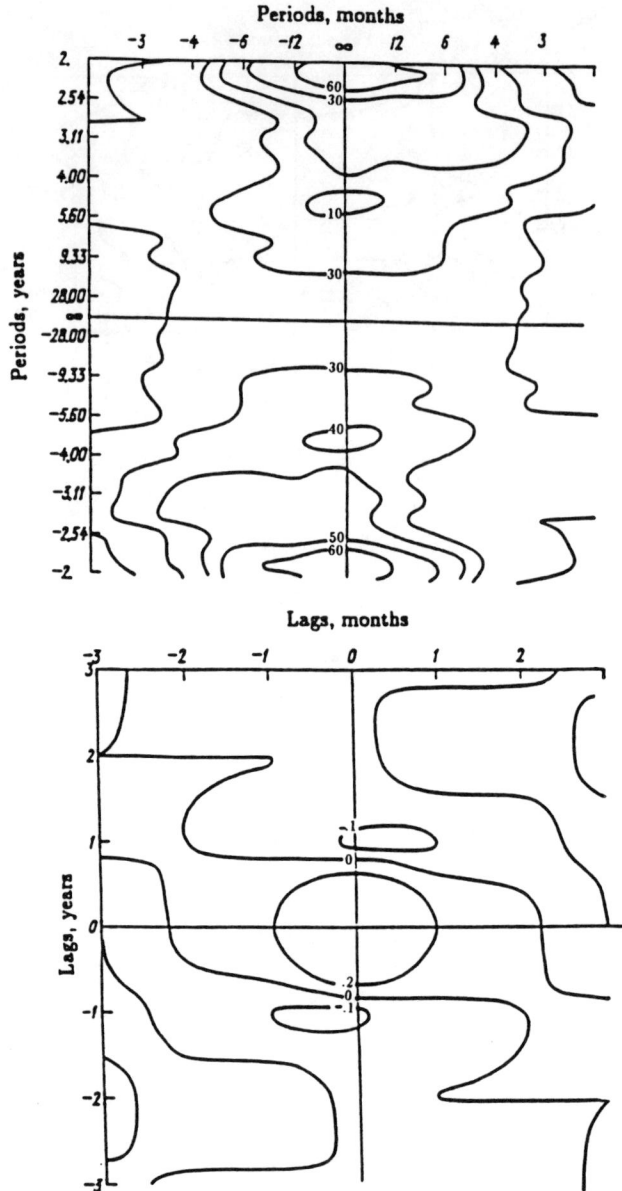

Figure 3.8: Central parts of two-dimensional spectrum (top) and correlation function (bottom) of the 500 mb surface monthly mean time series of geopotential.

3.8 Examples of Climatological Fields

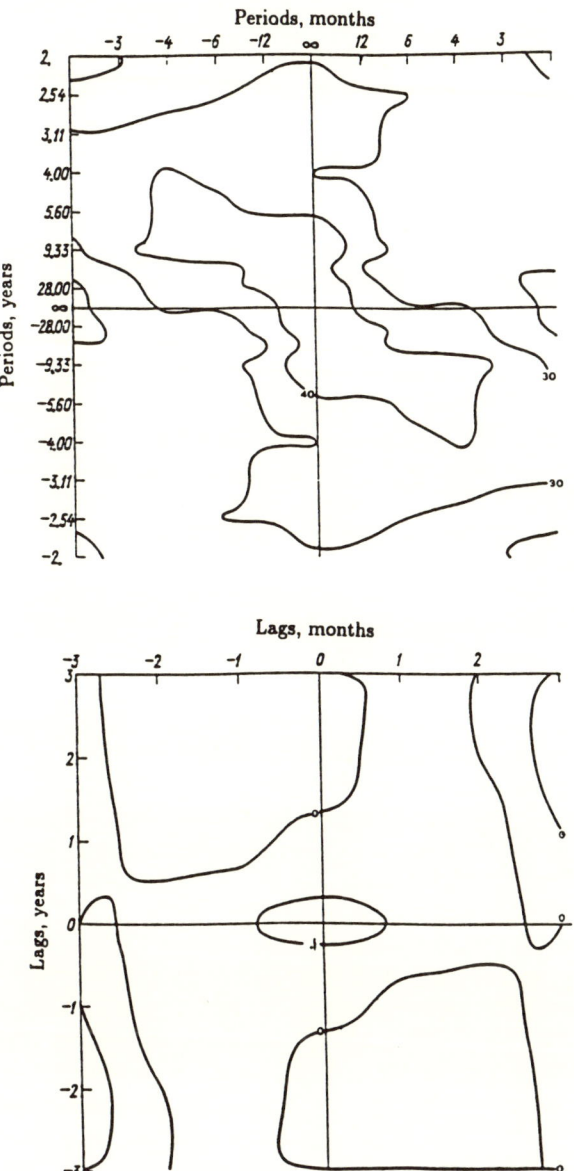

Figure 3.9: Central parts of two-dimensional spectrum (top) and correlation function (bottom) of the 500 mb surface monthly mean time series of geostrophic wind.

3.8.4 Altitude-Temporal Monthly Temperature Field Simulated by the Hamburg GCM

The next type of climatological field is the altitude-temporal field, in which ξ is altitude and η is time. This type of field presents the vertical cross-section of the three dimensional observed or simulated structure (latitude, altitude, time).

As Matrosova and Polyak (1990) have shown, such temperature fields are highly nonhomogeneous, or their first- and second-moment values vary with altitude. In particular, such fields of monthly data have a spatial trend (conditioned by the variation of the mean as a function of the altitude) as well as a seasonal cycle. For the example provided below, such trends were removed. Nonetheless, the results have a pure formal character because the correlations remained nonhomogeneous. To get a more accurate analysis, one must split this field along the altitude into at least four (Matrosova and Polyak, 1990) vertical subfields of a smaller size, simultaneously increasing their resolution. Such a procedure can be done with a specially designed simulation experiment.

Consider a vertical cross-section of the temperature data, simulated by one of the Hamburg GCMs (see Chapter 7). The size of the field is 10 (altitudes, from 1000 mb to 100 mb with a vertical step equal to 100 mb) × 372 (temporal points, 31 years of data with a one-month step). For each altitude level, data were averaged within a $10°$-width latitude band ($5°$ to $15°$) of the Northern Hemisphere. The seasonal cycle was removed.

The estimated spectrum (Figure 3.10, top) shows a concentration of power along the frequency axis with a significant maximum near the origin. This indicates that the major simulated variability is taking place in time and that the data from the model has a trend. The most powerful vertical fluctuations occur because of the long-length waves (greater than 350–400 mb).

The temporal statistical dependence of fluctuations (Figure 3.10, bottom) primarily indicates an intraannual character; the vertical correlation of data can be traced up to lags of 600 mb.

Of course, these examples do not exhaust all the field types, but this formal consideration demonstrates a diversity of climatological fields and their two-dimensional spectral and correlation characteristics. The estimates obtained were not accompanied with any physical analysis (as well as with accuracy estimation) because such insight requires special wording of climatological problems. Such problems will be considered in Chapters 7.

3.8 Examples of Climatological Fields

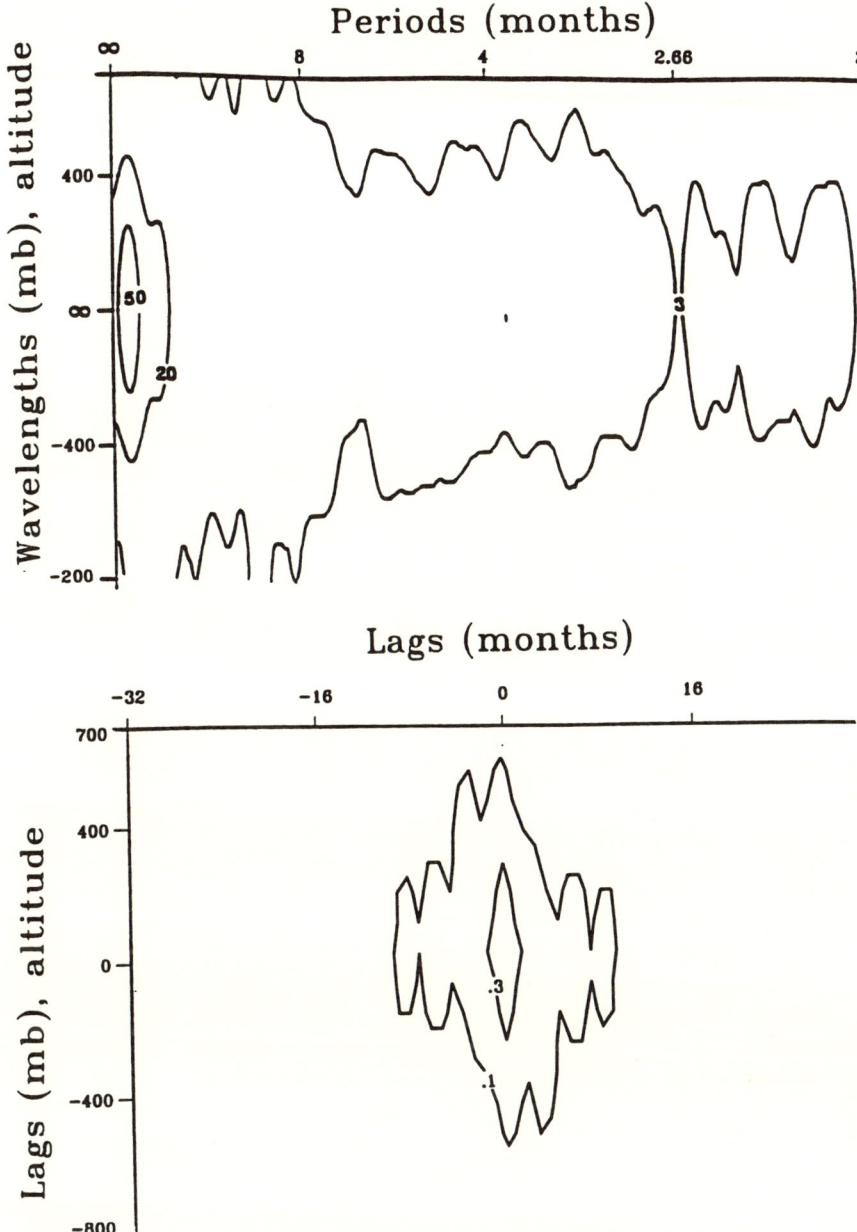

Figure 3.10: Spectrum (top) and correlation function (bottom) of the altitude-temporal simulated temperature field for 5° to 15° latitude band.

As has been mentioned, the principal obstacle to the two-dimensional spectral and correlation analysis of meteorological fields is in their nonhomogeneous statistical structure. One of the practical ways to overcome this obstacle is to reduce the field size while simultaneously increasing its spatial resolution. There are no principle difficulties in such procedures for the simulated data: a GCM must be integrated with a corresponding spatial grid. As for the observed data, it will require a new means of measurements, and especially the satellite observations.

3.9 Anisotropy of Climatological Fields

The spectral and correlation estimation schemes above were deduced for homogeneous random fields. As a rule, climatological fields are not homogeneous (and, of course, not isotropic). Processing their samples by the above estimation procedures leads to some approximations, although the accuracy varies for different elements and types of sampling. Corresponding reasoning about possible scales of distinction of the particular climate field under study from the homogeneous one can be done by analyzing its specific physical and statistical nature. In any case, as a rule, spectral and correlation estimates of climatological fields carry new information that sometimes helps to get approximate solutions to the problem considered or at least to make a step in a right direction.

However, for some meteorological variables and sampling types, the homogeneity assumption is nearly satisfied. Moreover, spatial correlation functions of the two-dimensional fields (of the above geopotential type) for many elements (air temperature, atmospheric, geopotential, and so on) were estimated under the assumption of isotropy and used in the objective analysis procedures (Gandin and Kagan, 1976).

Furthermore, some types of homogeneous random fields can be transformed to isotropic ones by rotating the coordinate system and scaling the axes differently. For this purpose, one can use, for example, modeling by the Laplace equation in accordance with the results of the paper by Jones and Vecchia (1993).

Consider the following two-dimensional stochastic Laplace equation:

$$\frac{\partial^2 \xi(x,y)}{\partial x^2} + \frac{\partial^2 \xi(x,y)}{\partial y^2} - a\xi(x,y) = \varepsilon(x,y), \qquad (3.100)$$

where a is positive and $\varepsilon(x,y)$ is two-dimensional white noise. The solution $\xi(x,y)$ for this equation is an isotropic random field. The

3.9 Anisotropy of Climatological Fields

correlation $\rho(r)$ between two points of this field at distance r apart is

$$\rho(r) = r\sqrt{a}K_1(r\sqrt{a}), \qquad (3.101)$$

where K_1 is the modified Bessel function of the second kind, order one. The corresponding standardized rational spectral density is

$$S(k_1, k_2) = \frac{1}{(k_1^2 + k_2^2 + a)^2}. \qquad (3.102)$$

If the new coordinate axes are u and v, the rotation of the coordinates by an appropriate angle α is the standard transformation

$$u = x\,\cos\alpha - y\,\sin\alpha$$

$$v = x\,\sin\alpha + y\,\cos\alpha.$$

The scaling of the two coordinate axes differently is carried out by an additional positive parameter λ^2. The scaled distance between observations i and j is

$$r = [\lambda^{-2}(u_i - u_j)^2 + \lambda^2(v_i - v_j)^2]^{1/2}. \qquad (3.103)$$

Therefore, to transform an anisotropic field to an isotropic one, two parameters, α and λ, must be estimated.

Practical applicability of this methodology can be evaluated by careful consideration of the isolines of the two-dimensional correlation function estimates to reveal that the anisotropic character of the field is not very complicated. As a result, it is possible to transform the field into an isotropic one by rotating and scaling coordinates. If the angle α, as well as the scaling parameter λ^2, is not constant and differs for different points of the field, this transformation cannot be done. The methodology (Jones and Vecchia, 1993) of transforming the anisotropic field into an isotropic field has many applications in climatology.

4

Variability of ARMA Processes

In this chapter, the numerical and pictorial interpretation of the dependence of the standard deviation of the forecast error for the different types and orders of univariate autoregressive-moving average (ARMA) processes on the lead time and on the autocorrelations (in the domains of the permissible autocorrelations) are given. While the convenience of fitting a stochastic model enables us to estimate its accuracy for the only time series under consideration, the graphs in this chapter demonstrate such accuracy for all possible models of the first and second order. Such a study can help in evaluating the appropriateness of the presupposed model, in carring out the model identification procedure, in designing an experiment, and in optimally organizing computations (or electing not to do so). A priori knowledge of the theoretical values of a forecast's accuracy indicates the reasonable limits of complicating the model and facilitates evaluation of the consequences of certain preliminary decisions concerning its application. The approach applied is similar to the methodology developed in Chapters 1 and 2. Because the linear process theory has been thoroughly discussed in the statistical literature (see, for example, Box and Jenkins, 1976; Kashyap and Rao, 1976; and so on), its principal concepts are presented in recipe form with the minimum of details necessary for understanding the computational aspects of the subject.

4.1 Fundamental ARMA Processes

Consider a discrete stationary random process z_t with null expected value $[\mathbf{E}(z_t) = 0]$ and autocovariance function

$$\mathbf{M}(\tau) = \sigma^2 \rho(\tau), \qquad (4.1)$$

where σ^2 is the variance and $\rho(\tau)$ is the autocorrelation function of z_t.

Let a_t be a discrete white noise process with a zero mean and a variance σ_a^2. Let us assume that processes z_t and a_t are normally distributed and that their cross-covariance function $\mathbf{M}_{za}(\tau) = 0$ if $\tau > 0$. Let us also assume that z_t can be presented as

$$z_t = \varphi_1 z_{t-1} + \ldots + \varphi_p z_{t-p} + a_t - \theta_1 a_{t-1} - \ldots - \theta_q a_{t-q}, \qquad (4.2)$$

where $\varphi_i (i = 1, 2, \ldots, p)$ and $\theta_j (j = 1, 2, \ldots, q)$ are parameters that satisfy the following conditions: The roots of the equations

$$1 - \varphi_1 x - \varphi_2 x^2 - \ldots - \varphi_p x^p = 0 \qquad (4.3)$$

and

$$1 - \theta_1 x - \theta_2 x^2 - \ldots - \theta_q x^q = 0 \qquad (4.4)$$

are outside the unit circle ($|x| > 1$). In this case, z_t is referred to as the autoregressive-moving average process of p, q order, and it is denoted as ARMA(p, q). Parameters φ_i are referred to as autoregressive parameters, and parameters θ_j are referred to as moving average parameters. The total number of parameters of the ARMA(p, q) process is equal to $p + q + 1$ (φ_i, θ_j, and σ_a^2). The condition ($|x| > 1$) determines the domain of the permissible values of parameters φ_i and θ_j.

Multiplying (4.2) by $z_{t-\tau}$ and taking the expectation of the product yields the following equations:

$$\mathbf{M}(\tau) = \varphi_1 \mathbf{M}(\tau - 1) + \ldots + \varphi_p \mathbf{M}(\tau - p) + \mathbf{M}_{za}(\tau)$$
$$- \theta_1 \mathbf{M}_{za}(\tau - 1) - \ldots - \theta_q \mathbf{M}_{za}(\tau - q) \quad \text{if } \tau > 0, \qquad (4.5)$$

$$\mathbf{M}(0) = \varphi_1 \mathbf{M}(1) + \ldots + \varphi_p \mathbf{M}(p) + \sigma_a^2 - \theta_1 \mathbf{M}_{za}(-1) - \ldots$$
$$- \theta_q \mathbf{M}_{za}(-q) \quad \text{if } \tau = 0. \qquad (4.6)$$

Letting $\tau = 1, 2, \ldots, p + q$ in (4.5) leads to a system of $p + q$ equations, which connect (in a single way) parameters φ_i and θ_j with the first

$p+q$ autocovariances and the first q cross-covariances. This system [together with conditions (4.3) and (4.4)] reveals that the domain of the permissible values of the parameters of the ARMA(p,q) process determines the domain of the permissible values of the autocorrelations.

Formula (4.2) enables us to forecast process z_t one step ahead. In such forecasting, the value of a_t is neglected and the forecast error variance is equal to the variance σ_a^2. By using (4.2) as a recursive expression for forecasting z_t for $l+1$ steps ahead, we obtain value $\bar{z}_{t+l}$. The variance $v_{p,q}^2(l)$ of the forecast error $(z_{t+l} - \bar{z}_{t+l})$ is determined by the formula (Box and Jenkins, 1976)

$$v_{p,q}^2(l) = \sigma_a^2(1 + \psi_1^2 + \psi_2^2 + \ldots + \psi_{l-1}^2), \tag{4.7}$$

where

$$\begin{aligned}
\psi_1 &= \varphi_1 - \theta_1, \\
\psi_2 &= \varphi_1\psi_1 + \varphi_2 - \theta_2, \\
&\ldots \\
\psi_j &= \varphi_1\psi_{j-1} + \ldots + \varphi_p\psi_{j-p} - \theta_j, \\
\psi_0 &= 1, \ \psi_j = 0 \ (\text{if } j < 0) \text{ and } \theta_j = 0 \ (\text{if } j > q).
\end{aligned} \tag{4.8}$$

Together with $v_{p,q}^2(l)$, the normalized (standardized) variance

$$\varepsilon_{p,q}^2(l) = v_{p,q}^2(l)/\sigma^2 \tag{4.9}$$

will be considered.

Later, when misunderstanding cannot arise, subscripts p and q will be omitted; that is, $\varepsilon(l)$ will be written instead of $\varepsilon_{p,q}(l)$ and $v(l)$ instead of $v_{p,q}(l)$.

Our goal is to study the theoretical limits of the forecast accuracy of the ARMA(p,q) processes in order to outline reasonable, practical situations for fitting a model and forecasting. Consequently, the main object of the numerical and pictorial interpretation will be the normalized standard deviation $\varepsilon(l)$ of the forecast error for $l+1$ steps ahead.

A fitting of the ARMA(p,q) process to a time series is provided by identifying the values of p,q and estimating the corresponding parameters φ_i, θ_j. The basis for such estimation is the system of $p+q$ equations [obtained from (4.5) for $\tau = 1, 2, \ldots, p+q$], where the values of the auto- and cross-covariance functions are replaced by their estimates.

4.2 AR Processes

If the stationary process z_t and white noise a_t are statistically independent, then all parameters θ_j are equal to zero. The process (4.2) is then called the autoregressive process of order p [abbreviated AR(p)]. It is presented by the formula

$$z_t = \varphi_1 z_{t-1} + \varphi_2 z_{t-2} + \ldots + \varphi_p z_{t-p} + a_t. \tag{4.10}$$

The number of the AR(p) process parameters equals $p+1$ (φ_i and σ_a).

From (4.5) and (4.1), the system of linear algebraic equations referred to as the Yule-Walker equations can be written as

$$\rho(\tau) = \varphi_1 \rho(\tau - 1) + \ldots + \varphi_p \rho(\tau - p), \ (\tau = 1, 2, \ldots, p). \tag{4.11}$$

This system connects parameters φ_i with the first p autocorrelations $\rho(\tau)$.

From (4.6), we have

$$\sigma_a^2 = \sigma^2(1 - r^2), \tag{4.12}$$

where

$$r^2 = \varphi_1 \rho(1) + \ldots + \varphi_p \rho(p) \tag{4.13}$$

is the squared multiple correlation coefficient of z_t and $z_{t-1}, z_{t-2}, \ldots, z_{t-p}$.

Therefore, all $p+1$ parameters of the AR(p) process and the values of its autocorrelation function are interconnected in a unique way.

Let $\varepsilon^2 = \varepsilon_{p,0}^2(1)$. Then (4.12) yields

$$\varepsilon^2 + r^2 = 1; \tag{4.14}$$

in other words, the normalized variance of the forecast error for one step ahead and the squared multiple correlation coefficients mutually determine each other.

By denoting the multiple correlation coefficient of z_{t+l} and $z_{t-1}, z_{t-2}, \ldots, z_{t-p}$ as $r^2(l)$, we get the relationship

$$\varepsilon^2(l) + r^2(l) = 1. \tag{4.15}$$

Formula (4.14) is a particular case of (4.15).

The AR(p) processes can be presented in matrix form in the following way:
Let us introduce the notation

$$\varphi = (\varphi_1, \varphi_2, \ldots, \varphi_p), \tag{4.16}$$

$$\mathbf{z}_0 = \begin{pmatrix} z_{t-1} \\ z_{t-2} \\ \ldots \\ z_{t-p} \end{pmatrix}, \tag{4.17}$$

$$\mathbf{A} = \begin{pmatrix} \varphi_1 & \varphi_2 & \varphi_3 & \ldots & \varphi_{p-1} & \varphi_p \\ 1 & 0 & 0 & \ldots & 0 & 0 \\ 0 & 1 & 0 & \ldots & 0 & 0 \\ \ldots & \ldots & \ldots & \ldots & \ldots & \ldots \\ 0 & 0 & 0 & \ldots & 1 & 0 \end{pmatrix}. \tag{4.18}$$

Then

$$\bar{z}_{t+l} = \varphi \mathbf{A}^l \mathbf{z}_0 \quad (l = 0, 1, 2, \ldots), \tag{4.19}$$

$$v_{p,0}^2(l) = \sigma^2[1 - \varphi \mathbf{A}^l \mathbf{R}(\mathbf{A}^T)^l \varphi^T] = \sigma^2 \varepsilon_{p,0}^2(l), \tag{4.20}$$

where

$$R = \begin{pmatrix} 1 & \rho(1) & \rho(2) & \ldots & \rho(p-1) \\ \rho(1) & 1 & \rho(1) & \ldots & \rho(p-2) \\ \ldots & \ldots & \ldots & \ldots & \ldots \\ \rho(p-1) & \rho(p-2) & \rho(p-3) & \ldots & 1 \end{pmatrix} \tag{4.21}$$

is the matrix of the Yule-Walker system; the T is the sign of transposition.

The algorithm of forecasting for l steps ahead can be presented in the following way:

$$l = 0, \ \mathbf{Q}_l = \varphi,$$

$$\begin{cases} \longrightarrow \bar{z}_{t+l} = \mathbf{Q}_l \mathbf{z}_0, \\ \\ v_{p,0}^2(l) = \sigma^2(1 - \mathbf{Q}_l \mathbf{R} \mathbf{Q}_l^T), \\ \\ \longleftarrow \mathbf{Q}_l = \mathbf{Q}_{l-1} \mathbf{A} \quad (l = 1, 2, \ldots). \end{cases}$$

As we will see in the next chapter, such a notation does not differ from the corresponding representation of the algorithm for the multivariate case.

4.3 AR(1) AND AR(2) Processes

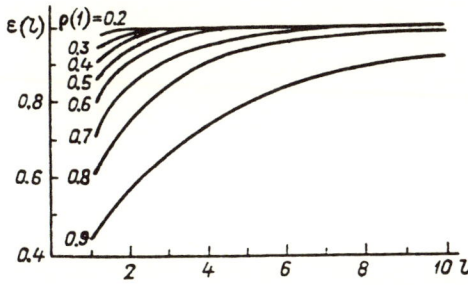

Figure 4.1: Dependence of the normalized standard deviation $\varepsilon(l)$ of the forecast error of the AR(1) process on autocorrelation ρ and on lead time l.

4.3 AR(1) and AR(2) Processes

If $p = 1$, formulas (4.10) to (4.12) yield

$$z_t = \rho z_{t-1} + a_t, \tag{4.22}$$

$$\varphi_1 = \rho(1) = \rho, \tag{4.23}$$

$$\rho(\tau) = \rho^{|\tau|}, \tag{4.24}$$

$$v^2(1) = \sigma^2(1 - \rho^2). \tag{4.25}$$

The normalized variance $\varepsilon^2(l)$ of the forecast error for l steps ahead is obtained from (4.6) and (4.7) by the simple rearrangement

$$\varepsilon^2(l) = 1 - \rho^{2l}. \tag{4.26}$$

The value of $\varepsilon(l)$ as a function of parameters ρ and l is presented in Figure 4.1.

We can see that, if $\rho \leq 0.3$, the standard deviation of the forecast error for any l is greater than 0.95 (or the standard deviation of the forecast error constitutes more than 95% of the standard deviation of process z_t). The practical usefulness of such a forecast is doubtful. If $0.3 < \rho \leq 0.5$, the value of $\varepsilon(1)$ varies from 0.87 to 0.95; that is, a forecast, even for one step ahead, has a very low accuracy; if l is greater than 1, such a forecast will likely have no practical value. If ρ is about 0.6, the forecast makes sense for not more than two steps ahead, and in this case $\varepsilon(l)$ is varied from 0.8 to 0.93. If ρ is about 0.7, the forecast can be considered for two or three steps ahead. It is only if ρ is greater than 0.8 that the forecast for four of more steps ahead is meaningful.

Because dependence of the variance on the autocorrelation (4.26) is continuous, the above statement about the boundaries, which separate a reasonable forecast from a senseless one, is purely subjective. However, from (4.26) and its numerical interpretation, it is clear that for a first-order process each value of $|\rho| < 1$ determines a theoretical limitation on the forecast accuracy. In particular, small autocorrelations ($\rho \leq 0.3$) (or their estimates, even statistically significant) mean that corresponding forecasts will be very close to the mean value (zero in our case) and that such forecasting makes no sense, even for one step ahead.

The above assertions become especially obvious when one considers the dependence of the lead time l on autocorrelation ρ and ε. In this case, (4.26) yields (for $\rho > 0$)

$$l = \frac{\ln[1 - \varepsilon^2(l)]}{2 \ln \rho}. \tag{4.27}$$

This dependence, presented in Figure 4.2, demonstrates a useful fact: that with ρ increasing from 0 to about 0.6, a possible lead time l increases very slowly, and a forecast for more than two steps ahead (as has already been shown) is senseless.

It is only for $\rho > 0.7$ (up to 1) that the possible lead time begins to increase rapidly, converging to the infinite value (when ρ is equal to 1 and the forecast is deterministic). The detailed consideration of the simplest AR(1) process is justified by the fact that, as will be shown shortly, the above numerical interpretation of the relationship between autocorrelation ρ and the forecast accuracy (or lead time) is approximately correct for the processes of any orders. Parameter $\varepsilon(l)$ is conveniently used in climatology as a primary characteristic of the forecast accuracy; therefore, it will be of constant interest to us.

The second-order autoregressive process

$$z_t = \varphi_1 z_{t-1} + \varphi_2 z_{t-2} + a_t \tag{4.28}$$

has the Yule-Walker system

$$\begin{aligned} \rho(1) &= \varphi_1 + \varphi_2 \rho(1) \\ \rho(2) &= \varphi_1 \rho(1) + \varphi_2, \end{aligned} \tag{4.29}$$

4.3 AR(1) AND AR(2) Processes

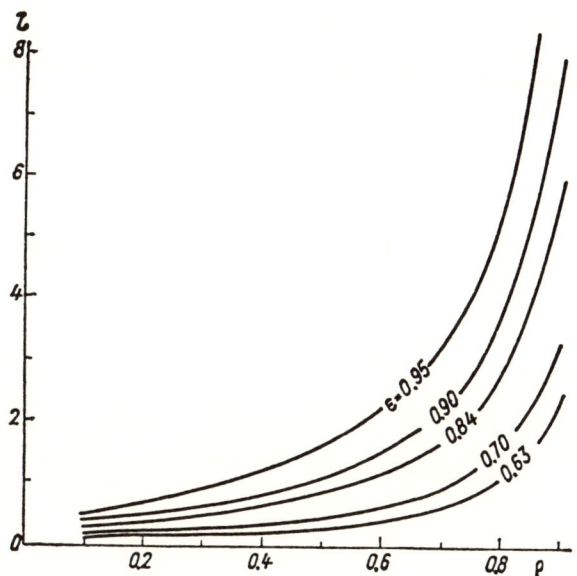

Figure 4.2: Dependence of the lead time of the AR(1) process on autocorrelation ρ and on standard deviation $\varepsilon(l)$.

the matrix of which

$$\mathbf{R} = \begin{pmatrix} 1 & \rho(1) \\ \rho(1) & 1 \end{pmatrix} \tag{4.30}$$

has the inverse matrix

$$\mathbf{R}^{-1} = \frac{1}{1 - \rho^2(1)} \begin{pmatrix} 1 & -\rho(1) \\ -\rho(1) & 1 \end{pmatrix}. \tag{4.31}$$

Using (4.12), the normalized variance ε^2 of the white noise a_t is presented as

$$\varepsilon^2 = 1 - \varphi_1 \rho(1) - \varphi_2 \rho(2) = \frac{1 - \rho(2)}{1 - \rho^2(1)} \left[1 + \rho(2) - 2\rho^2(1)\right]. \tag{4.32}$$

The AR(2) forecast accuracy is determined by the first and second autocorrelations [$\rho(1)$ and $\rho(2)$]. By defining a two-dimensional grid on the plane $[\rho(1), \rho(2)]$ with the domain of the permissible autocorrelations and performing the computations by (4.12) for each point of the grid, it is possible to obtain the fields of the normalized standard deviation $\varepsilon(l)$ for different l. The results of such computations are symmetrical relative to the plane $[\rho(2), \varepsilon(l)]$, and for this reason, they are presented in Figures 4.3 and 4.4 only for $\rho(1) > 0$. These figures show the shapes and corresponding isolines of the $\varepsilon(l)$ surface as a function of the autocorrelations; that is, they illustrate the character of the variations of the theoretical accuracy of the forecast corresponding to different lead times. If $l = 1$, the part of domain $[\rho(1), \rho(2)]$ where $\varepsilon(1)$ is close to 1 [i.e., $\varepsilon(l) > 0.95$] is not large. With increasing l, this part of the domain gradually increases, spreading over almost the entire area of the permissible autocorrelations. The closeness of ε to 1 means that variance $v^2(l)$ of the forecast error is close to variance σ^2 of process z_t and that the forecast is senseless in the sense that it does not differ markedly from the mean.

The greater $\rho(1)$ and $\rho(2)$ (i.e., the closer the point is to the boundary of the domain), the higher the accuracy of the forecast. If the statistical predictability of a time series is studied, one must evaluate the possible lead time of a forecast for the fixed accuracy values. In this case, together with the results presented in Figures 4.3 and 4.4, the lead time l field [as functions of $\rho(1)$ and $\rho(2)$ and parameter ε] can give a clearer picture of the predictability limits of the autoregressive processes.

Let us consider the numerical interpretation of the fields of l for the two fixed values of the parameter ε ($\varepsilon = 0.95$ and $\varepsilon = 0.8$). The results (Figure 1.5) show that the lead time is less than one (step ahead), if $\rho(1) < 0.3$ and $\rho(2) < 0.3$. Even for the minimal accuracy ($\varepsilon = 0.95$), one can speak about a forecast for only two steps ahead, if $\rho(1) > 0.5$ and $\rho(2) > 0.3$, and so on.

Therefore, the requirements for the autocorrelation values are approximately the same as those for the AR(1) process, and the forecast accuracy (and the lead time) of the second-order AR processes is also approximately the same as for the AR(1). Of course, a class of physical processes, which can be approximated by the AR(2) is extended: One can obtain a more detailed and accurate description of the approximated time series, including some subtle features of spectral density. But the significant increase in forecast accuracy and in the lead time cannot be expected in theoretical or practical terms. Later, this assertion will be illustrated more clearly.

Information presented in Figures 4.3 to 4.5 simplifies the preliminary analysis and identification of a model, as well as the accuracy evaluation and the building of a forecast confidence interval.

4.4 Order of the AR Process

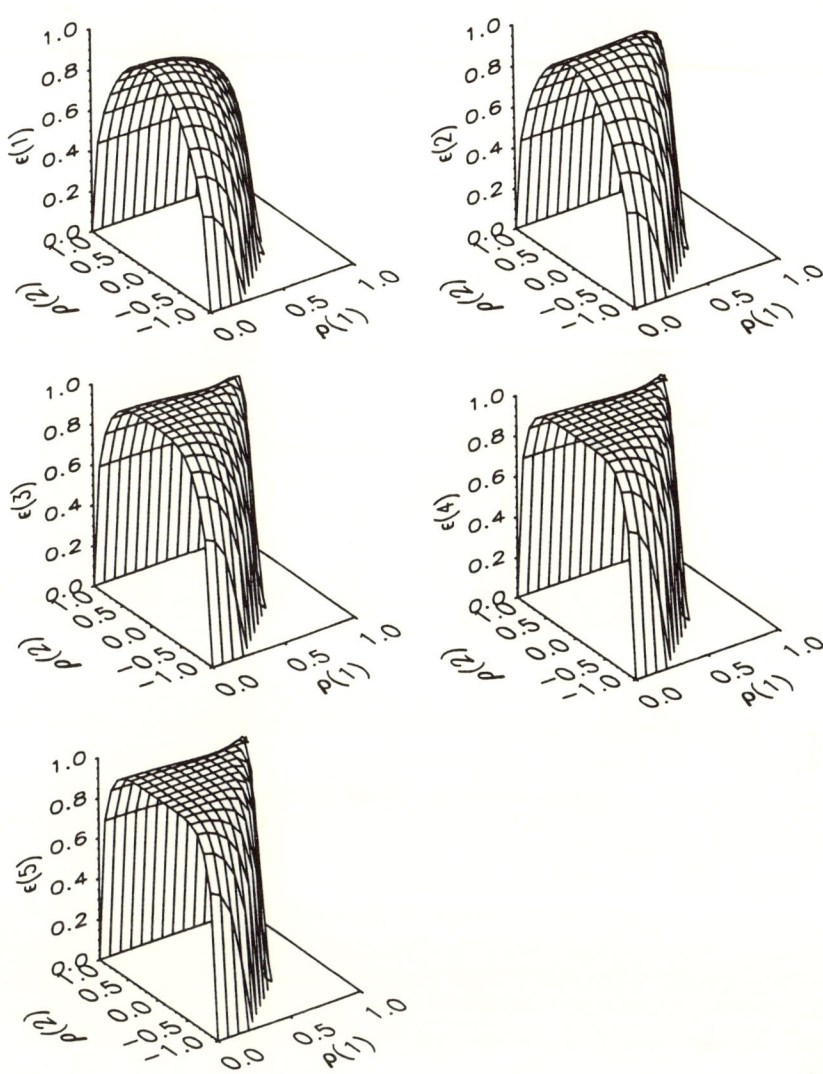

Figure 4.3: Normalized standard deviation $\varepsilon(l)$ of the forecast error of the AR(2) process for $l=1$ to 5.

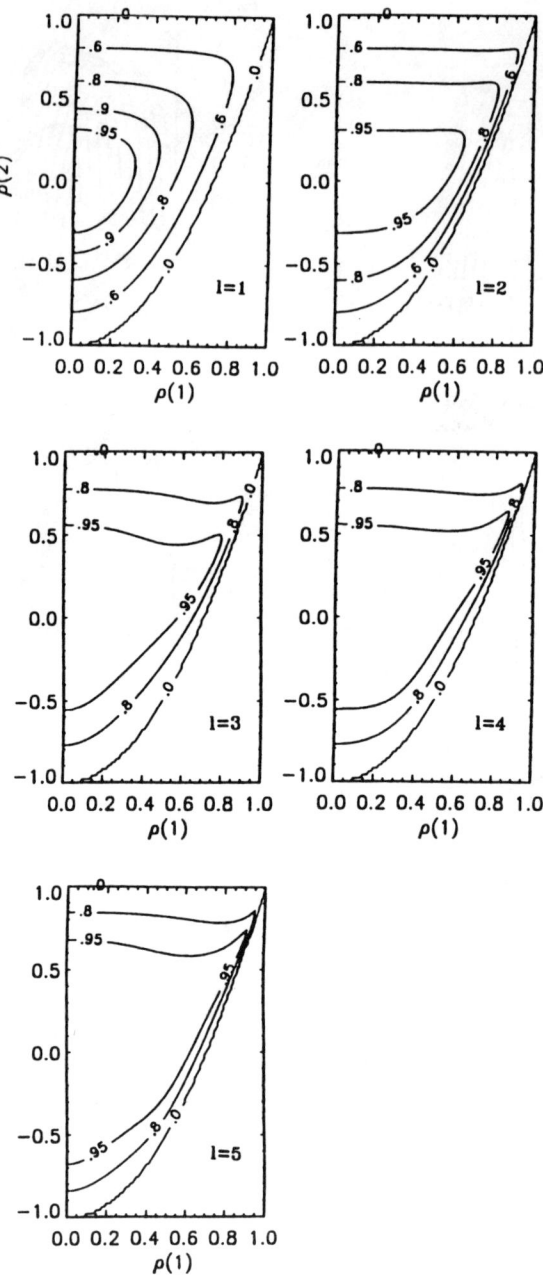

Figure 4.4: Contour lines of the structures in Figure 4.3.

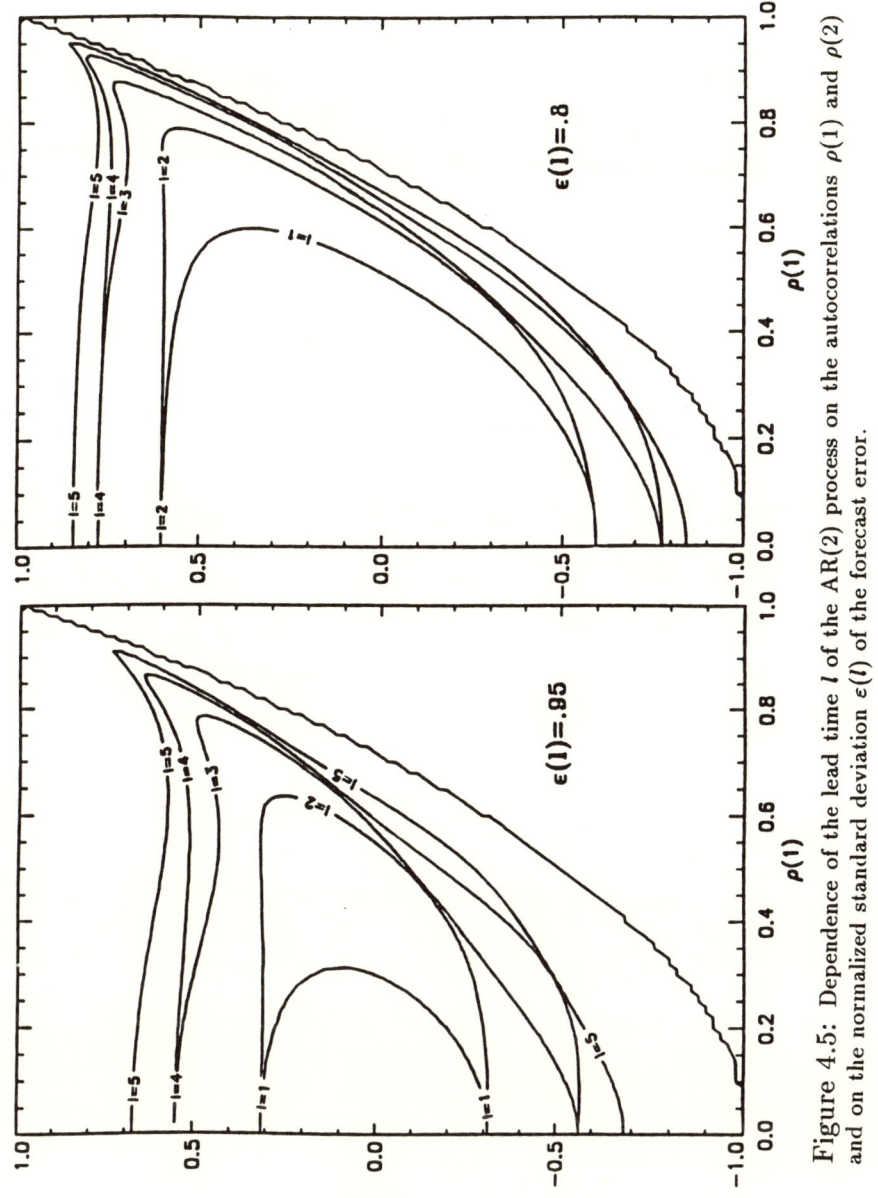

Figure 4.5: Dependence of the lead time l of the AR(2) process on the autocorrelations $\rho(1)$ and $\rho(2)$ and on the normalized standard deviation $\varepsilon(l)$ of the forecast error.

The advantage of such an approach is that it permits one to formulate the requirements to the statistical characteristics of the time series, which guarantee the desired forecast accuracy (or lead time). Such an analysis can also be helpful when working with a time series with little data. In this case, it is possible to find the confidence intervals of the autocorrelation estimates which determine the corresponding domain in lag space which, in turn, determines some region of the normalized standard deviations of the forecast error.

4.4 Order of the AR Process

The numerical and pictorial interpretation, analogous to those in Section 4.3, for the high-order (greater than 2) autoregressive process is cumbersome, because the elements of the large-size tensors must be computed and illustrated. Instead of obtaining a complete theoretical picture of the dependence of the forecast accuracy of any AR processes on possible autocorrelations, lead times, and orders, we simplify the problem and consider an example with a special type of autocorrelation function:

$$\rho(\tau) = \begin{cases} 1 & \text{if} \quad \tau = 0, \\ \rho & \text{if} \quad 0 < \tau < p. \end{cases} \qquad (4.33)$$

and $\rho(\tau) \to 0$ if $\tau \to \infty$. To determine parameters φ_i $(i = 1, 2, \ldots, p)$ of the corresponding autoregressive model

$$z_t = \varphi_1 z_{t-1} + \varphi_2 z_{t-2} + \ldots + \varphi_p z_{t-p} + a_t, \qquad (4.34)$$

it is necessary to find the solution of the Yule-Walker system

$$\begin{aligned} \rho &= \varphi_1 + \varphi_2 \rho + \varphi_3 \rho + \ldots + \varphi_p \rho, \\ \rho &= \varphi_1 \rho + \varphi_2 + \varphi_3 \rho + \ldots + \varphi_p \rho, \\ &\quad \ldots\ldots\ldots\ldots \\ \rho &= \varphi_1 \rho + \varphi_2 \rho + \varphi_3 \rho + \ldots + \varphi_p. \end{aligned} \qquad (4.35)$$

This solution

$$\varphi_i = \varphi = \frac{\rho}{1 + (p-1)\rho} = \frac{1}{1/\rho + p - 1} \quad (i = 1, 2, \ldots, p) \qquad (4.36)$$

4.4 Order of the AR Process

makes it possible to present (4.34) as

$$z_t = \frac{1}{1/\rho + p - 1}(z_{t-1} + z_{t-2} + \ldots + z_{t-p}) + a_t. \tag{4.37}$$

It is clear that the maximum weight φ in (4.37) obtained for ρ equals 1 when the process is a linear function.

The identity of all the autocorrelations (4.33) leads to the identity of all the parameters in (4.34) as well as to the identity (independence of lead time) of the standard deviations of the forecast errors. Substituting φ and ρ in (4.12) and (4.13) for (4.36) and (4.33) yields

$$\varepsilon^2 = 1 - \frac{p\rho^2}{1 + (p - 1)\rho}. \tag{4.38}$$

Formula (4.38) describes the behavior of the variance as a function of parameters ρ and p. If ρ is equal to zero, then $\varepsilon^2 = 1$, and the forecast is equal to the mean of the process. If $\rho = 1$, then $\varepsilon^2 = 0$, and the forecast is deterministic because separate values of z_t are linearly dependent.

If $p = 1$, then

$$\varepsilon^2 = 1 - \rho^2 \tag{4.39}$$

is the normalized variance of the forecast error of the $AR(1)$ process.

If p increases, the ε^2 monotonically decreases and

$$\varepsilon^2 \to 1 - \rho. \tag{4.40}$$

This fact shows that the normalized variance of the forecast error is limited from below (by the quantity $1 - \rho$), and building up the order cannot lead to the unlimited decrease of ε^2.

A more detailed study of the behavior of the normalized standard deviation ε can be done by computing values (4.38) for different $\rho > 0$ for some two-dimensional grid points of ρ and p.

The results of such computations (Figure 4.6) show that for small values of ρ ($\rho \leq 0.2$), the normalized standard deviation ε is greater than 0.9 independent of the order. In other words, it is equal to 90% or more of the standard deviation of the process.

Such a forecast is insufficiently deviated from the mean, and its estimation will not make any practical sense. In the band $0.2 < \rho \leq 0.3$ the normalized standard deviation gradually decreases (with increasing p) and (if $p > 5$) becomes very close to the value $1 - \rho$

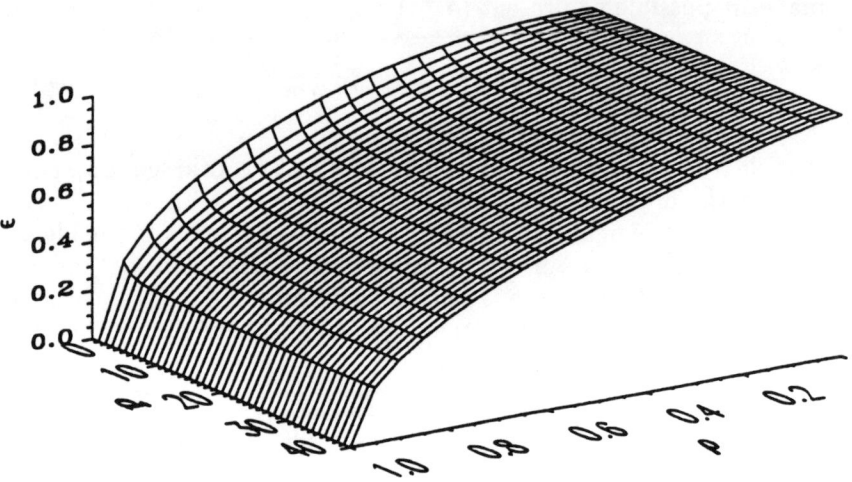

Figure 4.6: Normalized standard deviation ε of the forecast error as a function of autocorrelation ρ and order p.

(0.84–0.9); so the practical usefulness of increasing order ($p > 1$) is doubtful.

As will be shown in Chapter 6, which deals with the monthly and annual means of the historical records, the range (0.0–0.3) contains the values of the autocorrelation estimates of climatological time series of separate stations. Therefore, the above inferences characterize the possible accuracy of the forecasts of such a time series if they are approximated by the AR process. The results show that the benefit of forecasting by corresponding AR models is very low, if it has any benefit at all.

If $0.3 < \rho \leq 0.5$, the processes of the first through the fourth orders can provide relatively acceptable forecast accuracy ($\varepsilon = 0.84,\ldots,$ 0.71); however, building up the order further does not increase the accuracy. (For the first-order moving average process, the interval $|\rho| \leq 0.5$ presents the entire domain of the permissible autocorrelations; see the following section.) In this domain, the model type identification must distinguish between the autoregressive and the moving average model.

The values $\rho \geq 0.6$ can ensure a sufficiently high accuracy ($\varepsilon > 0.64$); therefore, fitting AR models can be of significant interest for the time series with such correlation values.

4.4 Order of the AR Process

The results (Figure 4.6) show that increasing the forecast accuracy by increasing the process order occurs very slowly, especially if $p > 3$. For this reason, if $p \geq 4$, inaccuracy in one or two units in the order identification cannot result in a significant increase of the forecast error.

From (4.39) and (4.40) one can derive the inequality

$$1 - \rho < \varepsilon^2 \leq 1 - \rho^2 < 1, \tag{4.41}$$

which determines the value band for the normalized variance of the forecast error of the autoregressive process independently of its order.

The above reasoning can be summed up with the following comments.

1. Evaluating the forecast accuracy of the AR processes, one can notice that the principal meaning presents not simply the non-zero autocorrelation between the terms of the time series but, primarily, its value. For instance, if $0.1 < |\rho| < 0.3$, the statistical dependence of the terms of time series is obvious but the corresponding forecast offers virtually no helpful information.

2. Large autocorrelations are the decisive factor for providing a high-accuracy forecast. Identification of the model's order can only slightly improve it. For this reason, sufficiently increasing the order (> 2) of the fitted model can be based mainly on very accurate physical reasoning. (Spectral analysis presents significantly higher requirements for the accuracy of the model order identification than forecasting does.)

3. The results obtained reveal the uncertainty of categorical statements, which can sometimes be found in climatological papers, about the predictability of various climate time series, when the latter differs from the sample of white noise. Because ε continuously depends on ρ, the autocorrelations and the forecast accuracy mutually determine each other. Therefore, reasoning about the predictability limits, as well as the practical ascertainment of any actual (fixed) value of ε as an acceptable normalized standard deviation of the forecast error, is subjective (arbitrary).

The main identification problem is determining an appropriate order p which guarantees that the AR(p) process fits the statistical and physical properties of the observations. Practically, such identification is carried out by a multi-step empirical-statistical procedure (see Anderson, 1971; Eikhoff, 1983; and Hannan, 1970); that is, by the multiple fitting to given data of the $AR(p)$ process by consequently increasing order p. For each step the two following problems (together with the parameter estimation) must be solved:

1. The evaluation of the statistical significance of the parameter φ_p estimate. If size of the sample is large, the simplest procedure for the evaluation is application of the asymptotic criterion, stating that the statistic

$$\sqrt{k} \cdot \bar{\varphi}_p \qquad (4.42)$$

has an approximately normal distribution with a zero mean and a variance equal to one.

2. Testing the hypothesis that the order is equal to p versus the alternative hypothesis that the order is $p+s$. Such testing can be carried out with the aid of the following statistic:

$$Q^2 = 1 - \frac{s_{p+\nu}^2(1)}{s_p^2(1)}, \qquad (4.43)$$

where $s_{p+\nu}^2(1)$ and $s_p^2(1)$ are the estimates of the variance σ_a^2 (4.12) for the $p+\nu$ and p order models respectively. For large k, quantity kQ^2 has χ^2 distribution with ν degrees of freedom.

Finding the parameter estimates φ_i by solving the system of Yule-Walker equations (4.11), where the theoretical autocorrelations are replaced by their estimates, is the first part of the process of fitting an AR model to the time series. After that, the estimates of $v(l)$ are found by (4.7), and a confidence interval of the forecast for $l+1$ steps ahead is constructed by the formula

$$\bar{z}_{t+l} \pm \nu v(l), \qquad (4.44)$$

where $\nu = 2$ or 3 respective to the 5% or the 1% level of significance.

4.5 MA(1) and MA(2) Processes

The moving average process z_t of order q [MA(q)] presents each value z_t as a linear combination of the uncorrelated random quantities. Such a process is denoted as

$$z_t = a_t - \theta_1 a_{t-1} - \theta_2 a_{t-2} - \ldots - \theta_q a_{t-q}. \tag{4.45}$$

The process MA(q) is determined by $q+1$ parameters $\theta_1, \theta_2, \ldots, \theta_q$, and σ_a^2. It is clear that (4.45) is the particular case of (4.2), if all the autoregressive parameters are zero.

The equation for the autocorrelation function of the MA(q) process is derived from (4.5) as

$$\rho(\tau) = \begin{cases} \dfrac{-\theta_\tau + \theta_1 \theta_{\tau+1} + \ldots + \theta_{q-\tau}\theta_q}{1 + \theta_1^2 + \theta_2^2 + \ldots + \theta_q^2} & \text{if } \tau = 1, 2, \ldots, q, \\ 0 & \text{if } \tau > q. \end{cases} \tag{4.46}$$

The equation for the variance σ_a^2 is derived from (4.6) as

$$\sigma_a^2 = \frac{\sigma^2}{1 + \theta_1^2 + \theta_2^2 + \ldots + \theta_q^2}. \tag{4.47}$$

Because the values of the MA(q) process are uncorrelated for lags greater than q, the corresponding forecasts do not make sense for more than q steps ahead.

The variance of the forecast error for l steps ahead is determined by the formula

$$v^2(l) = \sigma_a^2 \left(1 + \theta_1^2 + \theta_2^2 + \ldots + \theta_{l-1}^2\right). \tag{4.48}$$

The normalized variance of the forecast error for l steps ahead is obtained from (4.48) as

$$\varepsilon^2(l) = \frac{\sum_{j=0}^{l-1} \theta_j^2}{\sum_{j=0}^{q} \theta_j^2} \quad (l \leq q). \tag{4.49}$$

Let us consider normalized standard deviation of the forecast error for the MA(1) process.

Setting $q = 1$ in (4.45), the MA(1) process is presented as

$$z_t = a_t - \theta a_{t-1}. \tag{4.50}$$

The autocorrelation function and parameter θ are interconnected by the relationship (4.46) (if $q = 1$)

$$\rho(\tau) = \begin{cases} -\dfrac{\theta}{1+\theta^2} & \text{if } \tau = 1, \\ 0 & \text{if } \tau > 1. \end{cases} \tag{4.51}$$

From this formula, it follows that the estimator of θ can be found by solving the quadratic equation

$$\theta^2 + \theta/\rho + 1 = 0. \tag{4.52}$$

The forecast of the MA(1) process makes sense only for one step ahead. If $q = 1$, the normalized standard deviation (4.49) is

$$\varepsilon^2(1) = 1/(1+\theta^2). \tag{4.53}$$

The domain of permissible autocorrelations for an MA(1) process is determined by the inequality

$$|\rho| \leq 0.5. \tag{4.54}$$

Figure 4.7 enables us to compare the values of $\varepsilon(1)$ for MA(1) and AR(1) processes.

This reveals that the MA(1) process forecast has a slightly higher theoretical accuracy compared with the AR(1) process, but the small size of its domain of permissible autocorrelations limits the possibility of its practical application. If ρ increases, the distinction of the forecast accuracy of the MA(1) and AR(1) processes increases. This distinction is especially large in the subinterval (0.4–0.5), reaching a maximum at the endpoint of the domain (if $\rho = 0.5$).

Consider normalized standard deviation of the forecast error of the MA(2) process.

4.5 MA(1) and MA(2) Processes

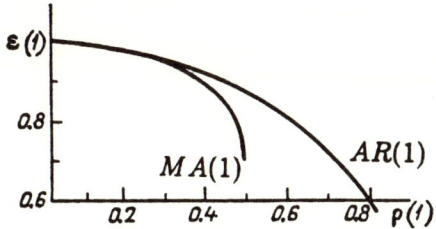

Figure 4.7: Dependence of the normalized standard deviation $\varepsilon(1)$ of the forecast error on the autocorrelation $\rho(1)$ for the MA(1) and AR(1) processes.

The domain of the permissible autocorrelations of the MA(2) process

$$z_t = a_t - \theta_1 a_{t-1} - \theta_2 a_{t-2} \tag{4.55}$$

is bounded by the lines

$$\rho^2(1) = 4\rho(2)(1 - 2\rho(2)),$$
$$\rho(2) + \rho(1) = -0.5,$$
$$\rho(2) - \rho(1) = -0.5. \tag{4.56}$$

This domain is presented in Figures 4.8 and 4.9.

The variance of process (4.55) is

$$\sigma^2 = \sigma_a^2(1 + \theta_1^2 + \theta_2^2) \tag{4.57}$$

and the relationships between the autocorrelation function $\rho(\tau)$ and parameters θ_i are obtained from (4.46) (setting $q = 2$) as

$$\begin{aligned}
\rho(1) &= -\frac{\theta_1(1 - \theta_2)}{1 + \theta_1^2 + \theta_2^2}, \\
\rho(2) &= \frac{-\theta_2}{1 + \theta_1^2 + \theta_2^2}, \\
\rho(\tau) &= 0 \text{ if } \tau > 2.
\end{aligned} \tag{4.58}$$

Our interest lies in analysis of the normalized variance of the forecast error. From (4.49), we have

$$\begin{aligned}
\varepsilon^2(1) &= \frac{1}{1 + \theta_1^2 + \theta_2^2}, \\
\varepsilon^2(2) &= \frac{1 + \theta_1^2}{1 + \theta_1^2 + \theta_2^2}.
\end{aligned} \tag{4.59}$$

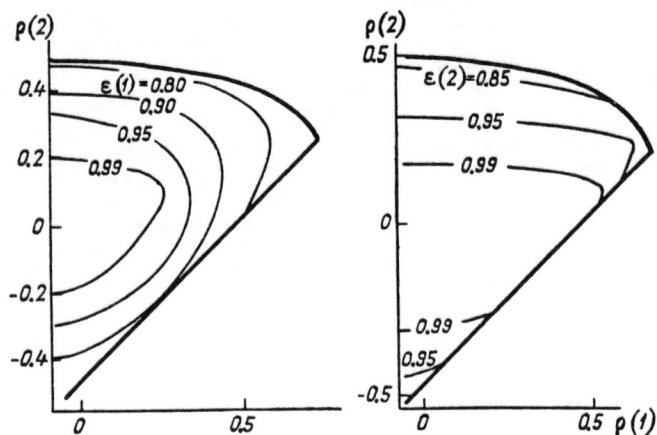

Figure 4.8: Dependence of the normalized standard deviation $\varepsilon(l)$ of the forecast error of the MA(2) process on autocorrelations $\rho(1)$ and $\rho(2)$.

The forecast for $l > 2$ coincides with the forecast for $l = 2$. The normalized standard deviations (4.59) are determined by the corresponding parameters which, in turn, depend on autocorrelations.

Having computed parameters θ_1 and θ_2 [with the aid of (4.58)] for some grid points of the $\rho(1)$ and $\rho(2)$ from the domain (4.56), one can use them to compute the values of $\varepsilon(1)$ and $\varepsilon(2)$ by (4.59) and to draw the isolines of the corresponding fields. The numerical outcome of such computations (Figure 4.8) demonstrates the sensitivity of the forecast accuracy on the autocorrelation values and on the lead time. The results show that, even for $l = 1$, forecasts with reasonable accuracy [for example, with $\varepsilon(l) < 0.9$], can be obtained only for the narrow strip of the points along the boundary of the domain of the permissible autocorrelations. If $l = 2$, such a strip is narrowed down to an even smaller size, which corresponds to the high possible values of the autocorrelations.

Together with the representation of the results in the form of the dependence $\varepsilon(l)$ on autocorrelations $\rho(1)$ and $\rho(2)$, the study of the dependence of the lead time l on $\rho(1)$ and $\rho(2)$ can be useful. The corresponding numerical results are presented in Figure 4.9. This figure shows that for very low requirements of forecast accuracy $[\varepsilon(l) = 0.95]$, there is a narrow region in the domain of the autocorrelations in which the forecast for two steps ahead can have practical value. If the requirement for accuracy is slightly high $[\varepsilon(l) \approx 0.8]$,

4.6 ARMA(1,1) Process

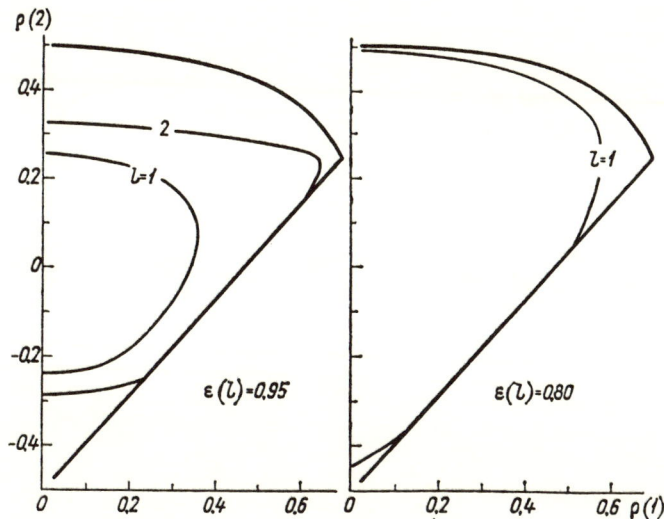

Figure 4.9: Dependence of the lead time l of the MA(2) process on autocorrelations $\rho(1)$ and $\rho(2)$ and on the normalized standard deviation $\varepsilon(l)$ of the forecast error.

the forecast for two steps ahead is senseless for almost all points of the domain.

A comparison of the results in Figures 4.8 and 4.9 with the corresponding results for the AR(2) process (see Figures 4.3–4.5) reveals that the MA(2) process can provide slightly higher theoretical accuracy than the AR(2) can.

However, the domain of the permissible autocorrelations of the AR(2) process is greater, making its practical use more convenient in spite of the slight disadvantage in the accuracy value. On the whole, the normalized variances of the forecast error of the MA(2) process are smallest on the coinciding parts of the domains of the permissible autocorrelations compared with other types of the above-mentioned processes. But this fact does not mean sufficient distinction of the requirements [obtained for the AR(1) and AR(2)] for the values of autocorrelations $\rho(1)$ and $\rho(2)$, which provide a reasonable forecast.

4.6 ARMA(1,1) Process

The ARMA(1,1) process

$$z_t - \varphi z_{t-1} = a_t - \theta a_{t-1} \tag{4.60}$$

is determined by parameters φ and θ, which satisfy the following relationships:

$$\rho(1) = \frac{(1-\varphi\theta)(\varphi-\theta)}{1+\theta^2-2\varphi\theta},$$

$$\rho(2) = \varphi\rho(1). \tag{4.61}$$

The domain of the permissible autocorrelations is determined by the following inequalities

$$\begin{aligned} |\rho(2)| &< |\rho(1)|, \\ \rho(2) &> \rho(1)(2\rho(1)+1), \\ \rho(2) &> \rho(1)(2\rho(1)-1), \end{aligned} \tag{4.62}$$

and it is presented in Figure 4.10.

The variance of the ARMA(1,1) process is

$$\sigma^2 = \frac{1+\theta^2-2\varphi\theta}{1-\varphi}\sigma_a^2. \tag{4.63}$$

For the ARMA(1,1) process, the expression for the normalized variance of the forecast error for $l+1$ steps ahead is obtained from (4.7) as

$$\varepsilon^2(l) = \frac{1-\varphi}{1+\theta^2-2\varphi\theta}\left[1+\frac{(\varphi-\theta)^2}{1-\varphi^2}\left(1-\varphi^{2(l-1)}\right)\right]. \tag{4.64}$$

The parameters φ and θ [and, consequently, the normalized standard deviation $\varepsilon(l)$] are determined by $\rho(1)$ and $\rho(2)$. The computation of $\varepsilon(l)$ can be done as described below.

Consider any mesh of two-dimensional grid points of $\rho(1)$ and $\rho(2)$ in the domain of the permissible autocorrelations (4.62). Substituting the values in the grid for $\rho(1)$ and $\rho(2)$ in (4.61), the sequence of the corresponding values of φ and θ is computed. And, finally, $\varepsilon(l)$ (for $l = 1, 2, 3, 4, 5$) is calculated by (4.64). The resulting fields of the normalized standard deviation $\varepsilon(l)$ are symmetric [as in the case of the AR(2) and MA(2) processes] relative to the plane $[\rho(2), \varepsilon(l)]$; for this reason, they are given in Figure 4.10 only for $\rho(1) > 0$. The isolines of these fields demonstrate the main features of the variation of the forecast accuracy as a function of the autocorrelations and of the lead time.

4.6 ARMA(1,1) Process

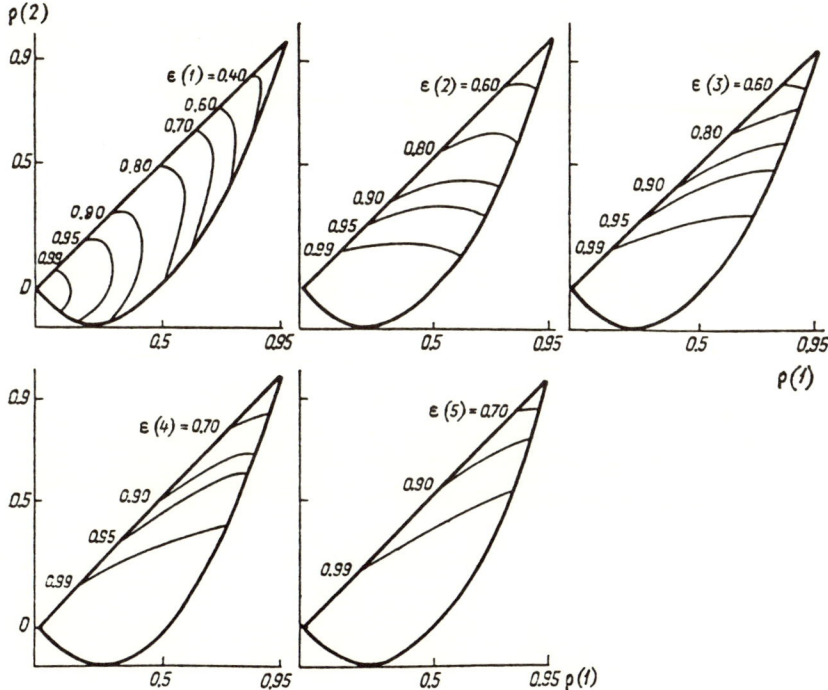

Figure 4.10: Dependence of the normalized standard deviation $\varepsilon(l)$ of the forecast error of the ARMA(1,1) process on the autocorrelations $\rho(1)$ and $\rho(2)$ and on the lead time l.

The domain of permissible autocorrelations of the ARMA(1,1) process is smaller than the corresponding domain of the AR(2) process, and its practical application is not as simple as the AR process. A detailed analysis reveals that, for the coinciding areas of the domains of the permissible autocorrelations, the accuracy of the ARMA(1,1) process is slightly worse than that of the MA(2) process, but it is slightly better than for the AR(2) process. The distinction in the values of $\varepsilon(l)$ can be clearly seen, especially on the periphery of the domain of the ARMA(1,1) process.

As we did in previous sections, let us consider the numerical fields of the lead time l of the forecasts as the function of autocorrelations $\rho(1)$ and $\rho(2)$ for the two fixed values (0.95 and 0.80) of the normalized standard deviation $\varepsilon(l)$. Figure 4.11 shows that it is possible to consider the forecast of the ARMA(1,1) process for three or four

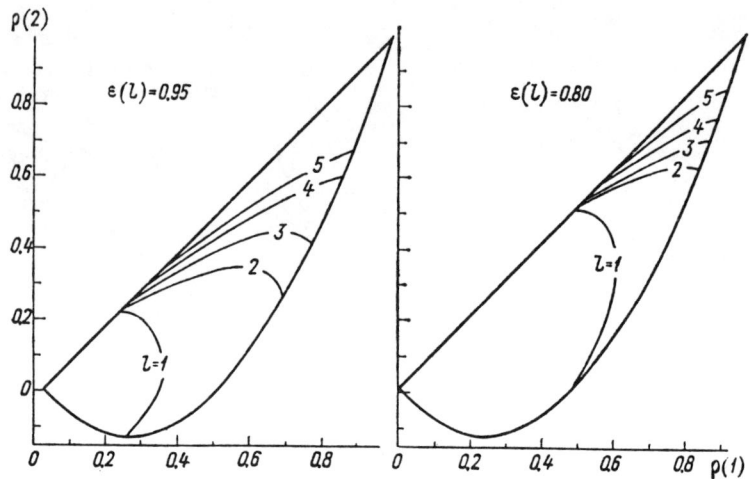

Figure 4.11: Dependence of the lead time l of the ARMA(1,1) process on the autocorrelations $\rho(1)$ and $\rho(2)$ and on the normalized standard deviation $\varepsilon(l)$ of the forecast error.

steps ahead only if the requirement for accuracy is minimal [for example, $\varepsilon(l) = 0.95$] and autocorrelations $\rho(1)$ and $\rho(2)$ are greater than 0.5–0.6.

If $\varepsilon(l) = 0.8$, such a forecast makes sense only for $\rho(1)$ and $\rho(2)$ greater than 0.7–0.8, but the corresponding area of the domain of the permissible autocorrelations is very small.

4.7 Comments

Carrying out the numerical analysis of the simplest $ARMA(p,q)$ processes enables us to consider the principal features of the forecast of the linear process.

1. The variance of the forecast error for l steps ahead is

$$v_{p,q}^2(l) = \sigma_a^2 \left(1 + \psi_1^2 + \psi_2^2 + \ldots + \psi_{l-1}^2,\right) \tag{4.65}$$

where $\sigma_a^2 = \sigma^2 \varepsilon^2$ is the variance of the white noise a_t and ε^2 is the normalized variance of the white noise.

The normalized variance of the forecast error for $l+1$ steps ahead $[\varepsilon_{p,q}^2(l) = v_{p,q}^2(l)/\sigma^2]$ is the principal theoretical characteristic of the model accuracy. If $\varepsilon(l)$ is close to 1, the forecast error variability is close to the natural variability of the process, and the forecast value is close to the mean. The closer the $\varepsilon(l)$ is to zero, the more accurate the forecast. For the standardized time series ($\sigma^2 = 1$), $\varepsilon^2(l)$ is the precise variance of the forecast error for l steps ahead.

2. The normalized variance $\varepsilon_{p,q}^2(l)$ of the forecast error of the ARMA process is functionally connected with the first $p + q$ autocorrelations and with the lead time l; the values of these parameters mutually determine each other. The numerical interpretation of the dependence of $\varepsilon(l)$ (and l) on the autocorrelations was done for several simple, frequently used processes that clarify the possible accuracy and lead time of the forecasts.

3. The value of the normalized variance $\varepsilon^2(l)$ converged to one with increasing lead time l or decreasing autocorrelations. From this, the conditional character of the forecast behavior inevitably follows: If $\varepsilon(l) \to 1$, the forecast smoothly approaches the mean. In principle, therefore, it is senseless to apply the stationary model for a forecast trajectory, which moves away from the mean with increasing l. Time series predictability is conveniently characterized by a fixed value of ε (or l). This fixed value always has a subjective meaning because the dependence between the autocorrelations, accuracy (ε), and lead time (l) is continuous.

4. The investigations carried out show that, when the autocorrelations are small, the variation of the fitted process type or its order value does not sufficiently affect the possible forecast accuracy or the lead time value. Therefore, the accuracy characteristics are stable, relative to the parameter values and the linear process types, as are the fluctuations of the samples of stationary processes. The numerical correspondence between the autocorrelation and the forecast accuracy, established for the AR(1), is approximately the same for other types of linear processes of any order. This correspondence can be expressed as follows: When the largest autocorrelation ρ of the process is less than 0.3, the forecast accuracy is very low ($\varepsilon > 0.95$), and the forecast has no practical value. If $0.3 < \rho < 0.5$, the forecast does not make practical sense for more than one step ahead; if $\rho \approx 0.6$, for more than two steps; and if $\rho \approx 0.7$, for more than three steps ahead. The practical consideration of the forecast for four or more steps ahead makes sense only when $\rho > 0.8$. Let us assume, for instance, that if $\varepsilon(l) > 0.95$, the forecast accuracy is unsatisfactory and its computation makes no sense. Then isoline $\varepsilon(l) \approx 0.95$ in the corresponding domains of the permissible autocorrelations (see Figures 4.4, 4.8, and 4.10) outlines the regions with the

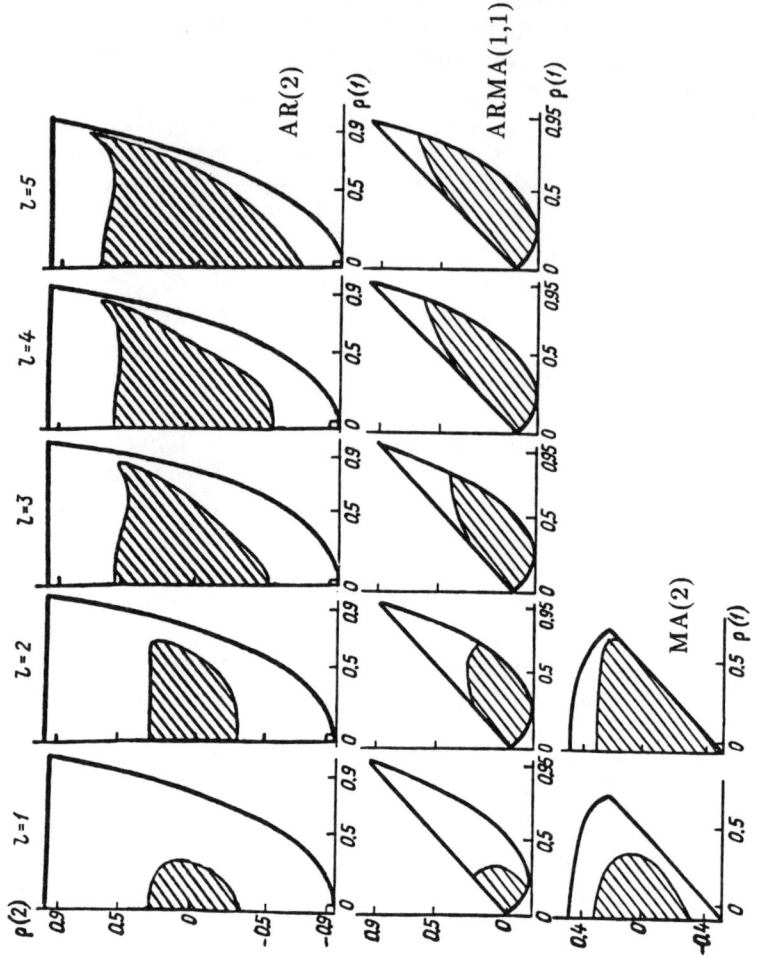

Figure 4.12: Increase of the domain $[\varepsilon(l) \geq 0.95]$ of the senseless forecasts with increasing l for all types of the second order ARMA processes.

4.7 Comments

senseless $[\varepsilon(l) > 0.95]$ forecasts. Such isolines, presented in Figure 4.12, clearly demonstrate the growth of the areas within these regions (with increasing l) for all types of second-order linear processes.

5. The results obtained enable us to evaluate the limits of potential predictability, corresponding to different autocorrelation values of the ARMA processes, and to get representation about the possibility of forecasting many physical processes of science and nature, the autocorrelation structures of which have already been estimated. One of the fields that benefits from this application is climatology. The pictorial materials presented can help to simplify the model identification process. While a particular model, fitted to a given time series, enables us to estimate its forecast and accuracy, Figures 4.1 through 4.11 allow us to obtain the theoretical accuracy and possible lead time for all first- and the second-order ARMA processes. The intuitive reasoning (that small autocorrelations yield bad forecasts and that, conversely, large autocorrelations yield good forecasts) finds its clear quantitative expression: Having fixed autocorrelations, the corresponding numerical value of the normalized standard deviation $\varepsilon(l)$ (or lead time) can be found from Figures 4.1 through 4.11.

6. The results presented can also be helpful when there is little data and when the confidence intervals of the estimates are wide. In this case, the numerical interpretation obtained helps to establish the agreement between the confidence interval of the autocorrelations and the corresponding region of the forecast error (or lead time); that is, some additional information can be obtained.

7. A comparative analysis of the accuracy characteristics of the second-order ARMA processes reveals that, in doubtful cases (when it is possible to fit any of the three considered models), the highest theoretical accuracy (and the largest lead time) is provided by the MA(2), then by the ARMA(1,1), and, finally, by the AR(2). However, the distinctions are not significant, and, because the AR(2) has the largest domain of the permissible autocorrelations, the corresponding models have the most practical application.

8. The ARMA processes make it possible to approximate a wide class of the observed atmospheric processes and to study their physical and statistical properties. The results of the theoretical analysis of the above five types of processes [AR(1), AR(2), MA(1), MA(2), ARMA(1,1)] present (as is shown in Section 6.5) virtually exhaustive information about their applicability to modeling the climate time series of the historical records (or to describing them with the aid of the corresponding first- and second-order differential equations).

9. Fitting the stochastic models is one of the approaches for estimating the spectral and correlation functions. This question

is considered, for example, by Anderson (1971), Box and Jenkins (1976), and Kashyap and Rao (1976), among others. Estimating the spectrum (and its physical analysis) is a more in-depth and subtle procedure than estimating the forecast. This is evident because a significant increase in the order of the ARMA process does not contribute significantly to improving the forecast accuracy (or increasing the lead time). However, the same increase in the process order can drastically change the spectrum shape, which, in turn, changes the understanding of the physical nature of a process.

10. The identification process is the key element in fitting the stochastic model to the observed time series. Some identification procedures, recommended in special publications (Kashyap and Rao, 1976; Eikhoff, 1983), are based on statistical criteria that evaluate the appropriateness of the presupposed model to the given time series. However, the final evaluation must be based not only on statistical properties but also on physical concepts. For instance, the identification of a model of climatological time series actually states the character (tendency) of climate change. But theoretical assumptions and conditions, which linear processes must satisfy, are mathematical abstractions. No natural processes strictly obey these abstractions, and *the identification is never absolutely precise. The conventionality and subjectivity of the model identification procedure result from the fact that there cannot be absolute adequacy between the process of nature and the stochastic (for example, ARMA) process.* Such inadequacy predetermines the nonuniqueness of climate modeling and the necessity of reasonable compromise, which must reconcile the difference between the reality and its approximate image contained in the fitted model. The above study also makes it possible to obtain an understanding of the extent of the distinction in the forecast accuracy, when the identification is not accurate.

4.8 Signal-Plus-White-Noise Type Processes

As will be shown shortly and in Chapter 6, a climatic time series can be modeled by signal-plus-white-noise type processes. Such processes allow an accurate theoretical interpretation in the framework of linear parametrical processes (Box and Jenkins, 1976; Parzen, 1966). In this case, the white noise is referred to as additive white noise.

The addition of a white noise to the general ARMA process does not change the order of the autoregressive part and affects only the moving average component. More accurately (Box and Jenkins, 1976), if the white noise is added to the ARMA(p, q) process, the

4.8 Signal-Plus-White-Noise Type Processes

result is the ARMA(p,Q) process, where Q is the maximum of p and q. For example, if $p \leq q$, then $Q = q$ and the order of the process with additive white noise is the same as the order of the original one; only the values of the parameters θ are changed.

Consider an example (see Kashyap and Rao, 1976) of a first-order ARMA process with additional white noise component. This example has application as an appropriate model for historical records of climate time series of temperature, pressure, precipitation, stream flow, and so on.

Let
$$z_t = S_t + N_t, \qquad (4.66)$$

where
$$S_t = \varphi S_{t-1} + a_t \qquad (4.67)$$

is the first-order autoregressive process (signal); N_t and a_t are the white noise processes; and σ_z^2, σ_S^2, σ_N^2, σ_a^2 are the variances of processes z_t, S_t, N_t, and a_t respectively.

Let z_t be the ARMA(1,1) process
$$z_t = \varphi z_{t-1} + \theta V_{t-1} + V_t, \qquad (4.68)$$

where parameter θ is the solution of the equation
$$\theta^2 + \theta \frac{(1+\varphi^2) + \sigma_a^2/\sigma_N^2}{\varphi} + 1 = 0, \qquad (4.69)$$

satisfying the condition $|\theta| < 1$. The variance σ_v^2 of the white noise V_t is
$$\sigma_v^2 = -\varphi \sigma_N^2/\theta. \qquad (4.70)$$

One of the informative approaches (with a clear physical interpretation) to the analysis of the signal-plus-white-noise process has been presented by Parzen (1966). Parzen's results match the well-known empirical procedure of "extrapolation to zero" of the autocorrelation function estimates used for many years in meteorology to estimate the noise level in observed data.

Let us briefly consider this simple approach. Let random process z_t be the sum of the signal S_t and the white noise N_t (4.66), where S_t is the AR(p) process. The signal ratio
$$\alpha = \frac{\sigma_S^2}{\sigma_z^2} = \frac{\sigma_S^2}{\sigma_S^2 + \sigma_N^2} = \frac{1}{1 + \sigma_N^2/\sigma_S^2} \qquad (4.71)$$

is a measure of the signal level (naturally, $1 - \alpha$ is the measure of the noise level). The ratio α shows the proportion of the variance of process z_t generated by signal S_t. Notice here that the ratio signal/noise, used in radio engineering, is connected with the ratio α by the relationship

$$\sigma_S^2/\sigma_N^2 = \alpha/(1-\alpha). \tag{4.72}$$

It is easy to show [for autocorrelation functions $\rho_z(\tau)$ and $\rho_S(\tau)$ of processes z_t and S_t] that

$$\rho_z(\tau) = \alpha \rho_S(\tau), \quad (\tau > 0). \tag{4.73}$$

Having estimated $\rho_z(\tau)$ and α, one can use (4.73) to compute autocorrelation function $\rho_S(\tau)$ of the signal. For estimating ratio α, let us notice that S_t is the AR(p) process, and its parameters φ_i must satisfy the Yule-Walker system

$$\rho_S(\tau) = \varphi_1 \rho_S(\tau-1) + \varphi_2 \rho_S(\tau-2) + \ldots + \varphi_p \rho_S(\tau-p),$$
$$(\tau = 1, 2, \ldots, p). \tag{4.74}$$

The autocorrelations $\rho_S(\tau)$ are unknown because the signal S_t is not observed; only the process z_t is observed. However, from (4.73) it follows that

$$\rho_S(\tau) = \rho_z(\tau)/\alpha, \tag{4.75}$$

where $\rho_z(\tau)$ can be estimated using the observations of z_t. Adding to the Yule-Walker system (4.74) the equation for $\tau = p+1$ and replacing $\rho_S(\tau)$ by (4.75), a system of $p+1$ algebraic equations (relative to $p+1$ unknown parameters $\varphi_1, \varphi_2, \ldots, \varphi_p$, and α) is obtained. These equations are nonlinear, but the analytical expressions for α can be found for small p.

If signal S_t is the AR(1) (4.67), parameters φ and α are estimated from the equations

$$\rho_S(1) = \varphi, \quad \rho_S(2) = \varphi \rho_S(1). \tag{4.76}$$

Taking into account (4.75) yields

$$\rho_z(1)/\alpha = \varphi, \quad \rho_z(2) = \varphi \rho_z(2), \tag{4.77}$$

which gives

$$\alpha = \rho_z^2(1)/\rho_z(2). \tag{4.78}$$

4.8 Signal-Plus-White-Noise Type Processes

Based on this formula, the least squares estimator for the ratio α can be found using the first $m+1$ estimates r_τ of autocorrelations $\rho_z(\tau)$. This estimator is

$$\hat{\alpha} = \frac{\sum_{\tau=1}^{m} r_\tau^2 r_{\tau+1}}{\sum_{\tau=1}^{m} r_{\tau+1}^2}, \qquad (4.79)$$

where the appropriate value of m depends on the particular time series considered. With the aid of $\hat{\alpha}$, it is possible to estimate autocorrelation function $\rho_S(\tau)$ by (4.75).

The absolute value of autocorrelation $\rho_S(\tau)$ is greater than the corresponding value of $\rho_z(\tau)$ (because $\alpha < 1$), which shows the distinction in the potential predictability of signal S_t as compared with observed processes z_t.

If the signal is the AR(2) process, parameter α can be found by solving the following equation

$$\alpha^2 - 2\alpha \frac{\rho_z(1)\rho_z(2)}{\rho_z(3)} - \rho_z(1) + \frac{\rho_z(1)\rho_z^2(2)}{\rho_z(3)} + \frac{\rho_z^2(1)}{\rho_z(3)} = 0,$$

where from the two values

$$\alpha = \frac{\rho_z(1)}{\rho_z(2)} \left\{ \rho_z(2) \pm \sqrt{\rho_z(1)[\rho_z(3) - \rho_z(1)]} \right\}$$

it is necessary to choose only the one that is in the interval (0,1).

The problem of finding α is complicated for the process with the signal that is presented by the autoregressive process of order greater than two.

Estimates of α will be systematically provided in this section and in Chapters 6 with climatological applications. In all these cases, we assume that S_t is the AR(1) process and that parameter α is estimated by (4.78).

4.8.1 Examples: Signal Ratio of Climate Time Series

Monitoring the modern climate requires evaluation of the potential statistical predictability of the historical records of different meteorological observations.

First we consider the longest temperature record in central England (the 315 years from 1659 to 1973), published by Manley (1974). Because the first 40 years of observations of this time series are not accurate enough, we use only the most accurate part of this record—the period from 1700 to 1973—for the evaluation of ratio α.

Another source of temperature data, for which the ratio α is estimated, is the data base collected in the World Data Center, Obninsk, Russia (Gruza, 1987). The Northern Hemisphere temperature data from these archives for the period from 1891 to 1985 are used to obtain the zonal and annual mean time series cited in Table 4.1.

Table 4.1 provides estimates of the signal ratio α for some climate time series. The results reveal that for the temperature data the smallest estimates of the first autocorrelations correspond to the time series of the stations (point gauges) [$\rho_z(1) = 0.18$ for the annual and $\rho_z(1) = 0.3$ for the monthly data of the central England temperature]. According to the results of Section 4.7, it does not make sense to use the ARMA model to forecast the time series with such small values of autocorrelations independently of the order and type of the fitted model. The value (0.37) of the first autocorrelation of the monthly temperature time series, obtained by averaging data within the ten-degree width latitude band ($50°$–$60°$) which includes the entire territory of England, can provide a low-accuracy forecast for one step ahead, according to the results of Section 4.7. Autocorrelation for the annual data, averaged over the entire Northern Hemisphere, is about 0.8. In fact, it makes sense to forecast such data for up to three steps ahead.

For the first three cases in Table 4.1, the model of the signal [the AR(1) process] plus white noise appears to match perfectly with the data. The lowest level ($\alpha = 0.25$) of the climate signal corresponds to the annual mean temperature time series of the separate station, and the highest level ($\alpha = 0.60$) to the monthly data averaged within the $50°$–$60°$ latitude band.

For the temperature time series of data averaged over the entire Northern Hemisphere, the α estimate [obtained by (4.78)] is greater than one, which means that this simple model [AR(1) process plus white noise] is not an appropriate approximation. The procedure of averaging all hemisphere observations that have sufficiently different level of autocorrelations in different regions of the Earth drastically changed the statistical structure of the resulting time series, compared with the original ones. Averaging of temperature observations with very diverse temporal statistical properties (for example, almost white noise in the polar regions and high autocorrelated time series near the equator) has led to the creation of a new time series with a new statistical nature. However, such averaging is widely applied in climatology without questioning its statistical legitimacy.

Spatial averaging reduces the noise component and increases the ratio α of the climate signal; therefore, it is one of the ways of detecting a climate signal when averaged time series have approximately identical statistical structures.

4.8 Signal-Plus-White-Noise Type Processes

Table 4.1: Estimates of signal ratio α for different time series.

Time Series	$\rho_z(1)$	$\rho_z(2)$	α
The central England monthly mean temperature time series (1891–1985)	0.30	0.16	0.56
The zonal (50°–60° latitude band of the Northern Hemisphere) monthly mean temperature time series (1891–1985)	0.37	0.23	0.60
The central England annual mean temperature time series (1700–1973)	0.18	0.13	0.25
The Northern Hemisphere annual mean temperature time series (1891–1985)	0.80	0.57	1.12
The Brazos River (Richmond gauge) monthly mean streamflow time series (1900–1984)	0.48	0.28	0.82
The Brazos River (Richmond gauge) annual mean streamflow time series (1900–1984)	0.08	0.06	0.11
Annual solar radiation budget in the USSR cities:			
Karadag (1934–1975)	0.20	0.10	0.40
Tashkent (1937–1975)	0.23	0.19	0.30
Tbilisi (1937–1975)	0.23	0.16	0.35
Irkytsk (1938–1975)	0.25	0.19	0.34
Vladivostok (1939–1975)	0.19	0.15	0.23

The next cases in Table 4.1 concern the normalized anomalies of the monthly mean streamflow time series (Richmond gauge, the mouth of the Brazos River), matching the above signal-plus-white-noise model with $\alpha = 0.82$. Analogous results obtained for the other 18 gages of the Brazos River basin gave the α estimates, which vary from 0.29 to 0.82. Corresponding annual streamflow time series are very close to the white noise samples.

The results in Table 4.1 for the annual solar radiation budget (the most accurate time series of several cities of the former USSR from the Pivovarova archives, Main Geophysical Observatory, St. Petersburg) also very closely follow this simple signal-plus-white-noise model with the α estimates in the relatively narrow interval of values from 0.23 to 0.40 (see also Polyak et al., 1979).

The fact that the variance proportion of the gauge annual time series corresponding to the signal is very small explains the difficulties connected with description and prediction. However, the results obtained provide evidence that such a signal does exist and can be considered as a long-period variation with a very small amplitude observed only on a short-term interval of instrumental observations.

Climate time series models with an additive white noise component clearly reveal the existence of the nonpredictable aspect of climate variability.

Consider more general results for the surface monthly mean temperatures obtained through the temperature time series mentioned above for the observational period from 1891 to 1985. The observations were spatially averaged over various geographical and political regions of the Northern Hemisphere, as cited in the first column of Table 4.2.

Before estimating the simple statistical characteristics, the seasonal cycle in the means and variances of the data was removed, and the time series of the normalized values were analyzed. It is assumed that the climatic signal can be presented by a first-order autoregressive model; so, after estimating $\rho_z(1)$ and $\rho_z(2)$, the signal ratio α was computed by formula (4.78). After analyzing the estimates given in Table 4.2, the following comments can be made:

1. The first autocorrelations of the spatially averaged time series are greater than the corresponding autocorrelations of the point gauges (see, for example, Polyak, 1975, 1979), which reveals the importance of spatial averaging procedure to detect the climate signal. The signal ratio for the time series of averages is also, as a rule, higher than that for the time series of the point gauges (see Polyak, 1975, 1979), for which the signal is responsible for not more than 30% of their variability.

4.8 Signal-Plus-White-Noise Type Processes

Table 4.2: Signal ratio and linear trend of the surface air temperature monthly mean time series averaged over different regions of the Earth. Observational interval 1891 to 1985.

Regions of averaging	$\rho_z(1)$	$\rho_z(2)$	α	β_1 10^{-4}°C per month	t
5-degree-width NH latitude bands with the center in:					
87.5°	0.28	0.11	0.71	9	3.5
82.5°	0.31	0.13	0.73	7	3.1
77.5°	0.37	0.16	0.85	4	1.8
72.5°	0.38	0.20	0.70	7	5.1
67.5°	0.35	0.22	0.57	5	3.1
62.5°	0.31	0.20	0.46	4	2.9
57.5°	0.27	0.18	0.43	1	1.0
52.5°	0.33	0.16	0.65	2	2.4
47.5°	0.34	0.16	0.73	3	3.8
42.5°	0.34	0.17	0.66	3	3.7
37.5°	0.37	0.25	0.56	0	0.3
32.5°	0.38	0.30	0.47	1	1.3
27.5°	0.40	0.30	0.62	3	6.3
22.5°	0.51	0.39	0.67	3	9.1
17.5°	0.61	0.46	0.80	3	8.2
12.5°	0.64	0.53	0.30	3	7.1
87.5°–67.5° NH	0.38	0.20	0.72	5	3.1
67.5°–37.5° NH	0.40	0.24	0.64	2	3.0
37.5°–27.5° NH	0.38	00.29	0.49	2	3.8
27.5°–12.5° NH	0.63	0.50	0.79	3	8.1
82.5°–12.5° NH	0.53	0.43	0.67	3	6.3
Eurasia	0.32	0.13	0.80	2	1.9
Europe	0.30	0.15	0.61	2	1.2
Asia	0.20	0.10	0.70	2	1.8
North America	0.22	0.10	0.35	4	3.1
North Africa	0.36	0.18	0.71	1	1.3
Atlantic Ocean					
67.5°– 2.5° NH	0.67	0.59	0.76	4	7.8
67.5°–37.5° NH	0.65	0.58	0.71	6	7.9
37.5°–12.5° NH	0.53	0.41	0.68	2	5.3
Pacific Ocean					
57.5°– 2.5° NH	0.63	0.55	0.71	7	12.1
57.5°–37.5° NH	0.49	0.40	0.59	7	8.9
37.5°–12.5° NH	0.63	0.54	0.74	7	12.7
USA	0.11	0.06	0.18	4	2.4
Canada	0.19	0.09	0.38	4	2.1
Russia	0.30	0.13	0.68	3	2.0

2. The greater the region of averaging (i.e., the greater the number of point gauges), the greater the corresponding autocorrelations.

3. The northward decreasing latitudinal trend of the first autocorrelations has a noticeable maximum in the equatorial region. For the high and middle latitudes (greater than 30°) of the Northern Hemisphere, the first autocorrelation of the temperature anomalies is less than 0.4. Therefore, according to the results of Section 4.7, the forecast of the corresponding time series makes sense for no more than one step ahead, and its accuracy will be very low. Indeed, the forecasting of the monthly or annual mean temperature time series by ARMA models for one to two steps ahead makes sense only for the tropical regions, where the first autocorrelations are greater than 0.5–0.6. The forecast accuracy of the temperature time series is higher for the equatorial regions than for the middle and northern latitudes.

4. Although the signal ratio value α in Table 4.2 varies within wide limits (from 18% to 85% of the variance), the appropriateness of the signal-plus-white-noise model in all the cases considered is unquestionable.

5. Finally, let us note the importance of the methodological study of the forecast accuracy of the different types of the ARMA models (see Sections 4.1 through 4.7). This study gave us the possibility of obtaining a complete representation of the statistical predictability of different climate time series without fitting any model.

Together with the signal ratio α, Table 4.2 gives the estimates of the linear trend parameter β_1 and its t-statistics values. Almost all estimates of β_1 are positive, and their range of variation is from 0 to about $(10^{-3})°C$ per month. In most cases, these estimates are statistically significant at any standard level of significance.

4.9 Process With Stationary Increments

Consider the nonstationary random process y_t, the finite differences

$$z_t = \nabla^\nu y_t \tag{4.80}$$

4.9 Process With Stationary Increments

of which are the stationary ARMA(p,q) process. Under some conditions, the process z_t can be presented as

$$z_t = \Phi_1 z_{t-1} + \ldots + \Phi_p z_{t-p} + a_t - \theta_1 a_{t-1} - \ldots - \theta_q a_{t-q}. \quad (4.81)$$

Replacing the finite differences by their definition (2.71–2.73) we obtain a difference presentation of process y_t as

$$y_t = \varphi_1 y_{t-1} + \varphi_2 y_{t-2} + \ldots + \varphi_{p+\nu} y_{t-p-\nu} + a_t - \theta_1 a_{t-1} - \ldots - \theta_q a_{t-q}. \quad (4.82)$$

The variance of the forecast error of y_t for l steps ahead (see Box and Jenkins, 1976) is determined by the formula

$$v^2(l) = \sigma_a^2 \left(1 + \psi_1^2 + \ldots + \psi_{l-1}^2\right), \quad (4.83)$$

where

$$\begin{aligned}
\psi_1 &= \varphi_1 - \theta_1, \\
\psi_2 &= \varphi_1 \psi_1 + \varphi_2 - \theta_2, \\
\ldots & \quad \ldots \quad \ldots \\
\psi_j &= \varphi_1 \psi_{j-1} + \ldots + \varphi_{p+\nu} \psi_{j-p-\nu} - \theta_j, \\
\psi_j &= 0, \text{ if } j < 0; \quad \theta_j = 0, \text{ if } j > q. \quad (4.84)
\end{aligned}$$

Therefore, in this case, the fitting of the linear model can be performed for the finite differences (4.80).

The simplest nonstationary process is the first-order autoregressive process

$$z_t = \rho z_{t-1} + a_t, \quad (\rho = \Phi_1) \quad (4.85)$$

of the first finite differences. It can be denoted as

$$y_t = (1 + \rho) y_{t-1} - \rho y_{t-2} + a_t \quad (4.86)$$

or

$$y_t = \varphi_1 y_{t-1} + \varphi_2 y_{t-2} + a_t, \quad (4.87)$$

where $\phi_1 = 1 + \rho$, and $\phi_2 = -\rho$. Substituting these parameter values in (4.83) and (4.84) enables us to compute $v^2(l)$ for any l. Following the denotation described in Section 4.1, let us consider the normalized standard deviation

$$\varepsilon(l) = v(l)/\sigma_y. \quad (4.88)$$

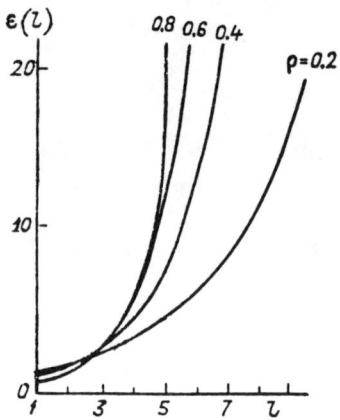

Figure 4.13: Dependence of the normalized standard deviation of the forecast error on the lead time l and on the first autocorrelation ρ of the AR(1) process of the first finite differences ∇y_t.

Figure 4.13 shows the dependence of $\varepsilon(l)$ on l for different values of the autocorrelation ρ of the first differences ∇y_t.

The normalized standard deviation (4.88) is less than 1 only for $l = 1$. If $l > 1$ and the lead time increases, the value of standard deviation ε increases infinitely (whereas the forecast value approaches a straight line), which means that the accuracy of the nonstationary forecast (for $l > 1$) is lower than the accuracy of the observations. An infinite growth of the standard deviation ε of the forecast error takes place for any ARMA process of the finite differences that is distinct from the stationary process, for which $\varepsilon(l) \leq 1$.

Evaluating the goodness of fit of a nonstationary model and its forecast accuracy for several steps ahead, one must be guided by the physics of the observed process rather than by statistical reasoning. The normalized standard deviation $\varepsilon(l)$ of the forecast error loses its comparative (with 1) value, and the forecast of the process with stationary finite differences asymptotically approaches (with increasing lead time l) one of the branches of the degree ν polynomial (straight line if $\nu = 1$).

Therefore, the closeness of the expected value of nonstationary process to the degree v polynomial is the decisive factor that determines the forecast accuracy. If the absolute values of the autocorrelations of the finite differences are small (for example, less than

0.3), the forecast for several steps ahead will not be distinct from the corresponding value of the polynomial.

4.10 Modeling the Five-Year Mean Surface Air Temperature

For many years, the time series of the annual mean surface air temperature, obtained by averaging gauge observations over the entire Northern Hemisphere has been studied as an indicator of climate change. Despite the doubtful statistical legitimacy of such averaging (over the observations with diverse spatial and temporal statistical structure), it is interesting to analyze the resulting time series, to build simple stochastic models, and to estimate the forecast accuracy. The objective of this study is to illustrate the dependence of the forecast scenarios on the assumption about stationarity or nonstationarity of the time series.

The time series of temperature anomalies under study was obtained with the aid of the Northern Hemisphere temperature data base of the World Data Center at Obninsk, Russia. This time series has a small linear trend (approximately 0.005°C per year), but it is statistically significant at the 95% level. To make the result more descriptive, the temporal scale of averaging was increased, and models of the five-year means of the Northern Hemisphere surface air temperature time series were built.

4.10.1 Stationary Model

The estimates of the first and second autocorrelations of this time series are 0.8 and 0.57, respectively; the standard deviation is about 0.2 C°; the standard deviation of the forecast error by AR(1) is 0.11; the predictability limit of this model for $\varepsilon(l) = 0.8$ is equal to about three steps (see Figure 4.2); and the model is $z_t = 0.8 z_{t-1} + a_t$. The results, presented in Figure 4.14, show that, if the lead time increases, the forecast converges to zero that was *a priori* clear for the stationary model. By looking at the picture (Figure 4.14) one can see some similarity in the forecast behavior with the cooling period observations from 1950 through 1970.

Of course, such a stationary model cannot be used for the modeling of any trend. In order to include any deterministic component in the description of climate, one must fit a nonstationary model.

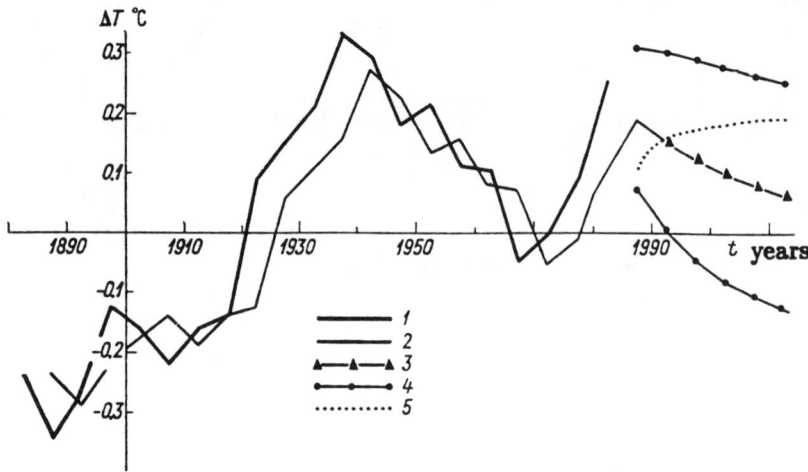

Figure 4.14: The Northern Hemisphere five-year mean surface air temperature anomalies (1); forecasts by the AR(1) model: for one step (five years) ahead (2), for 2 to 6 steps ahead (3); 70% confidence intervals (4); standard deviation of the forecast error (5).

4.10.2 Nonstationary Model

The principal motivation for fitting a nonstationary model to the above temperature time series of the Northern Hemisphere is that its least squares straight line $\beta_0 + \beta_1 t$ approximation has a statistically significant estimate of the parameter β_1 ($\beta_1 = 0.023 C°$ per five years; the Student's statistic t of this estimate is about 2; $\beta_0 = 0$ because we process the anomalies and the origin of the coordinate system was chosen in the center of the observational interval).

The ARMA models were fitted to the first finite differences of the above time series, and the value 0.023 of the β_1 estimate was taken as an estimate of the mean (μ) of these differences.

The best was the ARMA(1,1) model,

$$\nabla y_t - \mu = \varphi(\nabla y_{t-1} - \mu) - \theta a_{t-1} + a_t,$$

where the autoregressive parameter (φ) estimate is 0.056. The moving average parameter (θ) estimate is 0.13.

The obtained estimates provide the expression for the model,

$$y_t = 1.056 y_{t-1} - 0.056 y_{t-2} - 0.022 - 0.13 a_{t-1},$$

4.10 Modeling the Surface Air Temperature

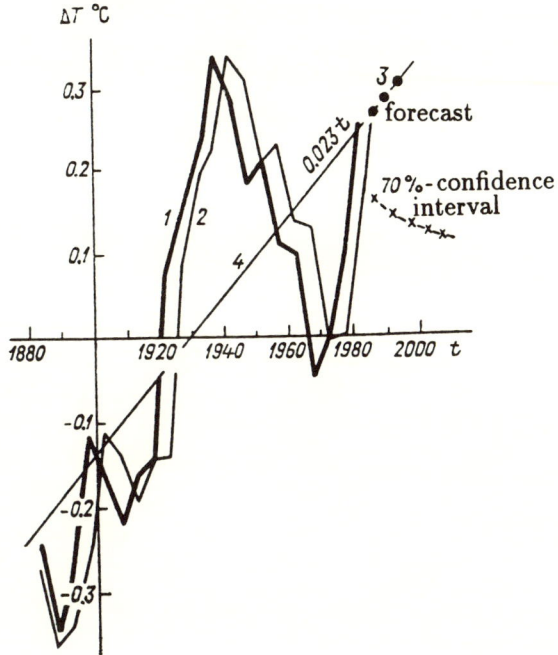

Figure 4.15: The Northern Hemisphere five-year mean surface air temperature anomalies (1); forecasts by the ARMA(1,1) model of the first finite differences: for one step (five years) ahead (2); for 2 to 4 steps ahead (3); and linear approximation (4).

the standard deviation of the forecast error of which is 0.1°C. The forecasts by this model and their 70% confidence interval are given in Figure 4.15.

The first autocorrelation estimate of the first finite differences time series is small (0.18), and for this reason the forecast, even for one step ahead, practically leads to the extrapolation of the straight line $0.023t$. In other words, the forecast shows that the mean temperature increases approximately 0.5°C per 100 years and gives zero value for the fluctuation with respect to this mean.

Comparing stationary (Figure 4.14) and nonstationary (Figure 4.15) models of the same time series, one can see that the results of forecasting are very different, although their standard deviations of forecast error are approximately the same (0.11 and 0.10).

It is interesting (methodologically) to attempt to increase the difference order and to build a model of the second finite differences. In this case, the differences $\nabla^2 y_t$ are proved to be statistically independent, and the forecast of the corresponding model coincides with the extrapolation of the corresponding second-degree polynomial

$$P(t) = 0.023t - 0.0026t^2. \tag{4.89}$$

The branches of this parabola are directed downward, unlike the forecasts of the previous model and the modern representation of global warming.

The simple stochastic models clearly reveal the conditional character of the forecasts, depending on the fitted model type and on the assumption of stationarity or nonstationarity. Fitting stationary and nonstationary models to the same climate time series makes it possible to obtain different climatic scenarios.

It is also important to notice that, since the period of instrumental observations is small, it is unrealistic to formulate an identification problem for the stochastic modeling of this climate temperature time series. In other words, it is unrealistic to determine the statistical law of climate variation with the aid of a time series with only 21 observations. For this reason, fitting low-order models is a necessary step in the right direction. Indeed, such investigations help to garner the experience that is necessary for building multivariate climate models.

4.11 Nonstationary and Nonlinear Models

As was shown, fitting a stochastic model to finite differences undermines the polynomial approximation of the trend and the modeling of the residuals.

The following two-step procedure provides a more direct empirical approach to the modeling of a long time series with the nonstationary expected value (trend). For the first step, a time series is approximated by the appropriate function, which is linear relative to the estimated parameters. For the second step, the ARMA model is fitted to the time series of the residuals.

In contrast to the stationary case, the variance of the forecast error of the nonstationary linear model increases infinitely with increasing lead time. For the polynomial trend, this variance is proportional to the value of $(k/2+l)^{2\nu}$, where l is the lead time and ν is the polynomial degree or the finite differences order (see Section 1.3). If

4.10 Nonstationary and Nonlinear Models

$l > 1$, the variance of the forecast error of the nonstationary model is greater than the variance of the observations, and, of course, it is greater than the corresponding variance of the stationary model.

If the statistical structure of the time series slowly varies in time and the observational interval is sufficiently long, one can use another empirical approach analogous to the smoothing moving average procedure. The interval of the observations can be divided into subintervals, possibly overlapping, for each of which the linear stationary model can be fitted. After that, one can study the temporal variation of the parameters' estimates and, if suitable, can approximate them by means of an appropriate function. In this simple method, monitoring a random process can be replaced by monitoring the model parameters.

Nonlinear modeling is, potentially, the most promising direction of climatology development. Fundamental concepts of nonlinear modeling are presented by Tong (1990). According to Tong, the general case of the nonlinear model can be denoted by

$$X_t = f(X_{t-1}, X_{t-2}, \ldots, X_{t-p}) + e_t, \qquad (4.90)$$

where e_t is the white noise.

The model, which has been intensively studied, is the threshold autoregressive model, the simplest form of which is

$$X_t = \begin{cases} aX_{t-1} + \text{white noise} & \text{if} \quad X_{t-1} \leq \text{threshold,} \\ bX_{t-1} + \text{white noise} & \text{if} \quad X_{t-1} > \text{threshold,} \end{cases} \qquad (4.91)$$

where a and b are the parameters.

Such nonlinearity can be associated with discontinuities of the observed process; for example, the climate time series of precipitation in the form of water and snow, solar radiation with and without cloudiness, and so on. Such discontinuities must be reflected in the structure of the fitted model. The intake of the threshold in the model is caused by the physics of the process under study and, in particular, by the possibilities of catastrophes (Tong, 1990).

Another type of nonlinear model is the class of bilinear models, the simplest example of which is

$$X_t = aX_{t-1} + be_{t-1} + ce_{t-j}X_{t-1} + e_t, \qquad (4.92)$$

where $a, b,$ and c are the parameters; e_t is independent of X_t white noise with a zero mean and a finite variance; and j is an integer.

The distinction of (4.92) from the ARMA model lies in the existence of the bilinear term $e_{t-j}X_{t-1}$. The important feature of the bilinear model is that the time series approximated by this model can have time-variable statistical structure, which is the case for some hydrometeorological processes.

Tong refers to (4.91) and (4.92) as nonlinear models of the first generation. A more general type of nonlinear model, the threshold model of the second order, is

$$X_t = \begin{cases} a_0 + a_1 X_{t-1} + a_2 X_{t-2} + e_t & \text{if} \quad X_{t-1} \leq C, \\ b_0 + b_1 X_{t-1} + b_2 X_{t-2} + e_t & \text{if} \quad X_{t-1} > C. \end{cases} \quad (4.93)$$

Fitting the nonlinear model makes it possible to decrease the standard deviation of the forecast error compared with the best linear models. For forecasting more than one step ahead, the situation is completely different; that is, the standard deviation of the forecast error becomes greater than the corresponding standard deviation of the linear model.

The necessity of moving toward considering nonlinear and nonstationary models has become especially clear in the process of identifying and agreeing on the structures of the stochastic and physical (hydrodynamic) models because, as a rule, the latter are nonlinear and nonstationary.

To conclude, the key points of the nonlinear modeling are as follows.

1. If the lead time increases, the nonstationary model forecast approaches the function (the polynomial), which approximates the time series trend.

2. If $l > 1$, the variance of the forecast error of the nonstationary and nonlinear models is greater than the corresponding variance of the observations and, naturally, is greater than the analogous variance of the forecast error of the stationary model.

3. Nonstationary and nonlinear models can describe a real climate change on the significant time intervals, whereas stationary models can characterize only the climate stability and invariability.

4. The fitting of stationary models on several subintervals of the entire observational interval can represent temporal variation of the model parameters.

4.10 Nonstationary and Nonlinear Models

5. Meteorological processes in the real-time scale are not stationary in principle. But in many cases, to meet the assumption of stationarity, the corresponding observations are transformed so that the resulting quantities do satisfy this condition. The standard transformations are: filtering out the annual cycle or the latitude trend, large-scale spatial and temporal averaging, and computing the principle components. Averaging additionally lowers the noise component and reduces the volume of information that must be analyzed.

5

Multivariate AR Processes

This chapter is devoted to different computational aspects of multivariate modeling. An algorithm for fitting such models with sequential increasing of the order is given. Several examples of approximation of multiple climatological time series by first-order AR models are considered. It is emphasized that such modeling can be used not only for forecasting, but also for the analysis of the parameter matrix as a matrix of the interactions and feedbacks of the observed processes. Some problems in identifying stochastic climate models are also discussed. It is shown that formulation of climatology problems within the strict frameworks of fundamental theory will facilitate natural progress along with the development of these methods.

5.1 Fundamental Multivariate AR Processes

Let us consider q discrete random processes $x_{1t}, x_{2t}, \ldots, x_{qt}$. Vector

$$\mathbf{X}_t = \begin{pmatrix} x_{1t} \\ x_{2t} \\ \ldots \\ x_{qt} \end{pmatrix} \quad (5.1)$$

is referred to as q-variate random process. We assume that this process is stationary and that $\mathbf{E}(x_{it}) = 0$; $i = 1, 2, \ldots, q$.

5.1 Fundamental Multivariate AR Processes

The auto- and cross-covariance functions of processes x_{it} can be presented in the form

$$\mathbf{V}(\tau) = \mathbf{E}(\mathbf{X}_t \mathbf{X}_{t+\tau}^{\mathrm{T}}) = \begin{pmatrix} v_{11}(\tau) & v_{12}(\tau) & \ldots & v_{1q}(\tau) \\ \ldots & \ldots & \ldots & \ldots \\ v_{q1}(\tau) & v_{q2}(\tau) & \ldots & v_{qq}(\tau) \end{pmatrix}, \qquad (5.2)$$

where T means transpose; $v_{ij}(\tau) = \mathbf{E}(x_{it} x_{jt+\tau})$ is the cross- (or auto- if $i = j$) covariance function of processes x_{it} and x_{jt}. The sequence of the matrices $\mathbf{V}(\tau)$ ($\tau = 0, 1, 2, \ldots$) is referred to as a q-variate covariance function or as the matrix covariance function of random process $\mathbf{X}_t$.

The q-variate covariance function satisfies the equation

$$\mathbf{V}^{\mathrm{T}}(\tau) = \mathbf{V}(-\tau) \qquad (5.3)$$

caused by the property of the stationarity

$$v_{ij}(\tau) = v_{ij}(-\tau). \qquad (5.4)$$

If $\tau \neq 0$, the matrix $\mathbf{V}(\tau)$ is not symmetric; however, the q-variate covariance function $\mathbf{V}(\tau)$ can be considered (and estimated) only for $\tau \geq 0$ because of (5.4) [$v_{ji}(\tau)$ is the continuation of $v_{ij}(\tau)$ to the domain of negative τ].

With the aid of the q-variate covariance function, one can define the q-variate correlation function (in the same way as for the univariate case), the elements of which are the correlations

$$\rho_{ij}(\tau) = \frac{v_{ij}(\tau)}{\sqrt{v_{ii}(0) v_{jj}(0)}}. \qquad (5.5)$$

Let

$$\mathbf{N}_t = \begin{pmatrix} n_{1t} \\ n_{2t} \\ \ldots \\ n_{qt} \end{pmatrix} \qquad (5.6)$$

be the q-variate process, which is defined in the following way:

1. Each n_{it} is the univariate white noise.

2. Processes n_{it} can be statistically dependent only at the same moment of time; for the nonzero lags, the covariances are zeroes.

The matrix covariance function $\Sigma(\tau)$ of process N_t is

$$\Sigma(\tau) = \begin{cases} \Sigma = \{\Sigma_{ij}\}_{i,j=1}^{q} & \text{if } \tau = 0, \\ \mathbf{0}_{q \times q} & \text{if } \tau \neq 0, \end{cases} \qquad (5.7)$$

where $\mathbf{0}_{q \times q}$ is the matrix of $q \times q$ size, all the elements of which are zeroes. This definition generalizes the notion of the white noise process for the multivariate case. The process N_t with the matrix covariance function (5.7) is referred to as the q-variate white noise.

If a sequence of square matrices $\boldsymbol{\alpha}_i$

$$\boldsymbol{\alpha}_i = \begin{pmatrix} \alpha_{i\,11} & \alpha_{i\,12} & \cdots & \alpha_{i\,1q} \\ \alpha_{i\,21} & \alpha_{i\,22} & \cdots & \alpha_{i\,2q} \\ \cdots & \cdots & \cdots & \cdots \\ \alpha_{i\,q1} & \alpha_{i\,q2} & \cdots & \alpha_{i\,qq} \end{pmatrix} \qquad (5.8)$$

(size $q \times q$; a_{ijk} are the numbers) is such that the value of process X_t in any moment of time is determined by the linear combination of its p previous values and the value of q-variate white noise N_t; that is,

$$X_{t+1} = \boldsymbol{\alpha}_1 X_t + \boldsymbol{\alpha}_2 X_{t-1} + \ldots + \boldsymbol{\alpha}_p X_{t-(p-1)} + N_{t+1}, \qquad (5.9)$$

then X_t is referred to as the q-variate autoregressive process of order p. Because the sequence of the matrices $\boldsymbol{\alpha}_i$ consists of $p \times q^2$ elements, and the white noise N_t is determined by the matrix Σ consisting of q^2 elements, the q-variate autoregressive process of order p is entirely determined by $(p+1) \times q^2$ parameters. Their estimation presents the principal problem of fitting the multivariate AR model to the multiple time series of the observed system.

The fundamental relationship, which is the basis for such an estimation, can be obtained by multipling the left and right parts of equation (5.9) by $X_{t+1-\tau}^{\mathrm{T}}$ and by taking the expectation:

$$E(X_{t+1} X_{t+1-\tau}^{\mathrm{T}}) = \boldsymbol{\alpha}_1 E(X_t X^{\mathrm{T}}{}_{t+1-\tau}) + \ldots$$

$$+ \boldsymbol{\alpha}_p E(X_{t-(p-1)} X_{t+1-\tau}^{\mathrm{T}}) + E(N_{t+1} X_{t+1-\tau}^{\mathrm{T}}). \qquad (5.10)$$

Then if $\tau > 0$ we have

$$\mathbf{V}^{\mathrm{T}}(\tau) = \boldsymbol{\alpha}_1 \mathbf{V}^{\mathrm{T}}(\tau - 1) + \ldots + \boldsymbol{\alpha}_p \mathbf{V}^{\mathrm{T}}(\tau - p); \qquad (5.11)$$

5.1 Fundamental Multivariate AR Processes

that is, the matrix covariance function satisfies the matrix difference equation (5.11).

Setting $\tau = 1, 2, \ldots, p$ in (5.11) gives the multivariate analog of the Yule-Walker equations (see Section 4.1)

$$\begin{aligned}
\mathbf{V}^T(1) &= \boldsymbol{\alpha}_1 \mathbf{V}(0) + \boldsymbol{\alpha}_2 \mathbf{V}(1) + \ldots + \boldsymbol{\alpha}_p \mathbf{V}(p-1), \\
\mathbf{V}^T(2) &= \boldsymbol{\alpha}_1 \mathbf{V}^T(1) + \boldsymbol{\alpha}_2 \mathbf{V}(0) + \ldots + \boldsymbol{\alpha}_p \mathbf{V}(p-2), \\
&\ldots \ldots \ldots \ldots \ldots \\
\mathbf{V}^T(p) &= \boldsymbol{\alpha}_1 \mathbf{V}^T(p-1) + \boldsymbol{\alpha}_2 \mathbf{V}^T(p-2) + \ldots + \boldsymbol{\alpha}_p \mathbf{V}(0).
\end{aligned} \tag{5.12}$$

This is a system of linear equations with $p \times q^2$ unknown scalar parameters. By replacing $\mathbf{V}(\tau)$ by their estimates and solving the system relative to the matrices of $\boldsymbol{\alpha}_i$, it is possible to obtain the parameter estimates.

Let us introduce the following notations:

$$\mathbf{U}_p^T = \left(\mathbf{V}^T(1), \mathbf{V}^T(2), \ldots, \mathbf{V}^T(p) \right), \tag{5.13}$$

$$\boldsymbol{\alpha} = (\boldsymbol{\alpha}_1, \boldsymbol{\alpha}_2, \ldots, \boldsymbol{\alpha}_p), \tag{5.14}$$

$$\mathbf{W}_p = \begin{pmatrix} \mathbf{V}(0) & \mathbf{V}^T(1) & \ldots & \mathbf{V}^T(p-1) \\ \mathbf{V}(1) & \mathbf{V}(0) & \ldots & \mathbf{V}^T(p-2) \\ \ldots & \ldots & \ldots & \ldots \\ \mathbf{V}(p-1) & \mathbf{V}(p-2) & \ldots & \mathbf{V}(0) \end{pmatrix}, \tag{5.15}$$

where $\mathbf{U}_p^T$ and $\boldsymbol{\alpha}$ are block matrices of size $q \times qp$; $\mathbf{W}_p$ is the block matrix of size $qp \times qp$.

Using these notations, the Yule-Walker system is

$$\mathbf{U}_p^T = \boldsymbol{\alpha} \mathbf{W}_p \tag{5.16}$$

or

$$\boldsymbol{\alpha} = \mathbf{U}_p^T \mathbf{W}_p^{-1}. \tag{5.17}$$

Equation (5.9) can be written in a more concise form by setting

$$\mathbf{X}_0 = \begin{pmatrix} \mathbf{X}_t \\ \mathbf{X}_{t-1} \\ \ldots \\ \mathbf{X}_{t-(p-1)} \end{pmatrix} \tag{5.18}$$

and

$$\mathbf{X}_{t+1} = \alpha\mathbf{X}_0 + \mathbf{N}_{t+1}. \tag{5.19}$$

The matrix $\mathbf{W}_p$ is the covariance matrix of vector $\mathbf{X}_0$.

These notations simplify the algorithmic structure and make it possible to use it in a recursive way when designing and developing computer routines.

The equation for the matrix $\mathbf{\Sigma}$ is obtained from (5.10), if $\tau = 0$:

$$\mathbf{V}(0) = \boldsymbol{\alpha}_1\mathbf{V}(1) + \ldots + \boldsymbol{\alpha}_p\mathbf{V}(p) + \mathbf{\Sigma}. \tag{5.20}$$

Therefore,

$$\begin{aligned}\mathbf{\Sigma} &= \mathbf{V}(0) - \boldsymbol{\alpha}_1\mathbf{V}(1) - \ldots - \boldsymbol{\alpha}_p\mathbf{V}(p) = \mathbf{V}(0) - \boldsymbol{\alpha}\mathbf{U}_p \\ &= \mathbf{V}(0) - \mathbf{U}_p^T\mathbf{W}_p^{-1}\mathbf{U}_p = \mathbf{V}(0) - \boldsymbol{\alpha}\mathbf{W}_p\boldsymbol{\alpha}^T.\end{aligned} \tag{5.21}$$

This equation is a multivariate analog of (4.12) for the variance of the white noise.

The elements on the main diagonal of matrix $\mathbf{\Sigma}$ are the variances of the components of vector $\mathbf{N}_{t+1}$. The estimation of $\mathbf{\Sigma}$ is necessary for constructing confidence intervals of the forecasts. If the model is fitted to the multiple time series, the estimated matrix $\mathbf{\Sigma}$ is referred to as the covariance matrix of the forecast error for one step ahead.

5.2 Multivariate AR(1) Process

The multivariate autoregressive process of the first order can be considered as the generalization of the concept of the scalar Markov random process.

The first-order multivariate AR process is

$$\mathbf{X}_{t+1} = \boldsymbol{\alpha}\mathbf{X}_t + \mathbf{N}_{t+1}, \tag{5.22}$$

where

$$\boldsymbol{\alpha} = \boldsymbol{\alpha}_1 = \begin{pmatrix} \alpha_{11} & \ldots & \alpha_{1q} \\ \ldots & \ldots & \ldots \\ \alpha_{q1} & \ldots & \alpha_{qq} \end{pmatrix} \tag{5.23}$$

is the parameter matrix, which must be estimated when a model is fitted to data.

The difference equation for the matrix covariance function is

$$\mathbf{V}^T(\tau) = \boldsymbol{\alpha}\mathbf{V}^T(\tau - 1). \tag{5.24}$$

5.2 Multivariate AR(1) Process

From (5.24) it follows that

$$\mathbf{V}^T(\tau) = \boldsymbol{\alpha}^\tau \mathbf{V}(0). \tag{5.25}$$

If $\tau = 1$, we get the Yule-Walker system

$$\mathbf{V}^T(1) = \boldsymbol{\alpha}\mathbf{V}(0), \tag{5.26}$$

which is the basis for the estimation of $\boldsymbol{\alpha}$:

$$\boldsymbol{\alpha} = \mathbf{V}^T(1)\mathbf{V}^{-1}(0). \tag{5.27}$$

The estimation of the elements of matrix $\boldsymbol{\Sigma}$ is based on the formula

$$\begin{aligned}\boldsymbol{\Sigma} &= \mathbf{V}(0) - \boldsymbol{\alpha}\mathbf{V}(1) = \mathbf{V}(0) - \boldsymbol{\alpha}\mathbf{V}(0)\boldsymbol{\alpha}^T \\ &= \mathbf{V}(0) - \mathbf{V}^T(1)\mathbf{V}^{-1}(0)\mathbf{V}(1). \end{aligned} \tag{5.28}$$

The expression for the variances $\{\boldsymbol{\Sigma}\}_{ii}$ can be used to determine the domain of the permissible correlations of the multivariate AR(1) process. Actually, the inequalities

$$\{\mathbf{V}(0) - \boldsymbol{\alpha}\mathbf{V}(1)\}_{ii} \geq 0 \tag{5.29}$$

must be satisfied because the process is stationary.

The forecast for l steps ahead is

$$\mathbf{X}_{t+l} = \boldsymbol{\alpha}^l \mathbf{X}_t + \mathbf{N}_{t+l}, \tag{5.30}$$

where $\mathbf{N}_{t+l}$ is the q-variate white noise. The covariance matrix $\boldsymbol{\Sigma}_l$ of $\mathbf{N}_{t+l}$ is

$$\boldsymbol{\Sigma}_l = \mathbf{V}(0) - \boldsymbol{\alpha}^l \mathbf{V}(l) = \mathbf{V}(0) - \boldsymbol{\alpha}^l \mathbf{V}(0)(\boldsymbol{\alpha}^T)^l \tag{5.31}$$

or

$$\boldsymbol{\Sigma}_l = \mathbf{V}(0) - \mathbf{R}_l, \tag{5.32}$$

where

$$\mathbf{R}_l = \boldsymbol{\alpha}^l V(0)(\alpha^T)^l \tag{5.33}$$

and $\boldsymbol{\Sigma}_1 = \boldsymbol{\Sigma}$.

If the forecasts are estimated step by step,

$$\bar{\mathbf{X}}_{t+i} = \boldsymbol{\alpha}\bar{\mathbf{X}}_{t+i-1} \quad (i = 1, 2, \ldots, l; \ \bar{X}_t = X_t), \tag{5.34}$$

then it is suitable to compute matrices $\boldsymbol{\Sigma}_i$ by means of the recursive equations

$$\boldsymbol{\Sigma}_i = \mathbf{V}(0) - \boldsymbol{\alpha}\mathbf{R}_{i-1}\boldsymbol{\alpha}^T \tag{5.35}$$

and

$$\mathbf{R}_0 = \mathbf{V}(0). \qquad (5.36)$$

Equations (5.31) and (5.35) are the multivariate analogs of the corresponding univariate relationships (4.12) and (4.20) for the Markov random process.

Estimating the elements of the matrix covariance function $\mathbf{V}(\tau)$ for zero and the first lags is the first step in fitting the model.

To demonstrate the dependence of the forecast accuracy of the multivariate AR(1) process on the correlations and the number of variates q, consider the following illustrative example.

Example

Suppose that the elements of the matrix covariance function $\mathbf{V}(0)$ and $\mathbf{V}(1)$ are determined by the formulas

$$v_{ij}(0) = \begin{cases} 1 & \text{if } i = j, \\ \rho > 0 & \text{if } i \neq j, \end{cases} \qquad (5.37)$$

$$v_{ij}(1) = r > 0. \qquad (5.38)$$

Let us approximate the process with this covariance function using the first-order multivariate autoregressive process. Replacing the values of v_{ij} in (5.27) and (5.28) with (5.37) and (5.38), we have

$$\alpha_{ij} = \frac{r}{1 + (q-1)\rho},$$

$$x_{it+1} = \frac{r}{1 + (q-1)\rho}(x_{1t} + x_{2t} + \ldots + x_{qt}) + n_{it+1},$$

and

$$\{\Sigma_1\}_{ij} = \begin{cases} \varepsilon^2 = 1 - \dfrac{qr^2}{1 + (q-1)\rho} & \text{if } i = j, \\ \gamma = \rho - \dfrac{qr^2}{1 + (q-1)\rho} & \text{if } i \neq j. \end{cases} \qquad (5.39)$$

The first equation of (5.39) determines the domain of the permissible values of the autocorrelations because

$$1 - \frac{qr^2}{1 + (q-1)\rho} \geq 0. \qquad (5.40)$$

5.2 Multivariate AR(1) Process

In spite of the particular character of this example, the analytic solution enables us to give some numerical interpretation of the results. It helps to understand the nature of the mutual stipulation of the correlations ρ and r for the zero and first lags, the number of variates q of the process, and the accuracy characteristics (ε and γ) of the corresponding forecast.

Let us study the behavior of standard deviation ε (5.39) for different r, ρ, and q. If $r = 0$, then $\varepsilon = 1$, and the forecast coincides with the mean value of the process. If the value of q is increased, then ε is monotonically decreased in the following way:

$$\varepsilon^2 \to 1 - r^2/\rho \quad (\text{if } q \to \infty). \tag{5.41}$$

Therefore, the accuracy of the forecast is limited by the value $1 - r^2/\rho$. In addition, beginning with some value of q, increasing the number of the mutually analyzed time series cannot lead to an unlimited reduction of ε. In general, the inequality

$$1 - r^2/\rho < \varepsilon^2 \leq 1 \tag{5.42}$$

determines the range of the normalized variance of the forecast error independently of the number of variates of the considered process. This inequality outlines the possible accuracy of the forecast of the process [with $\mathbf{V}(0)$ and $\mathbf{V}(1)$, given by (5.37) and (5.38)] and the expediency of fitting the corresponding model.

For a detailed numerical description of the behavior of the normalized standard deviation ε as a function of the correlations (ρ and r) and the number of variates (q), let us carry out the calculations by means of (5.39) for some fixed grid of ρ, r, and q values. The results are presented in Figures 5.1 and 5.2.

An analysis of Figure 5.1 enables us to make the following comments. First, the smaller the absolute value of the correlation ρ for zero lag, the smaller the domain (ρ, r) of the permissible correlations (and the higher the forecast accuracy). If q is increased, the largest growth of accuracy (decreasing of ε) occurs for small ρ ($\rho < 0.05$). Therefore, for this example, the statistical independence of the time series x_{it} at zero lag (and the maximum possible temporal correlation r) is ideal. For large ρ, the increase of the accuracy (with increasing q) occurs more slowly than for small ρ. Actually, if ρ is greater than 0.8, the standard deviation of the forecast error does not differ markedly from the corresponding standard deviation of the forecast of the univariate processes x_{it} (5.1) with the same value of r. This

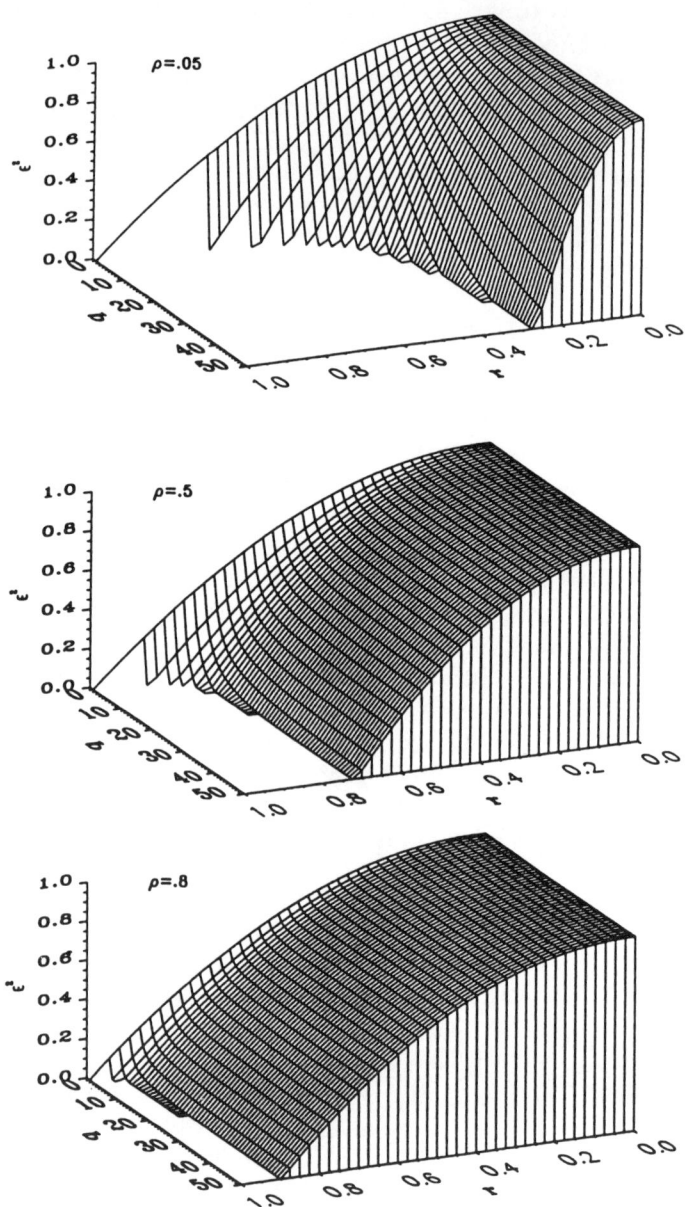

Figure 5.1: Normalized variance ϵ^2 (as a function of q, r, and ρ) of the forecast error of the multivariate AR(1) process determined by (5.37) and (5.38).

5.2 Multivariate AR(1) Process

Table 5.1: Correspondence between number of variates (q) and correlation (ρ) for zero lag.

ρ	0.0–0.2	0.2–0.5	0.5–0.7	0.7–0.9
q_s	8–10	6–8	3–6	2–3

fact is easy to understand; for, if ρ is close to 1, a statistical dependence of the variates is close to a linear (deterministic) dependence, and the cross-statistical analysis cannot provide a significant amount of new information. The growth of q leads to the decrease of ε, which asymptotically approaches the fixed (limit) value. Beginning with a certain value of q_s, some kind of saturation is taking place, and further increasing the number of variates does not affect the forecast accuracy. Approximate values of such q_s for corresponding ρ are presented in Table 5.1.

Second, an evaluation of the possibility of fitting the multivariate model and forecasting reveals that, exactly as for the univariate case, the principal meaning is not simply the existence of the temporal statistical dependence ($r \neq 0$), but the quantitative level of this dependence. For small r ($r \leq 0.2$), the standard deviation ε of the forecast error with any q and ρ is greater than 0.8; that is, it is equal to 80% (and more) of the standard deviation of the processes. The corresponding forecast does not significantly differ from the mean value of the process, and its estimation is of no practical value. Therefore, the interval $(0.0, 0.2)$ of the variation of r is outside the bounds of practical consideration.

If $0.2 < r < 0.3$ (the range of the values of the first autocorrelations for the monthly and annual mean time series of air temperature, radiation, streamflow, and so on) and q is increased, the standard deviation ε is gradually decreased to the values of 0.7–0.8, only if $\rho < 0.2$. Therefore, the possible accuracy of the multivariate models of the climate time series of point gauges is limited by these values (0.7–0.8); that is, the above calculations exhaustively characterize the possible accuracy of the statistical description of the local climate.

If $\rho \geq 0.3$, fitting the multivariate model with $r \leq 0.3$ is of no practical value because the standard deviation of the forecast error exceeds 90% of the standard deviation of the process (independent of the q value).

Analyzing Figure 5.1, it is possible to choose an appropriate region in the space of (r, ρ, q) in which the forecast error ε is so small that fitting the model makes sense. If the model with $\varepsilon < 0.8$ can be considered acceptable, then, as Figure 5.1 shows, the cross-correlation of the mutually analyzed time series must be greater than 0.3–0.4 at lag one.

From the above reasoning, it follows that one of the ways of building multivariate climate models is to prepare corresponding samples of spatially and temporally averaged time series (of modeling together climate characteristics), the cross-correlations of which satisfy the above requirements. More generally, it makes clear the necessity of developing different filtering methodologies. From application of such filters, the resulting time series of the means must be analyzed together in order to evaluate the possibility of building their multivariate model. If for such time series the estimated cross-correlations for the non-zero lags are greater than 0.3–0.4, then building the multivariate climate model makes sense. Of course, in this case it is necessary to conduct a theoretical study of the influence of the spatial averaging (transformation) procedure on the cross-correlation function of the samples of averages, analogous to those studies that were carried out in Section 2.4. The calculations and algebraic rearrangements in this case are more complicated than those presented in Section 2.4, and the interpretation of the results is not so obvious.

The character of the statistical dependence γ of the forecast errors on parameters ρ, r, q is demonstrated by Figure 5.2.

This example helps to represent some of the requirements of the statistical structure of the multiple time series, the multivariate climate model of which makes practical sense. In the general case of the first-order multivariate AR process, the analogous prior theoretical analysis is more complicated. However, we can notice that significantly increasing the number of the model's variates is necessary only for very accurate investigations. Having a small number of modeling variates, we do not lose much in terms of the forecast's accuracy; but, of course, we certainly can miss some of the subtle spectral features of the analyzed processes.

5.3 Algorithm for Multivariate Model

The above consideration of the theory of the multivariate autoregressive processes represents the standard approach because it directly follows the univariate case. When developing an algorithm for fitting a multivariate ARMA model to the system of the time series,

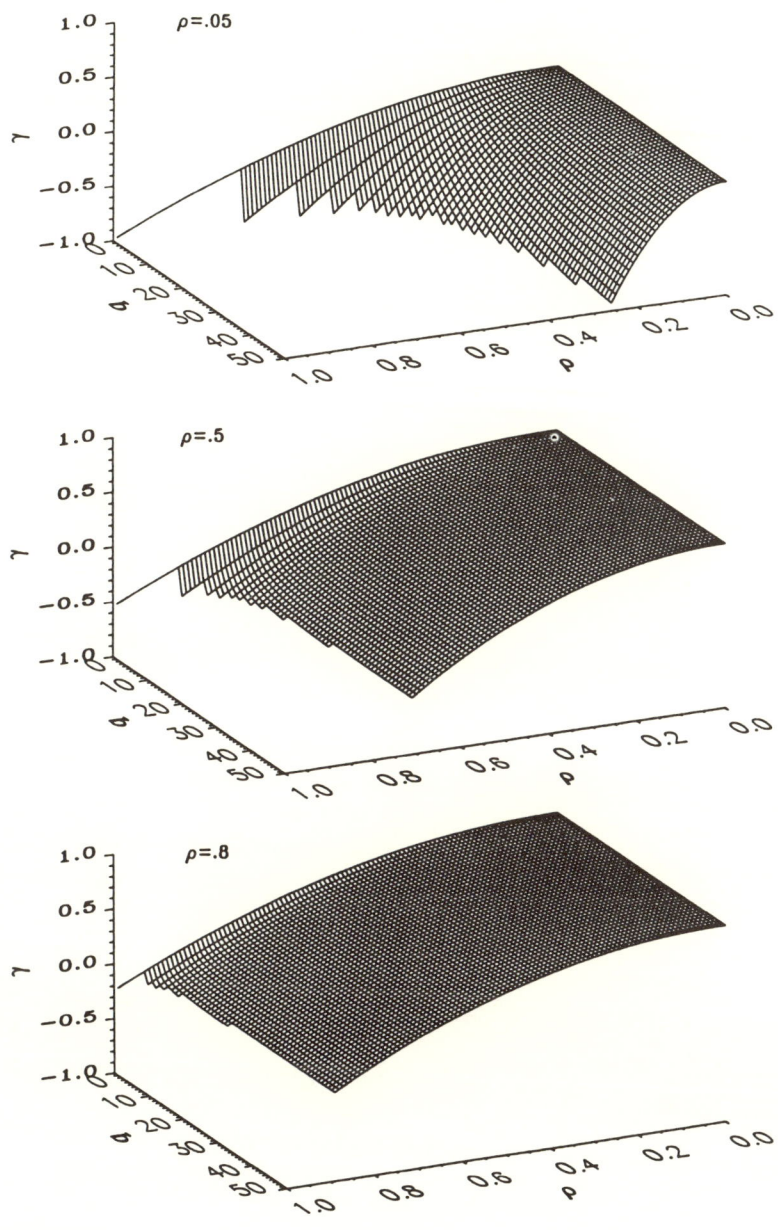

Figure 5.2: Correlation γ (as a function of q, r, and ρ) of the forecast errors of the multivariate AR(1) process determined by (5.37) and (5.38).

the most efficient methodology is the recursive scheme, which is analogous to that in Section 4.2.

The forecast for l steps ahead can be presented by formula

$$\bar{\mathbf{X}}_{t+l} = \alpha \mathbf{A}^{l-1} \mathbf{X}_0, \qquad (5.43)$$

where

$$\mathbf{X}_0 = \begin{pmatrix} \mathbf{X}_t \\ \mathbf{X}_{t-1} \\ \cdots \\ \mathbf{X}_{t-(p-1)} \end{pmatrix}, \qquad (5.44)$$

$$\mathbf{A} = \begin{pmatrix} \alpha_1 & \alpha_2 & \alpha_3 & \cdots & \alpha_{p-1} & \alpha_p \\ 1 & 0 & 0 & \cdots & 0 & 0 \\ 0 & 1 & 0 & \cdots & 0 & 0 \\ \cdots & \cdots & \cdots & \cdots & \cdots & \cdots \\ 0 & 0 & 0 & \cdots & 1 & 0 \end{pmatrix}, \qquad (5.45)$$

1 is the identity matrix of size $q \times q$, **0** is the matrix of size $q \times q$ with all zero elements. The covariance matrix Σ_l of the forecast errors for l steps ahead is denoted as

$$\Sigma_l = \mathbf{V}(0) - \alpha \mathbf{A}^{l-1} \mathbf{W}_p (\mathbf{A}^T)^{l-1} \alpha^T, \qquad (5.46)$$

where $\mathbf{W}$ is determined by (5.15). Formulas (5.43) and (5.46) yield the following recursive procedure:

$$l = 1, \quad \mathbf{Q}_l = \alpha, \qquad (5.47)$$
$$\to \bar{\mathbf{X}}_{t+l} = \mathbf{Q}_l \mathbf{X}_0 \qquad (5.48)$$
$$\Sigma_l = \mathbf{V}(0) - \mathbf{Q}_l \mathbf{W}_p \mathbf{Q}_l^T, \qquad (5.49)$$
$$\leftarrow \mathbf{Q}_l = \mathbf{Q}_{l-1} \mathbf{A} \quad (l = 2, 3, \ldots). \qquad (5.50)$$

Programming this algorithm in an algorithmic language, which permits the immediate use of the matrix expressions, is very simple.

Fitting the sequence of models to the given system of the time series by consecutive increase in the order p leads to the necessity of inverting the sequence of matrices $\mathbf{W}_p$ ($p = 1, 2, \ldots$). If the model order is increased by one (from $p - 1$ to p), then to obtain $\mathbf{W}_p$ the blocks of the last row and the last column must be added to the matrix $\mathbf{W}_{p-1}$. The inversion of $\mathbf{W}_p$ is significantly simplified by taking into account the fact that $\mathbf{W}_{p-1}^{-1}$ has already been obtained. The

5.3 Algorithm for Multivariate Model

most convenient algorithm for inverting matrix $\mathbf{W}_p$ is partitioning (Berezin and Zhidkov, 1965). Using this method one can obtain the recursive equation for $\mathbf{W_p}^{-1}$.

Let

$$\mathbf{W}_p = \left(\begin{array}{cccc|c} \mathbf{V}(0) & \mathbf{V}^T(1) & \ldots & \mathbf{V}^T(p-2) & \mathbf{V}^T(p-1) \\ \mathbf{V}(1) & \mathbf{V}(0) & \ldots & \mathbf{V}^T(p-1) & \mathbf{V}^T(p-2) \\ \ldots & \ldots & \ldots & \ldots & \ldots \\ \mathbf{V}(p-2) & \mathbf{V}(p-1) & \ldots & \mathbf{V}(0) & \mathbf{V}^T(1) \\ \hline \mathbf{V}(p-1) & \mathbf{V}(p-2) & \ldots & \mathbf{V}(1) & \mathbf{V}(0) \end{array} \right)$$

$$= \left(\begin{array}{cc} \mathbf{W}_{p-1} & \mathbf{B}^T \\ \mathbf{B} & \mathbf{V}(0) \end{array} \right), \qquad (5.51)$$

where

$$\mathbf{B} = [\mathbf{V}(p-1), \mathbf{V}(p-2), \ldots, \mathbf{V}(1)]. \qquad (5.52)$$

The matrices $\mathbf{W}_{p-1}^{-1}$ and $\mathbf{V}^{-1}(0)$ have already been computed in the previous steps.

The partitioning of matrix $\mathbf{W}_p^{-1}$, corresponding to the blocks of (5.51), is

$$\mathbf{W}_p^{-1} = \left(\begin{array}{cc} \mathbf{C}_{11} & \mathbf{C}_{12} \\ \mathbf{C}_{21} & \mathbf{C}_{22} \end{array} \right). \qquad (5.53)$$

According to the partitioning method (Berezin and Zhidkov, 1965), the expressions for the separate blocks of matrix $\mathbf{W}_p^{-1}$ (5.53) are

$$\mathbf{C}_{22} = \left[\mathbf{V}(0) - \mathbf{B}\mathbf{W}_{p-1}^{-1}\mathbf{B}^T \right]^{-1}, \qquad (5.54)$$

$$\mathbf{C}_{21} = \mathbf{C}_{12}^T = -\mathbf{C}_{22}\mathbf{B}\mathbf{W}_{p-1}^{-1}, \qquad (5.55)$$

$$\mathbf{C}_{11} = \mathbf{W}_{p-1}^{-1}\left(\mathbf{1} - \mathbf{B}^T\mathbf{C}_{21}\right), \qquad (5.56)$$

where $\mathbf{1}$ is an identity matrix.

Thus, in each step, a matrix of $q \times q$ size must be inverted (instead of matrix $\mathbf{W}_p$ of $pq \times qp$ size), thereby significantly saving time and increasing accuracy, compared with the direct inversion of $\mathbf{W}_p$.

Despite the efficiency of the algorithm considered, its representation in the form of a computer routine for arbitrary p and q requires a significant volume of computer memory.

The following comments are useful for software development and application of the algorithm considered.

1. In a multivariate case, an application of the (4.43)-type criterion leads to some difficulties because the scales (or the units) of measurement of the mutually analyzed processes can significantly differ in the numerical values. For example, eliminating the statistically insignificant estimates of parameters α_{ijk} according to (4.43) can lead to very different results, depending on whether the model is developed for the anomalies or for the standardized anomalies. Therefore, the computational aspects, in addition to the statistical nature of the data, can determine the final inferences. It is possible that processing the normalized anomalies is the most rational way of performing computations, but its theoretical foundation is not absolutely clear. The choice of the appropriate scales and units can be simplified significantly if there are physical relationships connecting the mutually analyzed processes.

2. In the case of normalization, the matrix autocovariance function in the above formulas must be replaced by the corresponding matrix autocorrelation function, and the estimation of parameters $\boldsymbol{\alpha}$ and matrices $\boldsymbol{\Sigma}_l$ must be carried out for the normalized time series. During the final stage of computations, to obtain the results in real scales or units, the forecast estimates and their confidence intervals must be multiplied by the corresponding estimates of the standard deviations. The analysis of variances $\{\boldsymbol{\Sigma}_l\}_{ii}$ of the normalized observations makes it possible to evaluate the accuracy of the results in exactly the same way as was done for the univariate processes (comparing the $\{\boldsymbol{\Sigma}_l\}_{ii}$ with 1). The range of variations of the forecast accuracy, which were found for the univariate processes, remain approximately the same for the multivariate cases.

3. The variances of the components of the q-variate white noise $\mathbf{N}_{t+l}$ are located along the main diagonal of matrix $\boldsymbol{\Sigma}_l$. Therefore, the estimation of $\boldsymbol{\Sigma}_l$ is necessary to construct a confidence interval of the forecast for l steps ahead. $\boldsymbol{\Sigma}_l$ is determined by the cross-statistical structure [the matrices $\mathbf{V}(\tau)$] of the processes x_{it} on the two intervals of lags: $\tau \in [-(p-1), p-1]$ and $\tau \in [l-p, l+p]$, which, of course, can overlap. However, by taking into account the relationship (5.11), it is clear that the forecast depends only on $\mathbf{V}(\tau)$ for $\tau = 0, 1, 2, \ldots, p-1$.

4. The identification process is developed by fitting the models with successively increasing order and by analyzing the following quantities:

a. values analogous to (4.43), $\sqrt{K}\alpha_{ijk}$, where K is the sample size, for each parameter (α_{ijk}) estimate to determine its statistical significance;

b. values analogous to (4.44) for the residuals of each equation of the consecutive models of the p and $p+1$ orders;

c. the proximity of the normalized variances, $\varepsilon_i^2 = \{\Sigma_l\}_{ii}/\mathbf{V}_{ii}(0)$, to 1. The estimation of the forecast accuracy is carried out by constructing the confidence intervals of the forecasts with the aid of the variances $\{\Sigma_l\}_{ii}$ of the forecast errors.

By making a decision on the basis of these results, one can identify the model's order and the number of variates, which must be jointly analyzed.

5. In climatology the general problems of identifying multivariate stochastic climate models have not been formulated. Even the appropriate models of the first order are not developed. But fitting such models and appraising their physical and statistical characteristics is an interesting area of the developing field of statistical climatology.

Of course, it is impossible to describe the climate only by AR processes. However, developing such models is a scientifically sound step in the right direction, which enables us to make accurate statistical inferences about the detectability of climate change. It seems that in the near future the first- and second-order AR multivariate climate models will have the most practical value.

5.4 AR(2) Process

The q-variate autoregressive process of the second order

$$\mathbf{X}_{t+1} = \boldsymbol{\alpha}_1 \mathbf{X}_t + \boldsymbol{\alpha}_2 \mathbf{X}_{t-1} + \mathbf{N}_{t+1} \qquad (5.57)$$

is determined by the parameter matrices $\boldsymbol{\alpha}_1$ and $\boldsymbol{\alpha}_2$ and by the covariance matrix Σ_1 of the white noise $\mathbf{N}_{t+1}$. Since each of these matrices consists of $q \times q$ numbers, the second-order process is totally determined by the $3 \times q^2$ parameters. Their estimation may be derived by the solution of the Yule-Walker system, with preliminary estimation of the elements of the matrix covariance function $\mathbf{V}(\tau)$,

$$\left.\begin{aligned}\mathbf{V}^T(1) &= \boldsymbol{\alpha}_1 \mathbf{V}(0) + \boldsymbol{\alpha}_2 \mathbf{V}(1), \\ \mathbf{V}^T(2) &= \boldsymbol{\alpha}_1 \mathbf{V}^T(1) + \boldsymbol{\alpha}_2 \mathbf{V}(0)\end{aligned}\right\} \qquad (5.58)$$

relative to matrices $\boldsymbol{\alpha}_1$ and $\boldsymbol{\alpha}_2$.

Then, Σ_1 can be computed by the formula

$$\Sigma_1 = V(0) - \alpha_1 V(1) - \alpha_2 V(2). \tag{5.59}$$

After estimating matrices α_1 and α_2, the forecast is made using the recursive formula

$$\bar{X}_{t+i} = \alpha_1 \bar{X}_{t+i-1} + \alpha_2 \bar{X}_{t+i-2}, \tag{5.60}$$

where the unknown quantities of $\bar{X}$ must be replaced by their estimates.

The computation is provided by (5.47) to (5.50), which are transformed into the following equations:

$$l = 1, \quad Q_l = \alpha = (\alpha_1, \alpha_2), \tag{5.61}$$

$$\rightarrow \quad \bar{X}_{t+l} = Q_l X_0, \tag{5.62}$$

$$\Sigma_l = V(0) - Q_l W_2 Q_l^T, \tag{5.63}$$

$$\leftarrow \quad Q_l = Q_{l-1} A \quad (l = 2, 3, \ldots), \tag{5.64}$$

where

$$A = \begin{pmatrix} \alpha_1 & \alpha_2 \\ 1 & 0 \end{pmatrix}, \tag{5.65}$$

1 is the identity matrix of order $q \times q$; 0 is the matrix of size $q \times q$ consisting completely of zeroes,

$$X_0 = \begin{pmatrix} X_t \\ X_{t-1} \end{pmatrix}, \tag{5.66}$$

and

$$W_2 = \begin{pmatrix} V(0) & V^T(1) \\ V(1) & V(0) \end{pmatrix}. \tag{5.67}$$

The main calculation is the inversion of the matrix W_2. If the first-order model has already been built, the matrix $V(0)$ has been inversed. Having $V^{-1}(0)$, matrix W_2 can be inverted by applying the scheme given in the Section 5.3. Let us consider

$$W_2^{-1} = \begin{pmatrix} C_{11} & C_{12} \\ C_{21} & C_{22} \end{pmatrix}, \tag{5.68}$$

where $\mathbf{C}_{ij}$ are matrices of order $q \times q$. Then, as we have already mentioned, the following equalities can be used:

$$\mathbf{C}_{22} = \left[\mathbf{V}(0) - \mathbf{V}(1)\mathbf{V}^{-1}(0)\mathbf{V}^{\mathrm{T}}(1)\right]^{-1}, \qquad (5.69)$$

$$\mathbf{C}_{21} = \mathbf{C}^{\mathrm{T}}{}_{12} = -\mathbf{C}_{22}\mathbf{V}(1)\mathbf{V}^{-1}(0) = -\mathbf{C}_{22}\bar{\boldsymbol{\alpha}}, \qquad (5.70)$$

$$\mathbf{C}_{11} = \mathbf{V}^{-1}(0)\left[\mathbf{1} + \mathbf{V}^{\mathrm{T}}(1)\mathbf{C}_{22}\bar{\boldsymbol{\alpha}}\right], \qquad (5.71)$$

where $\bar{\boldsymbol{\alpha}}$ is the matrix of the parameter estimates of the first-order model.

5.5 Examples of Climate Models

The statistical methodologies considered enable us to analyze the statistical properties and predictability of many kinds of hydrometeorological information. Naturally, to fit the stochastic models to climate data, the most voluminous and reliable sources of long-period climatological observations must be used. One such source is different atmospheric circulation statistics, obtained by Oort (1983), based on the radiosonde measurements of the atmosphere.

5.5.1 Modeling of Atmospheric Circulation Statistics

From Oort's data base, six mean monthly time series, averaged for the entire Northern Hemisphere, were selected (tables with data are given in Polyak, 1989). The list of statistics (in Oort's notations) includes:

1. TASFC—the 1000 mb level air temperature.

2. TAMN—the air temperature, obtained by averaging the observations over 20 standard isobaric levels from 1000 mb to 50 mb.

3. T850—the 850 mb level air temperature.

4. TS—the sea surface temperature.

5. QSFC—the 1000 mb level specific humidity.

6. Q850—the 850 mb level specific humidity.

Table 5.2: Estimates of the statistical structure of the Oort's atmospheric circulation statistics averaged over the entire Northern Hemisphere.

Characteristics	TASFC	TAMN	T850	TS	QFSC	Q850
Standard deviation of the time series	0.27	0.23	0.24	0.23	0.17	0.20
$\rho(1\ \text{month})$	0.49	0.74	0.58	0.88	0.76	0.86
$\rho(2\ \text{months})$	0.32	0.63	0.50	0.82	0.67	0.74
Lead time (month) for $\epsilon(l) = 0.8$	1	2	1	4	2	3
Signal ratio (4.78)	0.77	0.87	0.68	0.95	0.86	1.00
$\varepsilon(1)$ for univariate AR(1)	0.87	0.67	0.81	0.47	0.65	0.51
$\varepsilon(1)$ for six-variate AR(1)	0.86	0.62	0.76	0.46	0.64	0.49

The observational time span is 15 years, from May 1958 to April 1973, with 180 terms in each time series. The temperature unit is °C and the specific humidity unit is g/kg.

The seasonal cycle was removed by subtracting the mean estimates for the corresponding months from each term of the time series and dividing the anomalies by the standard deviation estimates for the corresponding months. Therefore, the normalized anomalies are analyzed.

The objective of this study is to compare the univariate and multivariate modeling of the same observations.

5.5.1.1 Univariate AR Models

As a first step, each time series was approximated by the first-order univariate autoregressive model. The normalized standard deviation $\varepsilon(l)$ of the forecast error for l steps ahead by the AR(1) of time series z_t with variance σ_z^2 and zero expected value was determined by formula (4.26). The computation of l as a function of $\epsilon(l)$ and $\rho(1)$ was done by formula (4.27).

The estimates of the basic statistical characteristics are collected in Table 5.2. The autocorrelations at one- and two-month lags reveal the forecasting value of each of the considered time series. As one might expect, the maximum value of the autocorrelation $[\rho(1) =$

5.5 Examples of Climate Models

0.88] corresponds to the time series of the ocean surface temperature. It is much less obvious that time series of specific humidity Q850 [$\rho(1) = 0.86$] and QSFC [$\rho(1) = 0.76$] are rarely inferior to TS. The autocorrelation coefficients of the temperature time series are smaller: $\rho(1) = 0.74$ for the TAMN; $\rho(1) = 0.58$ for the $T850$; and $\rho(1) = 0.49$ for the $TASFC$; i.e., the surface air temperature, often used as an indicator of climate change, is not the only or even best indicator of such change. The normalized standard deviations of the forecast error for the sea temperature $\varepsilon(1) = 0.47$ and for the specific humidity Q850 $\varepsilon(1) = 0.51$, which means that their forecast for one month ahead is about two times less variable than the anomalies.

In rows 5 and 6 of Table 5.2, the estimates of the lead time l (4.27) corresponding to the normalized standard deviations [$\varepsilon(l) = 0.8$] are presented. These values are the most informative characteristics of the time series under study. They show that, when the requirements of forecast accuracy are minimal [$\varepsilon(1) = 0.8$], the ocean temperature TS and the specific humidity Q850 can be forecast for about 3 to 4 months ahead. The air temperature observations can be forecast for not more than two months ahead.

The estimates of the signal ratio, given in Table 5.2, show that the climatic signal consists of about 65 to 100% of the variance of the corresponding time series; consequently, the white noise forms about 35 to 0% of that variance. The signal level is so high because the scales of spatial averaging are very large.

Figure 5.3 (a) presents the forecast estimates of the considered time series for six months ahead and the corresponding 70% confidence intervals. The forecasts smoothly converge to zero if the lead time increases. The standard deviations of the forecast errors converge to the corresponding standard deviations of the time series. The ocean temperature and specific humidity observations can be forecast most accurately.

Finally, we must notice that a more elaborative study, using zonal means time series of each of the elements, revealed that the possibility of modeling is a result of the high autocorrelations (and small variances) of the observations in the tropical regions of the Northern Hemisphere. For most of the above elements, the forecasts in the middle or higher latitude bands make sense for no more than one month ahead. Indeed, it is only for TS and Q850 that the forecasts in the middle and high latitudes make sense for a two-month lead time.

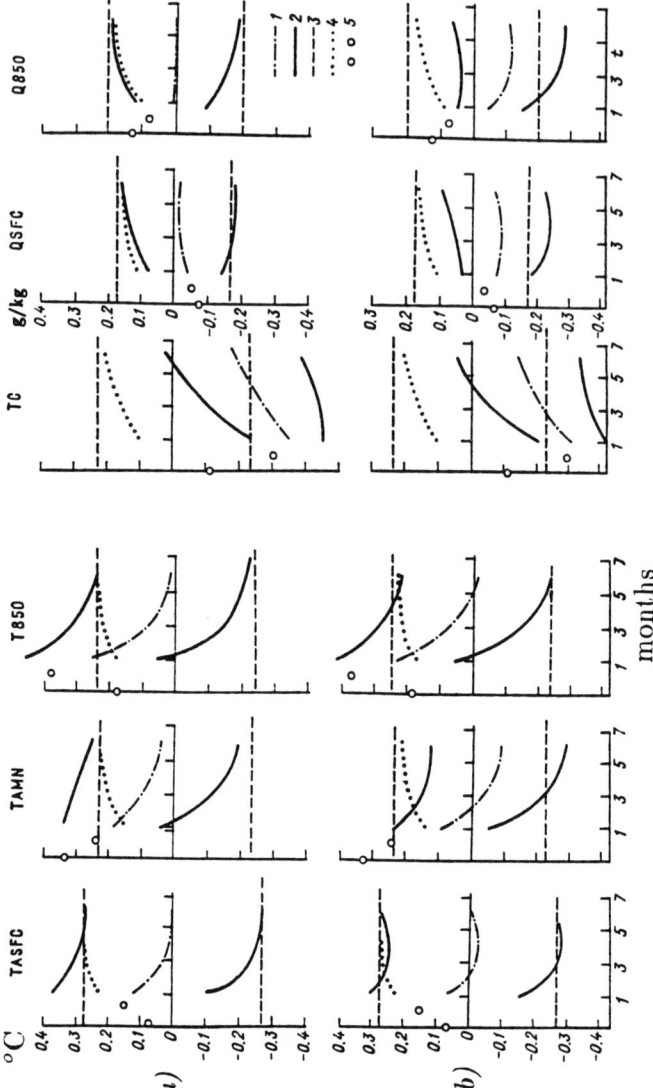

Figure 5.3: Forecasts of the mean monthly anomalies of the atmospheric circulation statistics by the univariate (a) and six-variate (b) AR(1) models. 1. Estimated forecasts. 2. 70% confidence intervals. 3. Standard deviations of the time series. 4. Standard deviations of the forecast errors. 5. The last observations of the time series.

5.5.1.2 Six-Variate AR Model

A need for a mutual statistical analysis of various meteorological data sets was clear at all stages of climatology development. However, this approach in the case of multivariate statistical modeling calls for a sufficiently large number of time series, which are to be sufficiently long. The number (180) of terms in each of the above time series is too small even to attempt to fit higher order multivariate models. Discussed below are the main results of the AR(1) model parameters estimation. Let us consider a six-variate AR model of the same time series that were studied in the previous subsection. The results of estimation are presented in Tables 5.3 and 5.4.

The diagonal elements (autocorrelations) of matrix $V(1)$ (Table 5.3) have already been discussed when we considered the univariate models. It is interesting to notice that some cross-correlations for the one-month lag are also sufficiently large (0.35 to 0.66). Thus, it is appropriate to build the multivariate model.

Table 5.4 gives the parameter estimates for the first order AR model. The main component, which forms fluctuations of each of the discussed processes, is the history of the process itself (diagonal elements α_{ii} of matrix α exceed all other elements of the line in terms of magnitude). This fact is a confirmation of the reality and physical significance of the model.

Along with its own history, the past of other processes influences the magnitude of the anomalies of each of the elements. Thus, TASFC(t+1) is determined directly by not only TASFC(t), but also by specific humidity QSFC(t) and Q850(t). Since the latter, to some extent, depend on TS(t), the TASFC(t+1) indirectly depends on the ocean surface temperature. Of course, both direct and indirect influences of the history of some processes on others are very small, which causes major difficulties in their estimation and tracing.

Actually, statistically significant estimates in Table 5.4 represent a matrix of interactions and feedbacks of observed processes. Having the analogous estimates for the time series simulated by a GCM, we could compare them, making statistical inference about the structure of such AR models and the closeness of the GCM to reality. The forecast estimates by this six-variate AR model presented in Figure 5.3(b) smoothly decline toward their mean values (to zero) with an increase of the lead time, but not as rapidly as in the case of the first order univariate autoregressive models.

The normalized standard deviations of the one-step-ahead forecast errors (last row of Table 5.2) show that the most accurate results

Table 5.3: Matrix correlation function for zero and first temporal lags.

τ		TASFC	TAMN	T850	TS	QSFC	Q850
0	TASFC	1.00	0.28	0.24	−0.03	0.42	0.01
	TAMN		1.00	0.69	0.38	0.45	0.47
	T850			1.00	0.12	0.11	0.17
	TS				1.00	0.38	0.67
	QSFC					1.00	0.52
	Q850						1.00
1	TASFC	0.49	0.10	−0.04	−0.04	0.35	0.02
	TAMN	0.13	0.74	0.52	0.37	0.36	0.39
	T850	0.08	0.42	0.58	0.13	0.01	0.13
	TS	−0.03	0.40	0.14	0.88	0.37	0.66
	QSFC	0.24	0.48	0.07	0.38	0.76	0.48
	Q850	−0.08	0.48	0.20	0.65	0.44	0.86

were obtained for the sea surface temperature and the 850 mbar level specific humidity.

The discussion above shows that the meaning of multivariate models is not limited to forecasts. Consideration of the parameter matrix α as a matrix of interactions and feedbacks can be a powerful tool for studies of different climate time series, their mutual influence, and mutual conditioning.

5.5.2 AR Models of Some Temperature Time Series

Further, the results of the approximation of different systems of monthly mean anomalies of the Northern Hemisphere temperature time series by the first-order multivariate AR models will be briefly discussed. We will try to find the main regularities of interdependences and feedbacks of temperature fluctuations in different regions of the troposphere and their mutual influence.

5.5 Examples of Climate Models

Table 5.4: Parameter estimates (exceeding 70% significance level) of six-variate model.

	TASFC$_t$	TAMN$_t$	T850$_t$	TS$_t$	QSFC$_t$	Q850$_t$
TASFC$_{t+1}$	0.44	—	—	—	0.21	−0.27
TAMN$_{t+1}$	−0.15	0.71	—	—	0.30	—
T850$_{t+1}$	−0.21	0.29	0.45	—	—	—
TS$_{t+1}$	—	—	—	0.80	—	0.11
QSFC$_{t+1}$	—	—	—	0.08	0.68	—
Q850$_{t+1}$	—	—	—	0.14	—	0.76

The first three models are built with the aid of radiosonde data (Oort, 1983); the observational time span is 15 years (1958–1973), with 180 terms in each time series.

5.5.2.1 AR Model of Zonal TAMN Temperature.

Consider Oort's data, which present the system of 9 time series of the troposphere air temperature obtained by averaging data within different 10° latitude zones and over the entire atmosphere from 1000 to 50 mb. For example, the time series corresponding to 80° was obtained by data averaging between the 85° and 75° latitudes and over the isobaric levels: 1000, 950, 900, ..., 50. The second time series (70°) was obtained by data averaging inside of 65° and 75° latitudes and over the same levels; and so on.

Table 5.5 shows the estimates of the parameters of the corresponding nine-variate model.

The main qualitative feature of this matrix is that most of the estimates, located below the main diagonal, are not statistically significant and can be considered as equal to zero. This means that climate fluctuations of zonal temperature are formed, on average, by the previous month's anomalies of this and southern latitudes. Dependence on the anomalies of the north regions (feedbacks or inverse influences, north to south) are practically undetectable with these time series.

Therefore, climate fluctuations at southern latitudes influence the

Table 5.5: Parameter estimates (exceeding 95% significance level) of nine-variate zonal TAMN temperature.

Lat. zones	80°	70°	60°	50°	40°	30°	20°	10°	0°	ε
80°	.13	.16	—	-.10	—	-.12	.71	-.15	—	.95
70°	—	.44	-.30	.33	-.23	.14	.63	-.20	—	.88
60°	—	.10	—	.41	-.19	.10	.48	-.24	—	.86
50°	—	—	—	.53	-.20	.44	-.15	—	—	.81
40°	—	—	—	.23	—	.51	-.19	.27	-.17	.84
30°	—	—	—	.16	-.19	.70	.11	—	—	.77
20°	—	—	—	—	—	—	.51	.19	—	.62
10°	—	—	—	—	—	-.22	.21	.62	.14	.53
0°	—	—	—	—	—	—	—	—	.77	.57

fluctuations of the North or, more generally, temperature anomalies of tropical regions are the main source of climate temperature anomalies of all other latitudes. This feature is related to the meridional atmospheric circulation.

In particular, as the coefficient estimates (of the first two equations) in Table 5.5 show, the climate fluctuations of the Arctic zone are formed by the temperature anomalies of almost all other latitude bands. According to the results of Polyak (1989), the point gauge time series of temperature anomalies of the Arctic regions are close to white noise. So, the first two rows of Table 5.5 show the process of formation of white noise samples (whitening) of Arctic temperature with the aid of a linear combination of, generally speaking, autocorrelated data of southern regions. For the surface air temperature this process will be discussed in Chapter 7.

It is interesting that approximately the same feature (that the observed fluctuations in the tropical regions are the main source of climate anomalies for the northern latitudes) were obtained for other time series (TASFC, T850, TS, QSFC, Q850) (Polyak, 1989).

5.5.2.2 AR Model of Troposphere Temperature

Here, we consider four troposphere temperature time series from the same Oort data base averaged horizontally on each level for the entire Northern Hemisphere and vertically in the four isobaric layers: 950–850 (T1), 800–400 (T2), 350–150 (T3), 100–50 (T4) mbar (see Matrosova and Polyak, 1990). The parameter estimates of the corresponding four-variate model are given in Table 5.6.

5.5 Examples of Climate Models

Table 5.6: Estimates (exceeding 95% significance level) of parameters of four-variate AR model.

Layers	950–850 T1(t)	800–400 T2(t)	350–150 T3(t)	100–50 T4(t)	ε
950–850, T1(t+1)	0.34	—	—	—	0.82
800–400, T2(t+1)	−0.24	0.83	—	—	0.63
350–150, T3(t+1)	—	—	0.81	—	0.62
100–50, T4(t+1)	−0.16	0.2	—	0.76	0.71

Table 5.6 shows that temperature fluctuations of the lowest level are determined only by its own history. Anomalies of the second layer depend on its own past and on the fluctuations of the first layer of the previous month.

Interrelation of the temperature fluctuations of the third layer, including tropopause, with others is statistically insignificant and not detected. So, the time series of the tropopause temperature can be considered and modeled independently, which corresponds to the understanding of the special physical nature of the tropopause.

As the last row of Table 5.6 shows, climate temperature fluctuations of the highest layer are formed by the temperature fluctuations of the previous month of all lower layers except the tropopause. We can see that the main source of the troposphere temperature fluctuations is fluctuation in its lowest level. In spite of the relatively small vertical climatic temperature variations, these fluctuations can be traced and estimated using multivariate stochastic models.

5.5.2.3 AR Model of Zonally Averaged Surface Air Temperature of the Midlatitude and Northern Regions

The next results were obtained using mean monthly surface air temperature anomalies averaged in three extra-tropical latitude bands: $27.5°–37.5°$ (Ts), $37.5°–67.5°$ (Tm), and $67.5°–87.5°$ (Tn). The observational interval is 95 years (1891–1985), and each time series has 1140 terms. One of the most complete surface air temperature data bases from the World Data Center, Obninsk (Russia), was used.

Table 5.7: Estimates (exceeding of 95% significance level) of parameters of three-variate surface air temperature AR model.

Time series	Tn(t)	Tm(t)	Ts(t)	ε
Tn(t+1)	0.37	0.14	0.29	0.92
Tm(t+1)	–	0.37	0.10	0.91
Ts(t+1)	–	–	0.34	0.90

The statistically significant parameter estimates, given in Table 5.7, are situated on and above the main diagonal of the parameter matrix. Therefore, in spite of a different source, nature, and time interval of these measurements (compared with Oort's data), qualitative consideration of the results are approximately the same as in Section 5.5.2; i.e., monthly mean anomalies of different latitude bands are formed on average by the fluctuations of the previous month in the band under consideration and in southern latitudes.

5.6 Climate System Identification

The AR models considered above appear not as a special physical wording of stochastic climate modeling problems but were basically evoked by the data bases at hand. Accurate physical formulation of such problems must be done. A stochastic model should have the basic statistical properties of the system under study and should represent our knowledge about this system in a convenient and compact form. Generally, the model must be a concentrated expression of modern achievements of the theory which facilitates better understanding of mutual conditioning of observed processes. The methodology of fitting the model and the estimation of its closeness to the real system forms a special branch of mathematical statistics, system identification. Contemporary identification methods (Eikhoff, 1983) of complicated stochastic systems show that for an adequate description of such systems, parallel building of their physical and stochastic models (SM) is necessary.

5.6 Climate System Identification

In climatology, the building of physical models (general circulation models, GCM) predominates. However, there are grounds (Hasselmann, 1976; North and Kim, 1991; Polyak, 1989) to suppose that SM, taught by qualitative observations, can offer an equally valid form of climate description and for this reason should be of great scientific and practical significance. Moreover, from an economical viewpoint, SM are very effective.

According to Eikhoff (1983), a general scheme of linear identification of climate system can be presented graphically (Figure 5.4). Neither the physical nor the stochastic approach alone can be the only method for model development. Figure 5.4 clearly shows that perfection of GCMs must be connected with developing the stochastic modeling and incorporating it into the process of building the GCM.

Of course, the identification procedure in Figure 5.4 presents an idealized picture because the physical climate models (GCM) were basically developed without any linearization and stochastic modeling. But this scheme may be helpful in the infinite process of diagnosis and perfection of GCMs.

Specifically, the identification of a linear stochastic model of climate includes the solution of the following problems:

1. climate data presentation;

2. determination of model order;

3. determination of dimensionality and the number of variates; and

4. determination of variates initial in nonstationary variations and selection of appropriate approximations for them.

Let us briefly discuss these questions.

1. Among the many possibilities (i.e., sampling, averaging, filtering, transformation) of presenting the multitude of global and local climate time series, we shall mention those that are given in two papers. Oort (1983) estimated many time series of global atmospheric circulation statistics by averaging aerological data over different latitude bands and altitude levels of the earth's atmosphere. Some of these data were used to build the above models. Another possibility was given by Matrosova and Polyak (1990), who showed that troposphere temperature observations of the Northern Hemisphere can be presented by

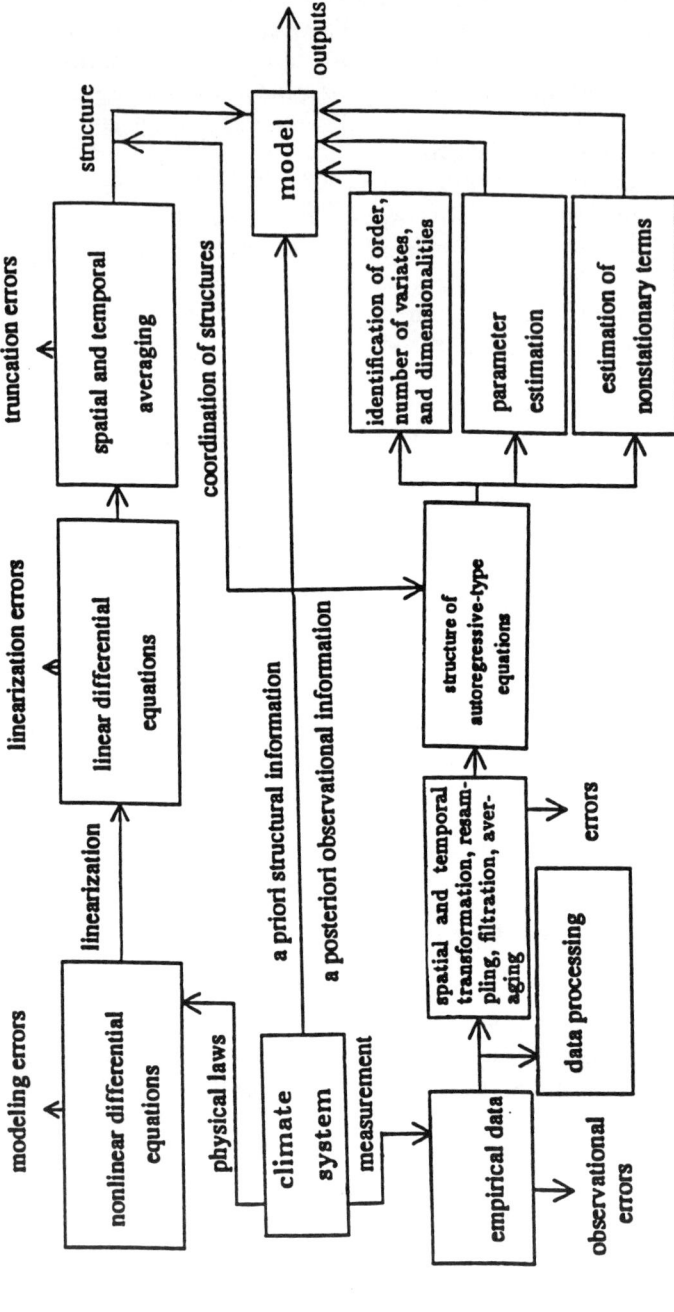

Figure 5.4: Linear identification of climate system.

5.6 Climate System Identification

monthly mean time series for 12 fixed spatial regions, inside of which statistical structures of data are approximately identical. Of course, a wide range of additional studies must be carried out to reveal spatial regions where observations have similar statistical structure and, thus, where averaging is permissible.

2. Determination of the appropriate order of a climate model may be based on physical and statistical concepts. As to the former, it is evident that basic laws of hydrodynamics (equations of continuity, movement, energy, etc.), used for developing the GCM, include time derivatives of different variables not greater than second order. Therefore, we should hardly expect that the stochastic climate model would be greater than the second order. Moreover, temperature, as a basic characteristic of climate system, is included in the equations of energy and diffusion as a first-order time derivative. This allows us to use first-order autoregressive models as the first step of climate statistical modeling. Since the volume of climate data is limited, the level of complexity of the stochastic model is limited, in principle.

3. Determination of a number of variates is based on the concept of causality, as defined by Kashyap and Rao (1976). Building the above climate models we used a simple empirical approach, choosing from Oort's data base the time series with the largest first autocorrelation coefficients.

4. Identification of the essentially nonstationary elements is based on statistical and physical analysis of tendencies of variation within a given observational interval. The problem of choosing an appropriate form of nonstationary stochastic models depends upon the development of methodology for solution of the classical problems of nonlinear modeling (Tong, 1990).

Concluding this chapter one can emphasize that multivariate SM is a powerful methodology, which can be used in a wide range of climatological studies, including diagnostic analysis and verification of GCM. The results in Chapters 7 and 8 and in Polyak (1989, 1992) show that the major advantage of multivariate AR models over the univariate case is in the possibility of interpretation of the parameter matrix as a matrix of interactions and feedbacks, even when the forecast has low accuracy or makes no sense at all. This model's core (matrix of parameters) may be used as a basis for physical analysis in the process of comparison with analogous estimates obtained using simulated data. Reliable estimation of each parameter α_{ij} allows us to determine the weight of fluctuation x_{jt} contributed to the variability of $x_{i(t+1)}$.

Analyzing sets of time series of different processes that characterize local or global climate makes it possible to estimate their mutual influence on the variability of each and, for example, to separate natural and man-induced variability.

For verification and evaluation of GCM, one must compare multidimensional probability densities of observed and simulated variables. In practice, different first- and second-order moments must be compared. The second moments of data generated by ARMA models are identical to corresponding moments of samples, which is impossible to guarantee for those simulated by GCM. Moreover, the matrix of the estimated parameters of the multivariate ARMA process (the matrix of interactions and feedbacks) contains all of the information about the second moments in a short and physically interpreted form, and it can be used for diagnostic analysis in the GCM verification problem as a standard for comparison with analogous results obtained using data simulated by GCM (see Chapter 7). In other words, the intercomparison of the stochastic models of the analogous systems of time series, both observed and simulated, gives the unique possibility of making a statistical inference about the identity or distinction of the observed and simulated climate processes.

Such an approach presents an opportunity to draw conclusions and to make recommendations concerning identification of the number of variates (and their types) and the order of models, for choosing appropriate regions of spatial averaging of time series, for comparing different stochastic or physical models (or results of different simulations by the same GCM), and for studying physical regularities and features of observations. Even the results obtained above with minimum observations can be used for qualitative comparison with analogous estimates found using data simulated by GCM. Therefore, multivariate ARMA models accumulate both physical and statistical qualities. This agrees with the requirements of the system identification theory.

This preliminary study demonstrates the great theoretical and practical possibilities of statistical modeling, which allows us to solve modern climatological problems on a new qualitative and quantitative level.

6

Historical Records

In this chapter, the historical records of annual surface air temperature, pressure, and precipitation with the longest observational time series will be studied. The analysis of the statistically significant systematic variations, as well as random fluctuations of such records, provides important empirical information for climate change studies or for statistical modeling and long-range climate forecasting. Of course, compared with the possible temporal scales of climatic variations, the interval of instrumental observations of meteorological elements proves to be very small. For this reason, in spite of the great value of such records, they basically characterize the climatic features of a particular interval of instrumental observations, and only some statistics, obtained with their aid, can have more general meaning.

Because each annual or monthly value of such records is obtained by averaging a large number of daily observations, the corresponding central limit theorem of the probability theory can guarantee their approximate normality. In spite of this, we computed the sample distribution functions for each time series analyzed below and evaluated their closeness to the normal distribution by the Kolmogorov-Smirnov criterion. As expected, the probability of the hypothesis that each of the climatic time series (annual or monthly) has a normal distribution is equal to one with three or four zeros after the decimal point.

6.1 Linear Trends

As seen in this section, the straight line least squares approximation of the climatic time series (see Table 6.1) enables us to obtain very simple and easy-to-interpret information about the power of the long period climate variability. Carrying out such an approximation, we assume that the fluctuation with a period several times greater than the observational interval will become apparent as a gradual increase or decrease of the observed values. Using only a small sample, it is impossible to determine accurately the amplitude and frequency of such long-period climate fluctuation. Consequently, the straight-line model is the simplest approach in this case.

Let us begin with an analysis of the annual surface air temperature time series, the observations of which are published in Bider et al., (1959), Bider and Schüepp, (1961), Lebrijn (1954), Manley (1974), and in the World Weather Records (1975). The units of the observations are °C. The stations' names, the observational intervals, and the linear trend parameter estimates are given in Table 6.1 (σ_0 and σ_1 are the standard errors for constant and straight-line approximations, respectively). The results in this table reveal that for seven (of the eighteen time series) the estimates of the parameter β_1 are statistically significant at the 99% level (the t statistic value is greater than 3). For ten of the time series, the estimates of the slope parameter are statistically significant at the 95% level (the t statistic value is greater than 2). Therefore, in such an approximation a stochastic model of the variations of the annual temperature time series analyzed is

$$Y(t) = \beta_0 + \beta_1 t + X(t), \qquad (6.1)$$

where β_0 is the annual mean for a given station, β_1 is the annual rate, and $X(t)$ is the random fluctuation.

Most of the estimates of β_1 in Table 6.1 are very small, but positive, which makes it clear that some kind of small global warming has taken place long before the industrial revolution and the subsequent increase of CO_2 in the atmosphere. The mean of the estimates of β_1 in Table 6.1 is about 0.002 (that is, two Celsius degrees per 1000 years); the mean of the corresponding standard errors σ_{β_1} is 0.001. Careful consideration of the results presented in Table 6.1 shows that, as a rule, the statistically significant β_1 estimates are obtained for the longer observational intervals. Therefore, the time series must have a considerable duration so that the parameter β_1 can be accurately estimated. This point is more fully illustrated in Table 6.2, where the linear trend parameter estimates and their accuracies are given for the different subintervals of the central England time series with a 315-year record of surface air temperature. The results reveal that approximately 90 years of observations were necessary to obtain a statistically significant estimate of the parameter β_1.

6.1 Linear Trends

Table 6.1: Linear trend parameter estimates of the annual mean surface air temperature (°C) time series.

Station	Observ. interval	$k+1$	σ_1	β_1	σ_{β_1}	t	σ_Y	σ_0
Strasbourg	1806–1955	150	.73	.002	.001	1.1	.06	.73
Prague	1775–1955	181	.87	−.001	.001	0.8	.06	.87
De Bilt	1755–1955	201	.70	.002	.001	2.4	.05	.71
San Bernard	1818–1955	138	.65	.002	.001	1.7	.06	.66
Berlin	1756–1955	200	.83	.002	.001	2.3	.06	.84
Trieste	1803–1955	153	.61	.002	.001	1.4	.05	.61
Hoenpisinberg	1781–1955	175	.78	−.001	.001	0.8	.06	.78
Basel	1755–1957	203	.69	.003	.001	3.6	.05	.71
Iena	1770–1955	186	.76	.003	.001	2.5	.06	.77
Vienna	1775–1955	181	.83	−.001	.001	1.0	.06	.82
Swanenburg	1735–1940	206	.73	.002	.001	1.9	.05	.07
Geneva	1755–1955	201	.64	.002	.001	3.0	.05	.65
Stuttgart	1792–1955	164	.81	.004	.001	3.3	.06	.83
Karlsruhe	1799–1955	157	.68	.008	.001	6.3	.06	.76
Paris	1757–1953	197	.70	.004	.001	4.4	.05	.73
Turin	1755–1911	157	.68	.001	.001	0.5	.05	.68
St. Petersburg	1752–1966	215	1.04	.004	.001	3.2	.07	1.08
Central England	1659–1973	315	.60	.002	.000	4.9	.04	.62

More elaborate studies of the annual temperature linear trend estimates for different stations and separate subintervals of the period of the instrumental observations reveal that the most noticeable increase in temperature has taken place during the first 35 to 40 years of our century, which corresponds to the growth of large cities such as Karlsruhe, Paris, and Berlin; naturally, the local climate was affected by the development of urban industry.

The results show that the long-period temperature fluctuations can be powerful enough for their rate to be statistically significant, or the model (6.1) can be approximately valid for the limited periods of time. This model is, of course, inappropriate for forecasting because the variance of such forecasts is greater than the variance of the observations (see, for example, Figure 1.2). Later, a more accurate

Table 6.2: Linear trend parameter estimates of the central England annual surface air temperature (C^o).

Observation interval	$k+1$	σ_1	β_0	β_1	σ_{β_0}	σ_{β_1}	t	$\sigma_{\hat{Y}}$	σ_0
1659–1973	315	.595	9.140	.002	.034	.0004	4.9	.037	.618
1664–1973	310	.599	9.142	.002	.034	.0004	4.8	.035	.621
1674–1973	300	.604	9.146	.002	.035	.0004	4.8	.036	.627
1684–1973	290	.609	9.166	.002	.036	.0004	4.1	.037	.626
1694–1973	280	.600	9.188	.001	.036	.0004	3.2	.036	.610
1704–1973	270	.584	9.217	.001	.036	.0005	2.0	.036	.588
1714–1973	260	.589	9.218	.001	.037	.0005	2.1	.037	.593
1724–1973	250	.596	9.218	.001	.038	.0005	2.2	.038	.601
1734–1973	240	.590	9.202	.002	.038	.0005	3.1	.039	.601
1744–1973	230	.564	9.199	.002	.037	.0006	3.6	.038	.579
1754–1973	220	.569	9.205	.002	.038	.0006	3.5	.039	.584
1764–1973	210	.568	9.204	.003	.039	.0006	3.8	.040	.587
1774–1973	200	.578	9.226	.002	.041	.0007	3.1	.042	.590
1784–1973	190	.566	9.219	.003	.041	.0007	3.7	.042	.585
1794–1973	180	.569	9.237	.003	.042	.0008	3.2	.044	.584
1804–1973	170	.569	9.244	.003	.044	.0009	3.2	.045	.585
1814–1973	160	.573	9.258	.003	.045	.0010	2.9	.046	.587
1824–1973	150	.557	9.281	.003	.046	.0011	2.4	.046	.566
1834–1973	140	.540	9.266	.004	.046	.0011	3.3	.047	.560
1844–1973	130	.534	9.288	.004	.047	.0012	2.9	.048	.550
1854–1973	120	.529	9.301	.004	.048	.0014	2.8	.050	.545
1864–1973	110	.516	9.316	.004	.049	.0015	2.7	.051	.531
1874–1973	100	.507	9.308	.006	.051	.0018	3.5	.054	.536
1884–1973	90	.496	9.340	.006	.052	.0020	3.1	.055	.520
1894–1973	80	.465	9.409	.003	.052	.0022	1.4	.052	.467
1904–1973	70	.474	9.420	.004	.057	.0028	3.1	.057	.476
1914–1973	60	.482	9.445	.002	.062	.0036	0.6	.062	.480
1924–1973	50	.443	9.490	−.002	.063	.0043	0.4	.062	.440
1934–1973	40	.461	9.520	−.010	.073	.0063	1.5	.074	.470
1944–1973	30	.476	9.497	−.016	.087	.0100	1.6	.089	.489
1954–1973	20	.486	9.390	−.001	.109	.0189	0.0	.106	.473
1964–1973	10	.231	9.390	.027	.073	.0255	1.0	.074	.233

6.1 Linear Trends

approximation based on the spectral characteristics of the time series will be presented.

Consider the linear trend parameter estimates (Table 6.3) of the longest historical records of the annual atmospheric pressure time series in ten European cities. For any data set, no β_1 estimates are statistically significant at the 95% level. The β_1 estimates are not statistically significant even for any ten-year (or greater) subinterval of observations of any of the above time series. Therefore, it is clear that atmospheric pressure had no sufficiently powerful long-period climatic fluctuations. Indeed, the results reveal that atmospheric pressure is the most stable characteristic of climate; moreover, as will be shown shortly, their annual time series can be used as a sample of a white noise process like a sample produced by any random number generator.

Considering the two longest historical records of precipitation in Swanenburg (Lebrijn, 1954) and in Paud Hole, East England (Craddock and Wales-Smith, 1977), one can see that there are no statistically significant estimates of the parameter β_1 (see Table 6.4). The standard errors are large, and the precipitation variability is greater than the variability of the air temperature or atmospheric pressure. The additional analysis of linear trends on separate (10-, 20-, and 30-year) subintervals in Swanenburg did not yield any statistically significant estimates of the parameter β_1.

Before concluding our consideration of the climatic trends, it is important to emphasize that the problem of their interpretation is significantly distinct from analogous problems in industry, astronomy, etc., where observations of certain variables are distorted by random errors. Of course, meteorological observations also contain errors; but when processing data which is a result of a large-scale averaging (such as annual values), one can assume that the fluctuations generally characterize the natural variability of the analyzed climatic element. This variability presents the unity of short- and long-period fluctuations; consequently, the notion of the trend (for which the approximation is obtained), and the random fluctuations (deviations from the trend) are very conditional. A variation that looks like a trend on the 10-year time interval can be a part of a random fluctuation on the 100-year time interval. Similarly, the secular variations of meteorological observations can be part of a random fluctuation on the 1000-year interval. Unity, the deep internal interdependence of the climate fluctuations of different time scales, reveals the conditional and subjective character of distinguishing between a trend and

Table 6.3: Linear trend parameter estimates of the annual atmospheric pressure (mm) time series.

Station	Observ. interval	$k+1$	σ_1	β_1	σ_{β_1}	t	$\sigma_{\hat{Y}}$	σ_0
Basel	1755–1959	205	0.83	−.000	.001	0.1	.06	0.83
Geneva	1768–1960	193	0.82	−.000	.001	0.1	.06	0.82
Copenhagen	1842–1960	119	0.97	.000	.003	0.1	.09	0.96
Happaranda	1860–1960	101	1.22	.000	.004	0.1	.12	1.22
Greenwich	1854–1949	96	1.37	.004	.005	0.8	.14	1.36
Valencia	1866–1960	95	1.25	.006	.005	1.3	.13	1.25
Aberdeen	1866–1960	95	1.63	−.007	.006	1.1	.17	1.63
Bordeaux	1868–1960	93	1.77	.001	.007	0.2	.18	1.76
Torshavn	1873–1959	87	1.25	−.003	.005	0.5	.13	1.25
St. Petersburg	1881–1950	70	1.32	−.002	.008	0.3	.16	1.31

a random component. But the distinction must be imposed methodologically when wording the climatological trend analysis problems because it can affect the selection of statistical methods and criteria.

In the study of climate variations, it is often presupposed implicitly that smaller microscale temporal fluctuations are a result of a large number of separate factors, and it is possible to consider their summary effect to be random. The larger fluctuations, such as the secular trend, are, supposedly, a result of only a few (or even one) physical phenomena, which can be monitored and explained. For example, one of the hypotheses about global warming during the first third of this century assumed that the refinement of the atmosphere's transparency resulted from a single factor: the lack of large volcanic eruptions at that time (Budyko, 1974). Today, there is an attempt to explain global warming by another single factor: the increase of CO_2 and other trace-gas concentrations in the atmosphere.

This reasoning shows that the way in which the climatic trend problem is studied represents an attempt to separate large- and small-scale climate fluctuations. In the framework of some problems, our main interest could be in analyzing long-period variations and their possible models, while in the framework of other problems, our primary interest could be in analyzing short-period fluctuations.

Table 6.4: The linear trend parameter estimates of the annual mean precipitation (mm) time series.

Station	Observ. interval	$k+1$	σ_1	β_0	β_1	σ_{β_1}	t	$\sigma_{\hat{Y}}$	σ_0
Swanenburg	1735–1944	210	124	743	.1	.1	.9	8.5	124
Paud Hole	1726–1975	250	501	2433	−.3	.4	.7	28.4	449

Where exactly, though, is the boundary between long- and short-period climate fluctuations? Does the use of a t-statistic criterion make it possible to impose a distinction that has any physical sense?

At any rate, the results of the approximation of the annual mean time series of meteorological elements can be considered as one of the characteristics of the power of long-period variations, part of which are observed on a given temporal interval.

6.2 Climate Trends over Russia

Let us begin by considering the annual surface air temperature time series for the 1891–1987 period of 104 stations, more or less evenly distributed over the territory of Russia (of course, the network is more dense in the European part than in Siberia). The data base was collected and maintained in the Main Geophysical Observatory (MGO), St. Petersburg, Russia. Citing the station names and corresponding estimates would take too much space, so only the summary results are given in Table 6.5.

Least squares straight-line approximations were carried out not only for the whole observational period, but for various subperiods (see the first line of Table 6.5), and the β_1 estimates were averaged over the stations ranging in accordance with the amount of population (see the first column of Table 6.5). It is clearly seen that there is a noticeable difference in the mean estimates of parameter β_1 between large cities with a population of more than 1,000,000 and smaller cities, especially when the observational period lasts more than 50 years. For such periods, the rate of temperature increase in the large cities (about 8°C/1000 years) is twice more that that of the

Table 6.5: Summary results of the temperature trend based on the observations of the 104 Russian stations for the period 1891 to 1987. The numerator presents the β_1 estimates in $°C$ per 1000 years; the denominator is the number of estimates with β_1 t-statistics greater than 2.

Averages over β_1 estimates of	Periods of observations							
	1891–1987	1901–1987	1911–1987	1921–1987	1931–1987	1941–1987	1951–1987	1961–1987
12 largest cities with population over 1,000,000	8/7	8/6	8/5	7/3	8/3	17/3	5/0	−7/0
21 cities with population from 0.3 to 1.0 million	4/10	3/3	5/4	4/4	5/3	11/5	0/3	−19/2
34 cities with population from 0.1 to 0.3 million	4/10	3/7	5/8	4/6	5/7	14/9	3/2	−14/1
37 cities with population less than 100,000	5/17	4/9	4/8	4/7	7/7	15/9	7/1	−12/0
all 104 stations	5/44	5/25	5/25	5/20	6/20	15/26	5/6	−12/3

smaller ones (about $4°$–$5°C/1000$ years).

For fewer than half of the stations, the parameter β_1 estimates are statistically significant (for the large cities the number of such estimates is more than half) at the 95% level (t-statistic is greater than 2) for the data on the entire observational interval; and only 20 to 25% of such estimates are statistically significant for the smaller subintervals. Since 1951, more than 95% of the β_1 estimates for the individual stations are not statistically significant.

We can conclude this brief consideration with the following remarks:

1. The results obtained show that, for about a century, surface air temperature increased in almost all Russian cities at an average rate of about $5°C$ per 1000 years.

2. Temperature increases in the large cities (with population larger than 1,000,000) were greater (about $8°C$ per 1000 years).

3. Since there are only a few large cities, a greater temperature increase there does not substantially influence the rate of warming over the large territories. When analyzing the global temperatures using spatial averaging of the entire global network, one probably should exclude data from the large cities because their weights are too great (with respect to the size of their territories) compared with the entire surface of the earth.

4. The greatest temperature increase was observed in the first third of our century.

Now consider the annual sums of precipitation for 104 time series of the most important agricultural regions of Russia for the same period of time. The precipitation data base was also collected and maintained in the MGO. The results, given in Table 6.6, show the tendency to decrease in the first third of our century and then to increase slightly. Statistically significant estimates of β_1 are obtained in about one fourth (or less) of the cases.

6.3 Periodograms

6.3.1 Central England Surface Air Temperature

Studying a periodogram of time series is really a deterministic approach since Fourier transform and computing amplitudes are not

Table 6.6: Summary results of the precipitation trend based on the observations of the 104 regions in Russia for the period 1891 to 1987. The numerator consists of averages over all β_1 estimates (mm per year); the denominator is the number of estimates with the β_1 t-statistics greater than 2.

Periods of observations				
1891–1987	1911–1987	1931–1987	1951–1987	1971–1987
−0.01/27	−0.15/11	0.20/7	0.12/2	1.17/13

accompanied by any statistical estimation procedure. In fact, a periodogram presents objective empirical information about the power of the harmonics with the various frequencies that compose the time series on the observational interval.

Let us begin our consideration with the central England annual mean surface air temperature (Manley, 1974). This time series spans 315 years (from 1659 to 1973), thus making it the longest temperature record known in climatology. The periodogram of this time series is shown in Figure 6.1 (the estimates are presented in the percentages of the sample variance value). Therefore, the periodogram values depend on the number of points in the time series, but the advantage of such a unit is in the possibility of comparing very explicitly the power of the fluctuations with different periods. Figure 6.1 shows that the periodogram values fluctuate significantly for different frequencies. To facilitate examination of Figure 6.1, the periods (105, 24, and 15 years) for the significant peaks are marked, and the percentages (8%, 5%, and 4%) of the variance corresponding to these peaks are also shown.

To understand whether or not these peaks have any physical meaning, it is necessary to answer the following two questions:

1. To what extent are these maxima stable in space? In other words, do other stations' periodograms of the temperature observations have exactly the same features?

6.3 Periodograms

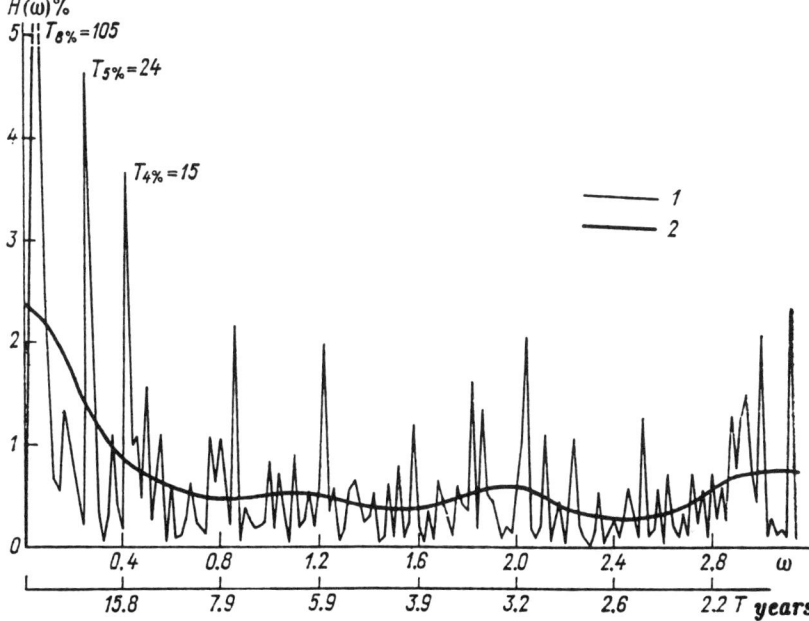

Figure 6.1: Periodogram (1) and spectral density estimates (2) of the central England annual mean surface air temperature time series (1659–1973).

2. To what extent are these maxima stable in time? In other words, are the peak periods and powers the same for any subinterval of this and other time series?

To answer the first question, it is necessary to compare periodograms that have been computed with data from a variety of stations. The results of such a comparison for the 18 longest temperature historical records, listed in Table 6.1, are given in Polyak (1975). These results show that in the low frequencies (with periods greater than 12 years) relatively large-power spikes are present on the periodograms of any station. But, in general, the frequency values of these spikes vary significantly for the various stations; or, in other words, the periodogram features do not have a regular spatial character.

Such irregular peaks on the periodograms of the various stations' temperature time series make it clear that, if the number of stations is sufficiently large, it is possible to find "cycles" at practically any frequency, each of which is inherent only to the data chosen. This phenomenon explains why so many cycles with a wide variety of periods were found in early publications of climatologists. A discussion of this question can be found, for example, in Monin (1969). There are several reasons for the development of the various long-term powerful fluctuations of the air temperature. For instance, volcanic eruptions lead to a decrease of air temperature with a gradual return to normal levels over several years; global processes connect with atmospheric carbon dioxide fluctuations; the earth orbit parameters change (astronomy climate theory); the intense growth in the industry of large cities affects temperature and leads to the gradual variation of the microclimate; and so on. Consequently, the cycles (large periodogram peaks) of the climatological temperature time series of different stations must be random and unstable in space.

To answer the second question about the spike temporal stability, it is possible to conduct a more detailed analysis of periodograms on the separate subintervals of the entire interval of observations and to compare the frequencies of their most powerful peaks with those presented in Figure 6.1. Such an analysis can help to locate in time both the most powerful fluctuations and the possible transfer of power from one frequency to another. This kind of analysis has been done, for example, for the St.Petersburg annual surface air temperature time series with the 1752–1966 observational period (Polyak, 1975). In that study, the periodograms were considered on the various overlapping 128-year subintervals. The results showed that the frequencies of the most powerful fluctuations are different for different subintervals, or, in short, the cycles mentioned above are temporally unstable.

Similar results were obtained for the historical temperature records of some other stations (Polyak, 1975). Consequently, separate powerful spikes on the periodograms (i.e., cycles) have a random character, and their appearance and disappearance in space and time is the manifestation of the sample variability. This means that it is senseless to use these cycles for long-period forecasts. A similar conclusion was obtained, for example, in the program "Understanding Climate Change" (1975).

Therefore, the primary feature of the periodograms of the climate temperature time series is a relatively significant part of the variance corresponding to some of the low frequency harmonics.

6.3.2 Atmospheric Pressure and Precipitation

This study of the interannual variability of atmospheric pressure and precipitation is based on the historical records of the atmospheric pressure for a 205-year span from 1755 to 1959 in Basel (Bider and

Schüepp, 1961) and of the precipitation for a 210-year span from 1735 to 1944 in Swanenburg (Lebrijn, 1954).

The periodogram of the atmospheric pressure (Figure 6.2, top) has three separate peaks with periods of 4, 3.5, and 2.5 years, respectively. The power of these peaks reaches about 4% of the sample variance. For periods of more than 4 years, there are no amplitudes with power exceeding 3% of the variance.

The periodogram of the precipitation time series (Figure 6.2, bottom) also has separate peaks, the maximum of which (6% of the variance) corresponds to a period of approximately 4.1 years. But generally, the distribution of the variances along the frequency axis has more or less even character for both time series. Multiple periodograms analogous to Figure 6.2 of many different time series of atmospheric pressure and precipitation (Polyak, 1975; 1979) reveal that the random cycles of pressure and precipitation with relatively large amplitudes are unstable in time and space. In other words, they appear as result of the sample variability.

6.4 Spectral and Correlation Analysis

6.4.1 Annual Mean Surface Air Temperature

In this subsection the spectral and correlation analysis of the Central England temperature time series will be carried out. The corresponding periodogram has already been studied

Figure 6.3 presents the cumulative spectral function of this time series and 99% confidence intervals for the white noise. The estimates exceed the confidence interval above. This means that the time series considered is not a sample of a white noise.

The spectral density estimates, given in Figure 6.1 (2), were obtained by smoothing the periodogram using a regressive filter (1.83) with $m = 3$ and a width $2r + 1 = 35$ points.

This spectrum's most noticeable feature is that it has a maximum at low frequencies (near the origin). This maximum has also been found in the spectral estimates of a variety of climate temperature time series (Landsberg et al., 1959; Monin and Vulis, 1969; Monin et al., 1971; Polyak, 1975; Madden, 1977; and so on). Most of the variability was caused by the white noise component of the fluctuations. It was shown (Polyak, 1975) that the shapes of the spectral density estimates for the longest temperature historical records of different stations and different observational intervals are almost identical.

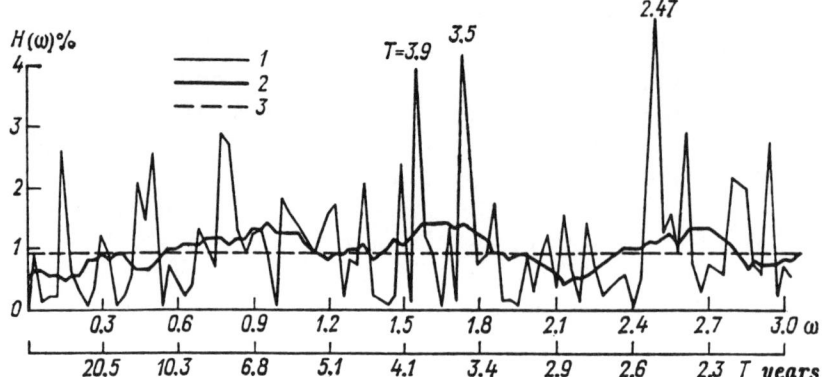

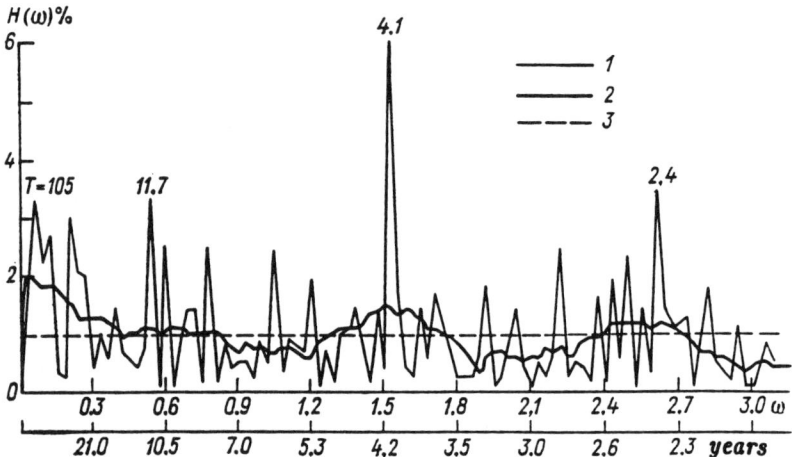

Figure 6.2: Periodogram (1) and spectral density estimates (2) of the Basel annual mean atmospheric pressure (top) time series (1755–1959) and of the Swanenburg annual precipitation (bottom) time series (1735–1944); (3) is white noise spectrum.

6.4 Spectral and Correlation Analysis

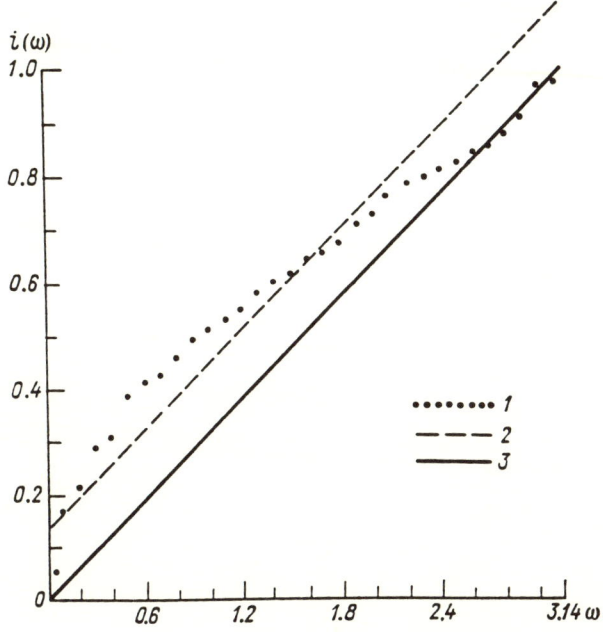

Figure 6.3: Cumulative spectral function (1) of the central England surface air temperature time series and the 99% confidence interval (2) for the white noise (3).

Figure 6.4 gives the central England time series autocorrelation function, which corresponds to the spectrum in Figure 6.1 (2). The first several autocorrelation estimates are small (less than 0.3), positive, and statistically significant at the 99% level. The white noise level, which is determined by the difference between one (value at the zero lag) and estimates 0.2–0.3 at lag equal to one year, is large (about 70 to 80 % of the variance). Approximately the same autocorrelation estimates were obtained for other temperature historical records listed in Table 6.1 (Polyak, 1975).

These results enable us to make the following comments:

1. The spectra of the climate air temperature time series have a small maximum in the vicinity of the origin, which corresponds

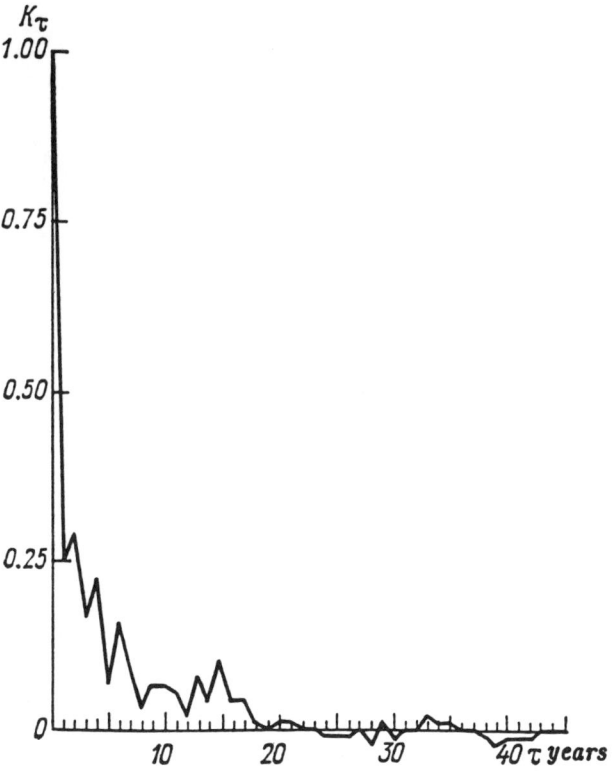

Figure 6.4: Autocorrelation function of the central England temperature time series.

to the periods of decades or centuries. Such "red spectrum character" reveals the possibility of the existence of relatively powerful low-frequency temperature fluctuations.

2. The temporal variability of climate air temperature primarily results from white noise, which makes up approximately 70 to 80% of the variance.

3. In contrast to the periodograms, the spectral and correlation estimates are spatially and temporally stable.

The above nonparametric consideration (together with the studies by Polyak, 1975; 1979) reveals that an approximate model for the

6.4 Spectral and Correlation Analysis

annual mean air temperature time series of a point gauge can be the random process

$$Y(t) \approx \bar{Y} + S(t) + N(t) \tag{6.2}$$

where $N(t)$ is a white noise, $S(t)$ is the Markov random process [independent of $N(t)$] with the autocorrelation function $e^{-\alpha|\tau|}$, and

$$E[N(t)] = E[S(t)] = 0. \tag{6.3}$$

The value of $\rho = e^{-\alpha}$ varies from station to station; for the middle latitudes the range of its variations is about 0.1 to 0.3. Each of the empirical approximations considered [(6.1) and (6.2)] has different features. Although (6.1) realistically describes the long-period temperature variation during the last century, it proves inadequate for forecasting even in the near future.

6.4.2 Atmospheric Pressure and Precipitation

Let us look at the spectral and correlation characteristics of the atmospheric pressure and precipitation time series, the periodograms of which are given in Figure 6.2. The parameters of the regressive filter (1.83) that was used for the spectra estimation are the same for both time series considered; the order equals three, and the width equals 25 points ($m = 3, r = 12$).

Figure 6.2 (top, 2) gives the spectral estimates of the Basel atmospheric pressure along with the white noise spectrum (3). These estimates are not smooth. However, by increasing the filter width, they can be made smoother and closer to the white noise spectrum. The minimum of the spectrum estimates in Figure 6.2 (top, 2) corresponds to the interval with periods greater than ten years, which reveals that powerful climatic fluctuations of atmospheric pressure are unlikely, just as one would expect. Nevertheless, it is clear that the spectrum estimates are very close to the white noise spectrum.

Figure 6.5 (top) compares a theoretical cumulative spectral function of the white noise with the estimated cumulative spectral function of the Basel atmospheric pressure.

The estimates are within the 99% confidence interval; that is, the analyzed time series presents a sample of white noise. This same conclusion is derived when one considers the autocorrelation function [Figure 6.6 (left)], whose values are very small and whose deviations from zero are statistically insignificant at any standard level of significance for all the non-zero lags.

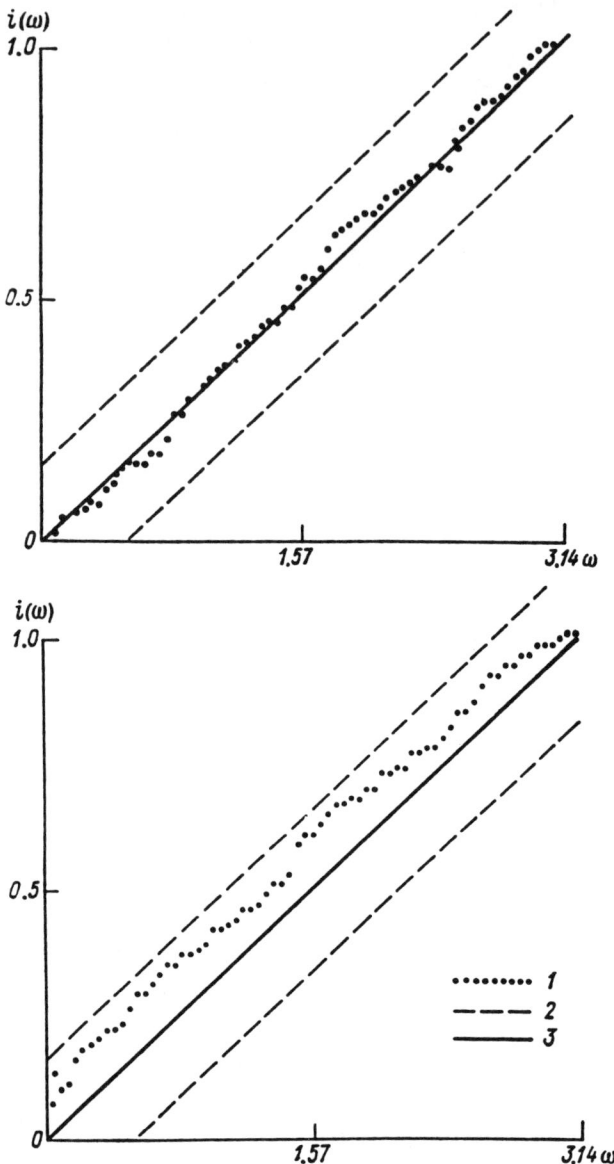

Figure 6.5: Basel atmospheric pressure (top) and Swanenburg precipitation (bottom) cumulative spectral functions (1) and the 99% confidence intervals (2) for the white noise (3).

6.4 Spectral and Correlation Analysis

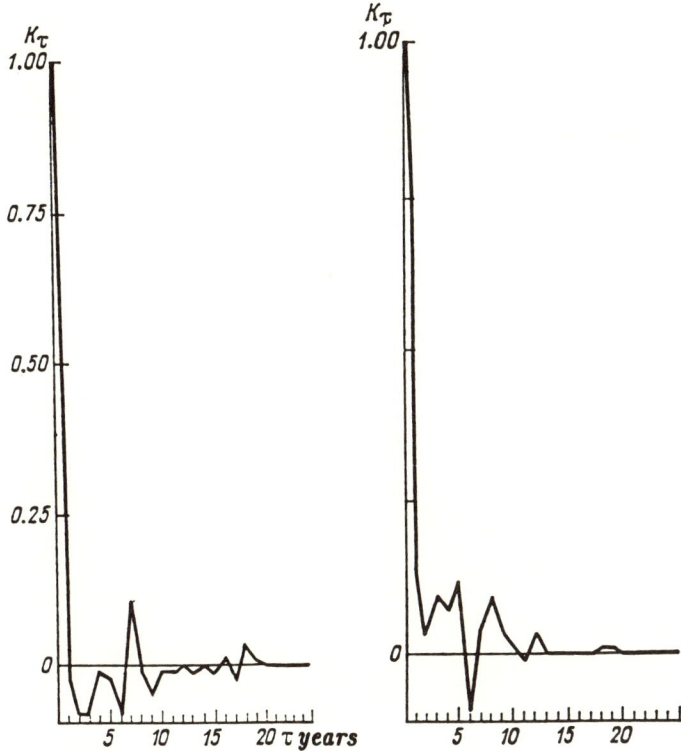

Figure 6.6: Basel atmospheric pressure (left) and Swanenburg precipitation (right) autocorrelation functions.

The analogous consideration of the Swanenburg precipitation time series reveals that all the estimates of the cumulative spectral function [Figure 6.5 (bottom)] are within the 99% confidence intervals for white noise. Consequently, the time series analyzed are close to the white noise sample. The same conclusion can be made analyzing the corresponding correlation function [Figure 6.6 (right)], the values of which for the non-zero lags are very small and statistically do not differ from zero.

The deviation of the spectral density estimates [Figure 6.5 (bottom)] from the white noise spectrum is slightly greater than the corresponding deviation for the atmospheric pressure, but the closeness of it to the white noise is indisputable.

Along with Basel and Swanenburg data, precipitation and atmospheric pressure time series of many other stations were also analyzed in Polyak (1975; 1979). Because the results are similar to those obtained above, it makes no sense to cite them here. We will only make the following comments:

1. The spectral and correlation characteristics of the climatic time series of the atmospheric pressure and precipitation are stable in space and in time.

2. It is quite probable that the annual mean atmospheric pressure and precipitation time series are samples of white noise. Even on the separate subintervals of each of these time series, the closeness of the observations to the white noise sample is obvious (Polyak, 1975; 1979). Therefore, any attempt to forecast these time series is meaningless.

3. Since the duration of the records considered is unique, the spectral and correlation estimates obtained not only carry the value of appropriate examples but also show the general characteristics of the atmospheric pressure and precipitation time series for the middle latitudes of the Northern Hemisphere.

The possibility of forecasting climate by statistical methods is determined by the character of the statistical and harmonic structure of the long-period fluctuations. The results in this section (together with those published in Polyak, 1975; 1979; 1989) reveal that the anomalies of the monthly and annual mean climate time series have a very high level of additive white noise. Based on these publications, Table 6.7 gives the range of estimated values of the autocorrelation coefficient for one month lag for the middle latitude Northern Hemisphere time series of different climatological variables.

The air temperature observations have the largest value of the autocorrelation coefficient. Table 6.7 provides information necessary for the temporal statistical modeling of corresponding climate data and also offers its forecasting value because the first autocorrelation is all of the information required for building the AR(1) model.

The autocorrelated component of the climate time series of many meteorological elements (with the exception of precipitation and atmospheric pressure) is conditioned by the small maxima in the low-frequency part of the spectra with periods of several decades or centuries. These maxima offer evidence of the possibility of the existence of long-period climate fluctuations with relatively powerful amplitude. Such fluctuations, which are usually referred to as cycles, have

6.5 Univariate Modeling

Table 6.7: Range of the autocorrelation coefficient estimates (for one month lag) of the normalized anomalies of the monthly mean climatological time series in the middle latitudes of the Northern Hemisphere.

Element	Autocorrelation coefficient
Air temperature on the tropopause level	0.20–0.40
Surface air temperature	0.20–0.35
Thickness (500–1000)	0.15–0.30
Atmospheric zone	0.10–0.30
Geopotential of 500mb level	0.10–0.20
Direct radiation	0.10–0.20
Heat budget	0.10–0.20
Precipitation	0.00–0.10
Zonal component of the geostrophic wind velocity of the 500mb level	0.00–0.15
Surface atmospheric pressure	0.00–0.10

a random spread in time and space. Considering only natural variability, it is important to recognize that the forecasting value of any cycle is equal to zero.

From the example in Section 2.5, it follows that temporal averaging of daily temperature observations with the first autocorrelation coefficient equal to about 0.8 must lead to a monthly time series with the first autocorrelation coefficient equal to about 0.07. The values in Table 6.7 are greater than 0.07, which shows that there are long-term variations beyond the synoptic scale fluctuations that affect climate variation.

6.5 Univariate Modeling

Application of the nonparametric methodology of spectral and correlation analysis to the climate (monthly or annual mean) time series of hydrometeorological elements has provided a way to understand the physical and statistical nature of long-period climate fluctuations and to explain such phenomena as multiple climate cycles.

By analyzing the graphs of the periodograms of the various hydrometeorological observations at the various points of the earth, one can see that each and every one of them has several separate, relatively powerful peaks corresponding (in its totality) to virtually any period ranging between several months to several hundred years. These peaks are unstable in space and in time; their frequencies are different for different stations or subintervals of the same time series. The part of the variability that corresponds to each maximum does not exceed 5 to 8% of the sample variance, which excludes any forecasting value. Such peaks clearly demonstrate the random character of the periodogram, or, in other words, the sample variability of the time series. The periodogram of the climate time series (and its peculiarities) does not carry general theoretical value. Rather, it represents a basis for the consequent estimation of the spectral density. Any attempt to give a profound meaning to the separate peaks (that is, cycles) is equivalent to attempting to build a law of behavior of a random variable based on only one observation. Smoothing (or averaging) of periodogram values in separate frequency subintervals is the key feature of spectral density estimation methodology.

Even the first attempts (for temperature data) to estimate climatic spectra have revealed the statistical nature of climate fluctuations from the viewpoint of the theory of probability: White noise, a random process with statistically independent separate values, dominated in the composition of such fluctuations.

The theoretical spectrum of white noise is constant for all frequencies; but, of course, the periodograms of its samples (for example, those generated by the computer random number generator) are similar to the periodograms of the climate time series. However, in spite of the high level of white noise in climate fluctuations, some spectral estimates of the separate stations' temperature time series enable us to establish with a high degree of confidence the existence of the single stable spectral feature: a small maximum in the vicinity of the zero frequency.

This spectral maximum at the origin is more pronounced for the time series of stream flow, air temperature, and radiative observations (see Table 6.7) and is significantly smaller (or does not exist at all) for the observations of the precipitation, the atmospheric pressure, the velocity and the direction of wind, the geopotential of the standard isobaric surfaces, and some other elements, the estimated spectra of which do not differ significantly from the white noise spectrum. Generally speaking, then, the assertion about the pure noisy

6.5 Univariate Modeling

character of the local climate's variations cannot be accepted. The more probable conclusion is that corresponding climatic observations are presented by the sum of a very small signal (autocorrelated component) and a significant portion of white noise, against the background of which the signal is often not detectable at all.

It is clear that slow variations of local climates during the earth's history must be associated with the low-frequency spectral maximum. Therefore, we must recognize that the study of this extreme has decisive meaning in the statistical description of climate change. This maximum provides the evidence for the possibility that there are climate fluctuations of a nonregular character on time scales ranging from decades to thousands of years.

The above analysis of meteorological data for the instrumental period of observations using the random process theory presents one possible description of the statistical nature of climate. Such a description shows that white noise assumes the primary role in the formation of climatic fluctuations. The autocorrelated component is very small, but the spectral maximum in the vicinity of the zero frequency demonstrates its existence.

However, in the framework of stationarity only the autocorrelated component can carry any information about climate change. Only this component determines the climate predictability (climatic memory). The white noise component is unpredictable, and the climate would also be unpredictable and unchangeable if the climatic fluctuations were entirely formed by the white noise.

White noise forms 70 to 100% of the monthly and annual mean anomalies of the point gauge climate time series in the middle latitudes. But in spite of the fact that the autocorrelated component is drowned out in the white noise, its small value (for some elements) reveals two peculiarities. First is the step from one for zero lag to the 0.0–0.3 values for the lag that is equal to one month or one year. This step demonstrates the possibility of presenting the climate time series as the sum of the signal plus the white noise.

Second, the closeness to zero of the autocorrelations for the lags greater than two, or sometimes three or four, months (or years for the annual data) enables us to interpret the signal not as a pure deterministic component, but as, in the first approximation, a random process with several non-zero autocorrelations. (If a signal were a pure deterministic component, the corresponding autocorrelation function would not approach zero). It is natural to think that the signal characterizes the physics of the observed process. Its reliable detection stands as one of the problems of statistical climatology.

The discussion above of spectral and correlation estimates, together with results of the theoretical study of the forecast error of different kinds of univariate ARMA processes presented in Chapter 4, enables us to draw certain conclusions about the possibility of identifying ARMA models of separate stations' climate time series and the limits of their statistical predictability without having to fit such models at all.

Indeed, if the maximum value of autocorrelations does not exceed 0.3, then according to the results of Section 4.7 it makes sense to forecast such a time series for, at most, one step ahead independent of the order and type of the fitted ARMA model. In the best case, the standard deviation of the forecast error would consist of 90% of the standard deviation of the observations, and the forecast estimates would be very close to the time series mean. The practical uselessness of such a forecast is quite clear.

Therefore, formulating the problem of univariate statistical ARMA modeling of the climate time series of point gauges presents no special interest, no deep physical matter. As for the type and order of such models, it is most likely that only the following three cases should be considered.

1. The white noise model, which matches the point gauges time series of precipitation, atmospheric pressure, wind, and perhaps some other variables.
2. The first order autoregressive model (annual river stream flow, air temperature, radiation), whose spectrum estimates have a small low-frequency maximum that characterizes the AR(1).
3. The signal plus white noise model, where it is possible to accept the AR(1) process for a signal. This latter model, ARMA(1,1), is considered the most general, since the white noise and the AR(1) model represent its particular cases.

Finally, nonparametric analysis of the historical records enables us to establish the reality of the signal-plus-white-noise character for the climatic time series of the point gauges. It also points out the difficulty of reliably detecting a signal that is masked by this noise. In other words, the climatic time series of point gauges must be further processed by an appropriate procedure of spatial and temporal averaging or filtering in order to evaluate the character of the long-period fluctuations (the signal) responsible for climate change.

6.6 Statistics and Climate Change

As we have already discussed, the periods of temporal fluctuations of meteorological elements range from a fraction of a second to many millennia.

However, climate change study concerns variations in periods that range from several weeks to many millennia. But by estimating the climate statistics (that is, the statistics of data with large-scale temporal averaging), we are actually analyzing the resulting state of the climate system, which is conditioned by the interactions and feedbacks of multiple physical factors and processes taking place in the atmosphere and ocean in real time. Such interactions and feedbacks, which are generally represented by the hydrodynamic laws, determine the character of fluctuations of the meteorological elements of any spatial and temporal scales. Estimated statistics provide a qualitative picture of this resulting state. Because for most of the instrumental period of meteorological observations the measurements have been done on the earth's surface, our knowledge of climate has been limited to the surface climate. The description of the local climate has had a relatively qualitative character based on estimates of climatic means (norms) and standard deviations for the separate months, seasons, and years of different regions of the earth. Unfortunately, these norms and variances do not give the entire representation about the large-scale spatial and long-period temporal variability of meteorological elements. In order to obtain a more complete description of climate, it is necessary to study the physical and statistical nature of the spatial and temporal climatic fluctuations in the form of multivariate stochastic models.

With the development of the system of radiosonde stations and corresponding data bases, several projects have been undertaken to describe the mean climate regime of the upper atmosphere (Oort, 1978; 1983; and others). These investigations, which are not yet completed, have significantly widened our knowledge of the atmospheric processes and their possible variations.

The primary subject under investigation is the cause of climate change. Various investigations have suggested several possible explanations (hypotheses), though none has become a generally recognized theory. The reason for this is that the climatic fluctuations are forced not only by separate powerful phenomena (for example, volcanic eruption, and so on) and processes, but also by interactions and feedbacks of multiple factors of a mechanical, physical, chemical, and biological character. Nevertheless, it is important to enumerate, at least briefly, the possible causes of climate change discussed by

Budyko, 1974; Hays et al. 1976; Monin, 1969, 1977; Monin and Vulis, 1971; and others.

One set of theories is based on elements external to the climatic system, which include the following:

1. The solar radiation power variation.

2. The amount of solar energy that reaches the earth surface due to variation of the concentration of intercelestial dust.

3. The variation of the seasonal and latitudinal distribution of incoming radiation due to variation in Earth's orbit geometry.

4. The content of volcanic dust in the atmosphere.

5. Earth's magnetic field and the tectonic motion of Earth's crust.

Another set of theories is based on elements internal to the climatic system. This set of theories assumes that there are physical mechanisms with sufficiently long response feedbacks to stimulate climatic fluctuations with periods of the order of thousands of years. These internal elements are:

1. The growth and decay of ice shields.

2. Snow cover variations.

3. Cloudiness variations.

4. Variation of the evaporation from the ocean surface near the equator.

5. Ocean deep-water circulation.

6. Deforestation.

There is a set of theories which assumes that the climatic system can have several different stable states and that it can move from one state to another (a scientific assumption otherwise known as the transitivity of climate). In the last few decades, anthropogenic causes, which result from the atmospheric CO_2 (and other gases) have increased, with the heat budget variation has dominated other theories.

Even this very brief list of possible causes of climate change reveals the complexity and multiplicity of climatological problems.

6.6 Statistics and Climate Change

Many papers (such as Lorenz, 1977; Monin, 1972; Flon, 1977; Hasselmann, 1976; and others) have considered the different physical aspects of climate change. There were also attempts (Mitchell, 1976) to build a spectrum of climatic fluctuations with periods that range from one hour to 10^9 years and to explain the spectrum's features. Mitchell differentiates between the deterministic and stochastic mechanisms of climate variations, estimates the temporal scales of different processes of nature, and presents them as features of his spectrum.

In many publications that analyze the nature of climate fluctuations, estimates of different climate statistics (for example, the spectral and correlation characteristics) are considered.

The continuous evolution of the living world for millions of years enables us to assume that earth's climate variations have taken place primarily evolutionarily and, therefore, have occurred relatively slowly (though, some regions, of course, experience catastrophes).

Indeed, the value of a trend (see Section 6.1) of any climatic process, such as surface air temperature, on the time intervals, compared with the period of instrumental measurements, is very small. It is several orders less than the climatic noise level or the accuracy of the instruments used today. For example, if for a period of 100,000 years the mean surface air temperature of a certain region has changed by about 10°C, the mean annual rate is equal to approximately 10^{-4} °C. It is clear that no observations with such accuracy currently exist. Therefore, the systematic variations of climate tend to be smaller on several orders than the accuracy of measurements. Consequently, it is very difficult to obtain a rigorous physical or statistical description of such long-period fluctuations.

7
The GCM Validation

In this chapter the observed and simulated (by the Hamburg GCM) Northern Hemisphere monthly surface air temperatures, averaged within different latitude bands, are statistically analyzed and compared. The objects used for the analysis are the two-dimensional spatial-temporal spectral and correlation characteristics, the multivariate autoregressive and linear regression model parameters, and the diffusion equation coefficients. A comparison shows that, generally, the shapes of the corresponding spectra and correlation functions are quite similar, but their numerical values and some features differ markedly, especially for the tropical regions. The spectra reveal a few randomly distributed maxima (along the frequency axis), the periods of which were not identical for both types of data. A comparative study of the estimates of the diffusion equation coefficients shows a significant distinction between the character of the meridional circulations of the observed and simulated systems.

The approach developed gives approximate stochastic models and reasonable descriptions of the temperature processes and fields, thereby providing an opportunity for solving some of the vital problems of theoretical and practical aspects surrounding validation, diagnosis, and application of the GCM. The methodology and results presented make it clear that formalization of the statistical description of the surface air temperature fluctuations can be achieved by applying the standard techniques of multivariate modeling and multidimensional spectral and correlation analysis to the data which have been averaged spatially and temporally.

7.1 Objectives

The idea of the statistical approach to the problems of GCM variability validation is contained in the comparison (observed vs. modeled) of the probability distributions of the different atmospheric and ocean processes and fields. At first, such a statement sounds like a standard statistical approach, and its solution would be obvious and simple if the number of climate processes taking place jointly were not huge and if they did not present a tremendously complicated (in its interrelationships and feedbacks) deterministic-stochastic system. As is known, the Stochastic System Identification Theory (see Eikhoff, 1983) deals mostly with the methodology for identifying linear systems. The interdependences of climatic processes and fields are not linear, and the application of this theory can give only highly approximate results. But it seems that such results can present very helpful information for many areas of climate studies and can also serve as an important step in the right direction for generalizing the approach to and solution of the GCM confirmation problems. Moreover, on the first stage of GCM variability study, even rough theoretical (statistical) approach is a principle way to refrain from some empirical validation schemes which are not recognized by the statisticians, and the results of which are impossible to compare statistically. Further developments in this direction will help us to define more accurately the details of such approaches and to identify appropriate observational structures.

This study is an attempt to compare the variability of one of the most important climate processes, surface air temperature, for the two complicated stochastic systems: real and modeled. As test statistics of such a comparison, the estimates of the corresponding second-order moments (or their functions) are used.

As is well known, some studies (e.g., Katz, 1992; Oort, 1983; Santer et al., 1990; Santer and Wigley, 1994) have provided a comparison of the means and standard deviations of different atmospheric characteristics derived from the observed and simulated data. In our study, the spatial-temporal climatic variability of the fluctuations (the deviations from the means) of the surface air temperature processes and fields is considered. Without studying and statistically modeling such deviations, it is impossible to make any statements about climate change (to forecast, to reason about the warming or cooling of the atmosphere, and so on) because in all these predictions one is reasoning about a deviation from the mean climatic regime.

In the lights of the results provided it will become clear that utilization of the GCMs for different physical experiments without any variability validation is not acceptable. But this is the situation on this moment: hundreds of papers have been published (and continue to release) where the GCMs with unknown variability are used for the analysis of the real climate variability (climate sensitivity, greenhouse effect, reconstruction of paleoclimates, El Niño study, and so on).

When comparing the second moments, one can use the estimates of the correlation matrices or the parameter matrices of the multivariate AR models as standard test statistics. Additionally, the parameter matrices as the matrices of interactions and feedbacks present interesting diagnostic information about both systems.

The methods in Chapter 3 and 5 have obviously illustrated ways in which to design and develop simple multivariate climate models, as well as ways in which to estimate the multidimensional spectra and correlation functions of different climatological structures. In this chapter these methodologies are used not only for estimation of the second-moment climate statistics, but also for conducting a sequence of their comparative studies by calculating some physically interpreted characteristics (the coefficients of the diffusion equation). Consequently, then, this study will help us to answer this important question: To what extent can GCM describe real large-scale climate variations?

In short, this study presents the estimates and the results of a comparison of a complex of the multiple second-moment statistics of the observed surface air temperature and the 1000 mb air temperature simulated by the Hamburg GCM.

7.2 Data

There are many ways in which to represent the air temperature in the form of time series and fields by spatial-temporal averaging and filtering. Because our first objective is to compare the climate statistics (or the statistics of the large-scale fluctuations), the following two different structures of data are used. The first is the 31-year zonally averaged monthly air temperature time series of the Northern Hemisphere (NH) at the 1000 mb isobaric level simulated by one of the Hamburg coupled global atmosphere-ocean general circulation models, see Cubasch et al. 1992, Max Plank Institute for Meteorology, Hamburg). The Hamburg GCM has described in details in many papers, for example, in Roeckner, 1989; Santer and Wigley, 1990; Santer et al., 1994. The temperature time series, available for various $5.625°$ width latitude zones, were additionally averaged for

7.2 Data

each pair of the adjacent latitude zones. As a result, eight time series were obtained for each of the eight latitudinal bands of $11.25°$ width.

This structure can be considered as eight variate time series (or two-dimensional latitude-temporal field)

$$T_i(t), \ i = 1, \ldots, 8; \ t = 1, \ldots, 372, \qquad (7.1)$$

where i and t are the latitude and temporal subscripts respectively.

The source of the observed data is the United Kingdom's global surface air temperature observations for 1891–1990 (Jones et al., 1986), archived at the NCAR The sample, used in this study, presents surface air temperature of the NH for 31 years (1959–1989) averaged within the same latitude bands of $11.25°$ width. The latitude-temporal field of these eight time series have the same structure as (7.1).

For ease of reference, Table 7.1 enumerates the time series of different latitude bands and identifies their boundary values.

Before the spectral analysis and multivariate modeling, the annual cycles $[\bar{T}_i(t) \text{ and } S_i(t)]$ in the mean and variances were removed and the time series of the normalized observations,

$$u_{it} = \frac{T_i(t) - \bar{T}_i(t)}{S_i(t)},$$

were analyzed.

The standard methodology (periodogram $\rightarrow$ smoothing $\rightarrow$ spectrum and periodogram$\rightarrow$ Fourier transform $\rightarrow$ correlation function) with a two-dimensional Tukey spectral window (see Section 1.11) was employed. Because the values of the periodograms (and the spectral estimates) are small, they were multiplied by the factor (1000) to be conveniently presented by illustrations. As a result of the estimation, the normalized sample spectral density and its standard deviations increased 1000 times will be analyzed systematically.

The observed and simulated sets compared are not completely identical climatic structures: The first is the surface observations and the second is the 1000 mb simulated data. As will become clear shortly, these facts are of no import, at least at this stage of the study.

But in principle, this discrepancy restricts the possibility of making the quantitative statistical inferences (by hypotheses testing based on corresponding test statistics) about parameters of the compared distributions, and it constrains us to carry out a basically qualitative study. If this approach and the form of the results representation

Table 7.1: Estimates of the linear trends $\beta_1(°C/\text{year})$ and t-statistic values for the observed and simulated data.

Time series number	Latitude band (degrees)	Observed data				Simulated data	
		100 years		31 years		31 years	
		β_1	t	β_1	t	β_1	t
1	90.00–78.75	.024	10.4	.005	0.8	.008	0.9
2	78.75–65.50	.009	8.4	.009	1.7	.010	1.7
3	67.50–56.25	.007	8.5	.012	2.6	.004	0.9
4	56.25–45.00	.006	9.9	.010	3.0	−.003	−1.1
5	45.00–33.75	.005	13.6	.000	0.2	−.007	−4.0
6	33.75–22.50	.005	24.1	.001	1.3	−.004	−3.0
7	22.50–11.25	.005	23.9	.006	4.6	−.003	−2.9
8	11.25–00.00	.004	16.7	.006	4.7	.000	0.4

proves acceptable to the scientific community involved in climate analysis and modeling, it will be possible to carry on a special experiments for purposefully recollecting both kinds of data on the identical level to apply more rigorous statistical inferences in multiple comparison and hypotheses testing.

To get a comparative representation about the long-term variability of the two 31-year data structures with respect to the entire interval of 100 years of the surface air temperature observations, the estimates of the parameter β_1 (of the simple linear regression lines $\beta_0 + \beta_1 t$, which approximate each of the latitudinal time series) were found together with corresponding t-statistic values (Table 7.1). Although for the observed data almost all of the β_1 estimates are positive, the linear trends for 31 years of observations are not as persistent as for the entire 100-year interval. For the latter, each of the eight β_1 is statistically significant (t-statistic values are from 8 to 24), while for the former, at most four of them are statistically significant, with smaller t-statistic values lying in the interval (2.6, 4.7). As for the simulated data, four of the β_1 estimates are negative, and at most three of them are statistically significant. So quantitatively as well as qualitatively, there is no strict consistency in these simple characteristics of the long-period fluctuations. But one remark must be made here: The scales of variations of the β_1 estimates are approximately the same for both types of data (less than about 1°C per 100 years).

7.3 Zonal Time Series

When one has a monthly time series of observations, it is possible (in contrast to the traditional univariate technique) to apply a two-dimensional methodology of spectral and correlation analysis. The time series has to be presented in table (field) form with 12 observations in each row and with the number of rows equal to the number of years (31) of observations (see Subsection 3.8.3). In our case, the time series for each latitudinal band is presented as a two-dimensional field of the 31×12 size. Such an approach makes it possible to separate the interannual and intraannual fluctuations and to obtain their description more expressively.

Before considering the results, the following remark must be made: We will begin with a descriptive consideration of the estimates, mentioning their peculiarity (e.g., possible spectral maxima) as a reflection of the sample variability. After that the speculation about the possible character of the spectra will be provided. Then, the statistical inferences about the equality of the correlation coefficients will be done.

Figure 7.1 gives estimates of the two-dimensional spectral densities of the observed and simulated time series for each latitudinal band.

Visual comparison of the pictures can lead to the preliminary conclusion that there is some general similarity in their shapes. The intraannual parts of them for the monthly frequencies have specific bell-shape graphs with maxima corresponding to the zero monthly frequency, which are expressed by the crest spreading along the annual frequency axis. The heights of these maxima (the heights of the crests) vary for different annual frequencies. The most powerful monthly fluctuations correspond to the periods of more than six months. Fluctuations with periods of less than 3 or 4 months have the least power. So the shape of the intraannual part of the spectra seems to be very close to the shape of the spectra of the first-order univariate autoregressive process. It clearly shows that for periods of more than a year, the long-scale spatial fluctuations predominate

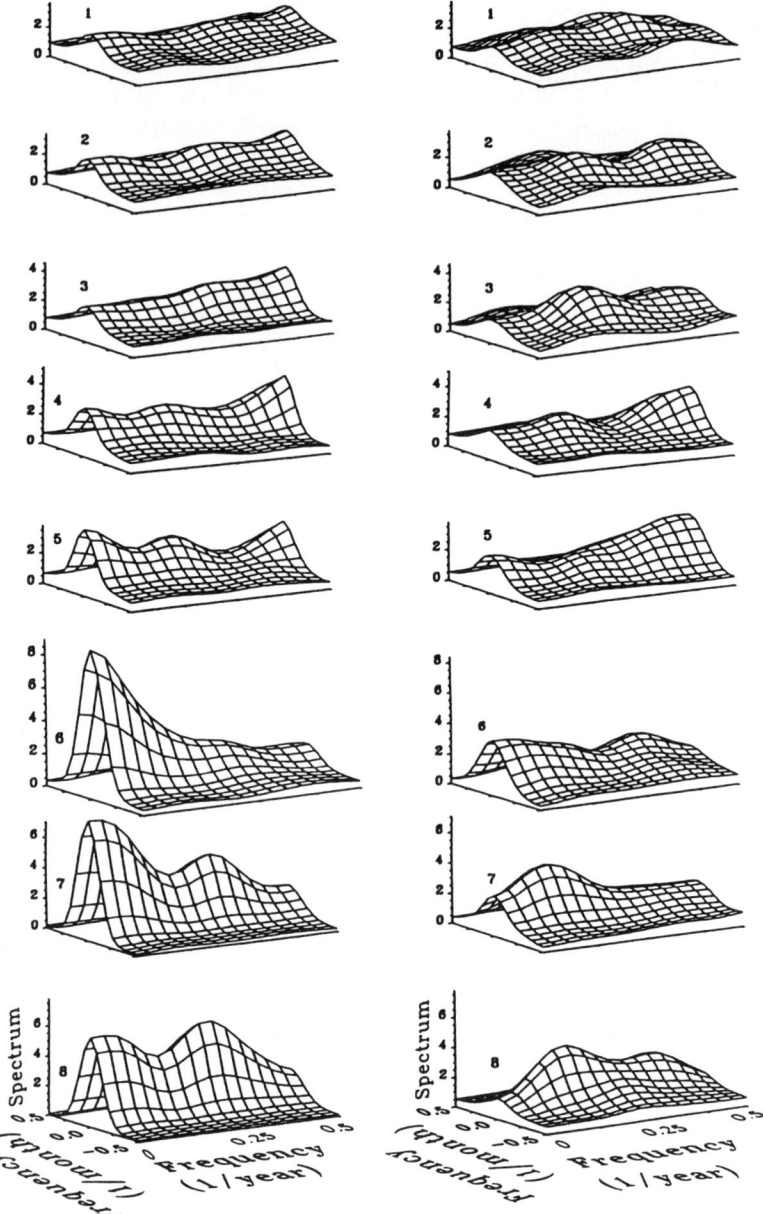

Figure 7.1: Two-dimensional spectra of the observed (left) and simulated (right) surface air temperature time series of different latitude bands. (Correspondence between the picture number and the latitude band boundary is given in Table 7.1.)

7.3 Zonal Time Series

Table 7.2: Standard deviations of the spectral estimates and periods (years) of the spectral maxima within the annual frequency interval (see Figure 7.1) of the time series of different latitudinal bands.

Time series number	Observed data			Simulated data		
	Standard deviations		Periods of maxima	Standard deviations		Periods of maxima
	$s1$	$s2$		$s1$	$s2$	
1	0.10	0.41	3.2	0.11	0.53	3.2; 8.
2	0.12	0.52	3.6	0.12	0.58	2.4; 5.3
3	0.14	0.48	3.2	0.12	0.68	5.3
4	0.15	0.49	5.3	0.12	0.40	6.2
5	0.17	0.90	5.3	0.14	0.72	2.3
6	0.31	0.44	∞	0.15	0.38	2.9
7	0.31	0.93	3.6	0.15	0.59	8.0
8	0.31	1.95	3.6	0.17	0.80	2.9; 5.9

in the variability of the surface air temperature. Therefore, the most powerful part of the spectra is the crest (for zero monthly frequency), with several maxima. This is a brief description of the similarity of the spectra.

The details of the spectra are quite different. First, the interannual maxima periods are different for different latitude bands (their numerical values are given in Table 7.2).

For the observed data there are about three clearly distinguishable latitudinal regions, with the periods of the maxima being about 3.6 years for the tropics, 5.3 years for the midlatitudes (35° to 57°), and 3.2 to 3.6 years for the northern and Arctic regions. Roughly speaking, this means that within each of these three regions, the spectral structure of the observed temperature fluctuations can be considered as approximately identical.

The specific frequency composition of the simulated data differs from that of the observed data: The periods of the maxima (see Figure 7.1 and Table 7.2) vary randomly for the time series of different

Table 7.3: Correlations (see Figure 7.2) for the first two lags (months) and corresponding N statistic values.

Time series number	Observed data		Simulated data		N	
	Correlations		Correlations			
	1	2	1	2	1	2
1	0.16	0.02	0.15	0.12	0.1	1.2
2	0.20	0.12	0.25	0.05	−0.7	0.9
3	0.30	0.19	0.28	0.10	0.3	1.1
4	0.35	0.20	0.32	0.12	0.4	1.0
5	0.32	0.15	0.41	0.16	−1.4	−0.1
6	0.55	0.41	0.40	0.13	2.5	3.8
7	0.76	0.65	0.44	0.23	6.8	6.7
8	0.82	0.68	0.43	0.25	9.1	7.1

bands. There is no noticeable separation of the statistical structure of the atmosphere into three regions, as was seen for the observed data.

The principal question which arises in the process of interpreting the maxima is that of their statistical significance. As is known (3.9), in general, the accuracy of the spectral estimates is different for different frequencies because the variance of such estimates is proportional to the corresponding squared value of the spectra. The greater the spectral value, the greater the error of its estimate can be. To get some representation about accuracy, approximate values of the standard deviations were found with the aid of the residuals: periodogram−spectral estimates. The value of such a standard deviation depends on the region of the frequency domain over which the squared residuals are summed up. For each spectrum in Figure 7.1, the two standard deviations, $s1$ and $s2$, are given in Table 7.2: $s1$ was computed with the aid of the residuals for all the points of the frequency plane, while $s2$ was found only with the aid of the residuals along the major crest of the spectrum estimates. Like the other characteristics analyzed here, these standard deviations are

7.3 Zonal Time Series

latitude dependent, with greater values for the tropical bands where the spectral estimate variability for the observed data is about two times greater than that for the simulated data. By comparing the standard deviations in Table 7.2 with the heights of the maxima in Figure 7.1, it is easy to notice that the estimates of the crest values are certainly statistically significant (for instance, at the 95% significance level). The question about the statistical significance of the separate maxima with respect to the mean level of the corresponding crest, however, is more complicated. Although some low-frequency maxima of the tropical bands are certainly significant, at least at the 70% level, it is likely that for the rest part of the interannual frequency interval and for the time series of the nontropical regions, they are simply the result of the sample variability. This means that for the considered period of observations, the variability in the vicinity of these maxima was certainly slightly higher than for other frequencies; however, the possible theoretical spectral density of the corresponding processes exhibits no such features. Therefore, the estimates in Figure 7.1 demonstrate two kinds of features: The first is the specific shape of the random process spectrum under study with only small maxima in the vicinity of the zero frequency; the second is the specific character of the observed sample Fourier decomposition.

Of course, it is possible to increase the width of the two-dimensional filter used and to get rid of the most of the maxima that we can see in Figure 7.1. But the empirical study of some spectral features of the analyzed samples, which represent real and simulated 31 year climate, can be useful for the comparison. For example, having the results in Figure 7.1, one can conclude that the separate observed climate fluctuations are significantly more powerful than the simulated ones, especially in the tropical regions.

In both cases, the above-mentioned periods of the spectral maxima are randomly distributed along the frequency axis. But it seems that for the observed data, the low frequency maxima are the results of some systematic long-period forcing (may be of the CO_2 or El Niño type), whereas for the simulated data the maxima are the results of the nonlinear transformation [of the amplitude modulation type (Polyak, 1975)], which occurs in the process of the numerical integration of the hydrodynamic equations. The most important distinctions of the spectra for the two types of data are in the white noise component level, which can be seen in the illustrations as the distance between the minimal spectral estimates and the plane with zero values. (Because normalized data is considered, none of the discrepancies in the level of this component are absolute values; they are respective to the values of the corresponding standard deviations.) For both systems, the white noise component increases as the latitude increases, but the starting level (at the equatorial region) is significantly greater for the simulated data.

Therefore, Figure 7.1 shows the "whitening" of the temporal temperature fluctuations in the process of spatial northward transport. The value of the white noise component and the character of its variation from one latitude band to another are different for both types of data, but the observed spectra have one very specific additional feature: a dramatic change of the frequency structure on the boundary ($35°$) between the tropics and the middle latitudes (see Figure 7.1, 6), where the heat balance is reached. For this latitude band (Figure 7.1, 6), the powerful maximum formed in the origin shows that something unusual abruptly occurred in the frequency composition of the temperature fluctuations on the boundary of this region. Nothing like this appears in the simulated spectra, where the latitudinal variation of the spectral structure is gradual and smooth.

The shapes of both types of the two-dimensional correlation functions corresponding to the spectra in Figure 7.1 do not change markedly from one latitude band to another, but their values vary significantly (see Table 7.3). These shapes are approximately similar to those in Figure 7.2, which provides examples of such correlation functions in the $56.25°$ to $45°$ latitude band. The intraannual correlations are non-zero only for up to 4 monthly lags, while the interannual statistical dependence of the monthly fluctuations is close to zero. Therefore, the main feature of the shape of such correlation functions is that they are almost flat (nonzero correlations are concentrated in the narrow strip about the origin of the annual lag axes).

The distinction between the correlation functions is reflected in the different character of their latitudinal variation, which is illustrated by the trend of the first two correlations, given in Table 7.3. For the observed data, the first correlations decrease from 0.82 in the equatorial region to 0.16 in the Arctic, while for the simulated data such a decrease is from about 0.43–0.44 to 0.15. There is a dramatic change (from 0.76 to 0.32) in the observed correlation value in the sixth ($33.75°$ to $22.5°$) latitude band for the transition from the tropics to the midlatitudes. This change in the span of the narrow $11.25°$ latitude area is greater than the total latitudinal variations of the simulated correlations.

For the observed data, the noise level changes from almost zero for the tropics (where there are especially powerful low frequency maxima) to the significant level of about 80% of the total variability

7.3 Zonal Time Series

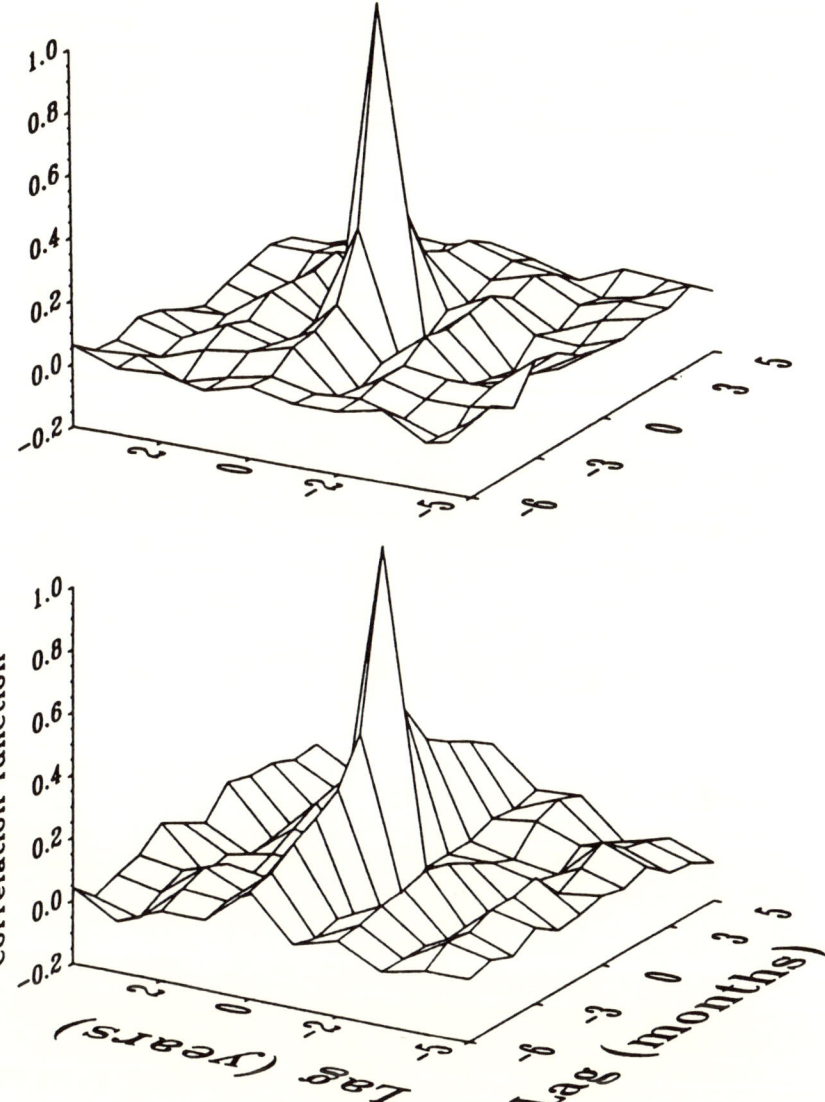

Figure 7.2: Two-dimensional correlation functions of the observed (bottom) and simulated (top) time series of the fourth latitude band.

of the fluctuations in the Arctic area. The noise component for the simulated data is significantly high for the tropical regions (about 40% of the variability) and approximately the same (as for the observed data) for the Arctic. Therefore, the character of whitening (that is, increasing the white noise component in the data of the northern regions) of the long-period temperature fluctuations, expressed as a latitudinal trend of the statistical structure, is the most distinctive feature of the temperature fluctuations, and it is quite different for both kinds of data.

The abrupt change in the statistical structure of the observed fluctuations on the boundary between the tropics and the middle latitudes corresponds to the theoretical understanding of the meridional types of the atmospheric circulation (see, for example, Palmen and Newton, 1969, where Figures 1.5, 1.6, 1.11, and some others show a specific character of the tropical and midlatitude mass circulation in the NH and the boundaries between them for different seasons). Thus, it is not unusual that this character has been reflected in the observed second-moment statistics; what is really unusual is that it is not clearly distinguishable in the simulated data.

To compare the correlations more accurately let us consider a test statistic (Rao, 1973)

$$N = \frac{log\frac{1+r_1}{1-r_1} - log\frac{1+r_2}{1-r_2}}{2\sqrt{\frac{1}{n_1-3} + \frac{1}{n_2-3}}}$$

(where r_1 and r_2 are the sample correlation coefficients and n_1 and n_2 are the sizes of the samples), of a simple asymptotic criterion for testing the equality of the two correlation coefficients. If n_1 and n_2 are large, then N has an approximately normal distribution.

The N values for the first two correlations (Table 7.3) show that the difference in the correlations compared are statistically significant (at 95% level) for the three tropical bands (values of N are from 2.5 to 9.1). As for the other five bands, such differences are not statistically significant.

These results show that the temporal variability in the middle and northern regions of the observed and simulated data is quite similar or, more accurately, the spectral and correlation structures of these two types of temperature fluctuations for these regions are statistically undistinguished. But such variability in the tropics and on the boundary between the tropics and midlatitudes is different. In this region the abrupt change of the observed temporal statistical structure is accompanied by the slow and gradual variations of the analogous simulated statistics.

7.4 Multivariate Models

Evaluation of the similarities of or distinctions between both types of second-moment climate statistics is also possible through the intercomparison of the parameter estimates of the multivariate autoregressive models fitted to their multiple time series. The AR stochastic models are an approximation of the stationary time series by a system of linear differential equations (more exactly, by corresponding difference equations); or, put another way, they are a numerical linearization of a system under study. This approach proves very interesting because the linear approximations of the analogous observed and simulated data structures must be very close if the samples are from the same distribution. But to build a multivariate AR model, one needs sufficiently long multiple samples. The volume of the data (372 points in each monthly time series) is enough to consider only the first-order multivariate AR models. It is also important to study the multivariate AR climate models, for they offer an alternative approach to the climate system description and to the analysis of the interactions and feedbacks of the atmospheric processes.

A description of an algorithm used for building first-order multivariate autoregressive models is given in Section 5.2. The dataset of the normalized anomalies for the eight fixed latitude bands (see Table 7.1) is assumed to be a multiple (eight-variate) time series with 372 temporal terms.

The estimates of the autocorrelation and cross-correlation coefficients for the first two temporal lags of the multiple time series are presented in Tables 7.4 and 7.5. For the observed data, the matrix correlation function estimates for $\tau = 0$ (the spatial correlation field) show that when the lags are equal to about 1260 km (one latitude zone separation) there is a noticeable spatial statistical dependence with correlations from 0.54 to 0.83.

If spatial lags greater than 2520 km, most of correlation estimates are not statistically significant. For a temporal lag equal to 1 month (Table 7.5), it is possible to see the crest along the main diagonal with the correlation maximum for the equator (in the lower right corner). The smallest value (0.15) of the autocorrelations is in the Arctic region, while the largest (0.93) is in the equator. As for the analogous simulated data estimates (Tables 7.4 and 7.5), spatial correlations for the 1260 km lag are sufficiently fewer (from 0.22 to 0.39) as well as autocorrelations for the one-month lag in the tropical region.

To make accurate statistical inferences, one has to compute the N statistic values for the corresponding correlations in Tables 7.4 and

Table 7.4: Matrix correlation functions for the zero lag ($\tau = 0$).

	u_1	u_2	u_3	u_4	u_5	u_6	u_7	u_8
				Observed data				
u_1	1.00	0.68	0.31	−0.10	−0.21	−0.16	−0.03	−0.04
u_2	0.68	1.00	0.79	0.14	−0.13	−0.07	0.07	−0.01
u_3	0.31	0.79	1.00	0.61	0.07	−0.03	0.05	0.00
u_4	−0.10	0.14	0.61	1.00	0.61	0.13	−0.03	0.00
u_5	−0.21	−0.13	0.07	0.61	1.00	0.54	0.03	−0.02
u_6	−0.16	−0.07	−0.03	0.13	0.54	1.00	0.67	0.41
u_7	−0.03	0.07	0.05	−0.03	0.03	0.67	1.00	0.83
u_8	−0.04	−0.01	0.00	0.00	−0.02	0.41	.083	1.00
				Simulated data				
u_1	1.00	0.22	−0.01	−0.04	−0.07	0.03	0.03	−0.03
u_2	0.22	1.00	0.39	−0.01	−0.20	0.03	0.09	0.00
u_3	−0.01	0.39	1.00	0.27	−0.17	0.00	0.05	−0.05
u_4	−0.04	−0.01	0.27	1.00	0.27	0.09	0.06	0.00
u_5	−0.07	−0.20	−0.17	0.27	1.00	0.31	0.02	0.01
u_6	0.03	0.03	0.00	0.09	0.31	1.00	0.24	0.12
u_7	0.03	0.09	0.05	0.06	0.02	0.24	1.00	0.25
u_8	−0.03	0.00	−0.05	0.00	0.01	0.12	0.25	1.00

7.5. Such values show that the difference in the spatial correlations (see the first matrix in Table 7.6) when the lag is about 1260 km are statistically significant at any standard level of significance. The largest value (12.7) of N corresponds to the equatorial band. This comparison shows that the spatial variability of the observed and simulated climate are sufficiently different.

As for the temporal-spatial correlations, the N statistic values (the second matrix in Table 7.6), show the statistically significant difference for the three tropical bands as well as for some of the midlatitude bands.

In spite of a certain qualitative similarity between the latitudinal trend of the estimated correlations (for the one-month lag) for both kinds of data, it is obvious that the values of the corresponding correlations in Tables 7.4 and 7.5 differ markedly, which means that the observed and simulated samples considered are from different distributions.

Figure 7.3 (bottom) illustrates with special clarity the fundamental distinction in the observed statistical structures (matrix correlation function from Table 7.5 for the one-month lag) of the tropical,

7.4 Multivariate Models

Table 7.5: Matrix correlation function for the first lag ($\tau = 1$ month).

	u_1	u_2	u_3	u_4	u_5	u_6	u_7	u_8
Observed data									
u_1	0.15	0.08	0.06	0.06	−0.02	−0.02	−0.01	−0.05
u_2	0.12	0.17	0.20	0.20	0.11	0.07	0.06	−0.05
u_3	0.09	0.19	0.30	0.38	0.24	0.13	0.05	−0.04
u_4	0.08	0.16	0.23	0.41	0.38	0.17	0.02	0.01
u_5	0.01	0.09	0.08	0.19	0.36	0.27	0.06	0.01
u_6	−0.07	0.10	0.11	0.05	0.18	0.61	0.56	0.38
u_7	−0.09	0.03	0.05	0.00	0.03	0.56	0.84	0.74
u_8	−0.08	−0.03	0.02	0.06	0.01	0.42	0.80	0.93
Simulated data									
u_1	0.17	0.15	0.00	0.04	0.04	0.02	−0.06	−0.07
u_2	0.11	0.23	0.14	0.11	−0.14	−0.02	0.04	−0.03
u_3	0.06	0.17	0.28	0.25	−0.04	0.06	0.06	0.00
u_4	0.02	0.01	0.04	0.32	0.26	0.09	0.11	0.08
u_5	−0.05	−0.06	0.01	0.00	0.42	0.18	0.06	−0.02
u_6	0.02	0.00	−0.01	0.08	0.20	0.43	0.16	0.07
u_7	−0.07	−0.01	0.08	0.09	−0.02	0.09	0.49	0.23
u_8	0.03	0.03	0.05	0.06	−0.03	0.11	0.18	0.49

midlatitude, and polar regions with abrupt and significant changes on the boundaries, which separate three major zones of the meridional atmospheric circulation (Oort and Rasmussen, 1971; Oort, 1983; Palmen and Newton, 1969).

An analogous picture for the simulated data in Figure 7.6 (top) shows the gradual trend of the correlation structure, whose values in the tropical region are about two times less.

The parameter estimates (exceeding the 99% significance level) of eight-variate AR and linear regression models for both kinds of data are given in Tables 7.7 and 7.8. Each row in these tables presents an equation in finite differences (for the autoregressive model) or a linear dependence (for the regression model) for the normalized anomalies $u_i(t)$.

The standard error ε_{t+1} for the autoregressive equation presents the normalized standard deviation estimate of the forecast for one month ahead as a proportion of the standard deviation of the original anomalies.

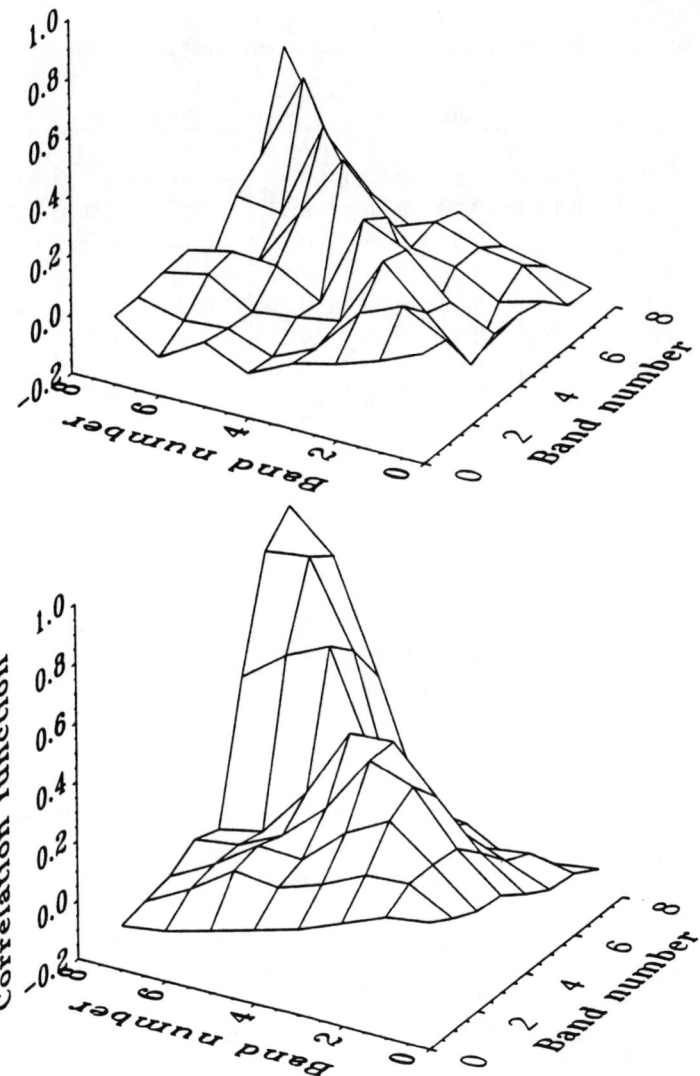

Figure 7.3: Matrix correlation functions for the one month lag of the observed (bottom) and simulated (top) temperature time series.

7.4 Multivariate Models

Table 7.6: N statistic values for the comparison of the correlations in Tables 7.4 and 7.5.

	u_1	u_2	u_3	u_4	u_5	u_6	u_7	u_8
\multicolumn{9}{c}{$\tau=0$ (month)}
u_1	—	8.2	4.5	−0.8	−1.9	−2.6	−0.8	−0.1
u_2	8.2	—	9.0	2.1	1.0	−1.4	−0.3	−0.1
u_3	4.5	9.0	—	5.9	3.3	−0.4	0.0	0.7
u_4	−1.9	2.1	5.9	—	5.9	0.6	−1.2	0.0
u_5	−1.9	1.0	3.3	5.9	—	3.9	0.1	−0.4
u_6	−2.6	−1.4	−0.4	0.6	3.9	—	7.7	4.3
u_7	−0.8	−0.3	0.0	−1.2	0.1	7.7	—	12.7
u_8	−0.1	−0.1	0.7	0.0	−0.4	4.3	12.7	—
\multicolumn{9}{c}{$\tau=1$ (month)}
u_1	−0.3	−1.0	0.8	0.3	−0.8	−0.5	0.7	0.3
u_2	0.1	−0.8	0.8	1.3	3.4	1.2	0.3	−0.3
u_3	0.4	0.3	0.3	2.0	3.9	1.0	−0.1	−0.5
u_4	0.8	2.1	2.6	1.4	1.8	1.1	−1.2	−1.0
u_5	0.8	2.0	1.0	2.6	−1.0	1.3	0.0	0.4
u_6	−1.2	1.4	1.6	−0.4	−0.3	3.4	6.4	4.5
u_7	−0.3	0.5	−0.4	−1.2	0.7	7.4	9.3	9.7
u_8	−1.5	−0.8	−0.4	0.0	0.5	4.6	12.4	15.2

Even a brief examination of the estimates in Table 7.7 of the model parameters for the observed data reveals that the number of statistically significant values above the main diagonal is greater than the number of those below the main diagonal. In other words, the temperature anomaly values are mainly determined by their own past and by the anomaly values of the more southern bands (northward transport of heat). We have already found this feature, while considering the example in Subsection 5.5.2. The observed temperature fluctuations in the moment of $t+1$ of the middle and northern latitude bands depend on almost all southern bands' anomalies (the spreading of the atmospheric temperature anomalies takes place, on average, northward; the southward transport is weaker because most of the corresponding parameter estimates are not statistically significant). The more or less regular change of the signs of the parameter estimates in Table 7.7 shows that some kind of stable spatial waves are responsible for the northward temperature fluctuation transport. Therefore, the approximation of the multiple time series by the multivariate AR model has the potential to detect some regularities in the second-moment structure in spite of the significant variability and noise associated with the fluctuations.

For the simulated data in Table 7.7, the temporal dependence is determined primarily by the past of the time series itself. This means, firstly, that the spatial-temporal statistical dependence of the temperature fluctuations of different latitude bands is very low, and, secondly, their statistical structure is simple and can be presented by a univariate first order autoregressive model.

In tropical regions, the approximation accuracy (ε_{t+1}) for the observed data is much higher than the analogous accuracy for the simulated data, and the corresponding forecasts for a couple of months ahead proves relatively accurate, while such forecasts for the corresponding simulated time series do not make sense for more than one month ahead. The forecast in the polar and midlatitude regions proves virtually useless for both types of data.

The linear regression model parameters in Table 7.8 also indicates a significant difference in the character of the spatial dependence of the compared anomalies. For the observed data, each anomaly is determined by the values on two to four closest latitudinal bands from both sides, while for the simulated data such spatial dependence can be seen mainly on the values of one adjacent band from both sides. For any of the latitudinal bands, the accuracy of the linear regression approximation of the observed data is significantly higher (ε is from 0.23 to 0.55) than for the simulated data (ε is from 0.85 to 0.97). The observed zonal monthly temperatures considered can be accurately interpolated by the linear combination of the observations for the same month of several adjacent latitude bands, while the spatial interpolation of the analogous simulated data cannot be accurately done at all.

Therefore, comparison of the estimated parameters of the multivariate models showed that the temporal and spatial statistical dependence of the observed data is greater than that of the simulated data, especially in the tropical regions. This discrepancy clearly shows a fundamental distinction between the statistical qualities of the temperature fluctuations for the two kinds of data, which means that both multiple samples are from different distributions.

Table 7.7: Estimates (exceeding the 95% significance level) of the parameters of the eight-variate autoregressive models for the latitude-temporal data on the 31-year interval (ε is the normalized standard error).

	u_{1t}	u_{2t}	u_{3t}	u_{4t}	u_{5t}	u_{6t}	u_{7t}	u_{8t}	ε
Observed data									
u_{1t+1}	—	0.24	−0.31	0.34	−0.17	—	−0.12	—	.98
u_{2t+1}	—	0.44	−0.34	0.42	−0.28	0.33	−0.17	—	.96
u_{3t+1}	—	—	0.15	0.25	−0.28	0.41	−0.32	—	.93
u_{4t+1}	—	—	0.27	0.28	—	0.19	−0.33	0.25	.89
u_{5t+1}	—	—	—	0.19	0.19	—	—	—	.91
u_{6t+1}	—	—	—	0.19	−0.12	0.54	0.12	—	.75
u_{7t+1}	—	—	—	—	—	0.13	0.45	0.37	.50
u_{8t+1}	—	—	—	—	—	—	−0.16	1.03	.35
Simulated data									
u_{1t+1}	0.16	—	—	—	—	—	—	—	.98
u_{2t+1}	—	0.17	—	—	—	—	—	—	.96
u_{3t+1}	—	—	0.30	—	—	—	—	—	.95
u_{4t+1}	—	—	0.14	0.29	—	—	—	—	.92
u_{5t+1}	—	—	—	0.16	0.34	—	—	—	.88
u_{6t+1}	—	—	—	—	—	0.41	—	—	.89
u_{7t+1}	—	—	—	—	—	—	0.46	—	.86
u_{8t+1}	—	—	—	—	—	—	—	0.46	.86

7.5 The Diffusion Process

The observed temperature fluctuations are most strongly dependent (temporally) upon their own past as well as upon the fluctuations of the southern regions' past since the computed matrix of the parameters has a special form with mostly non-zero elements above the main diagonal (Table 7.7). This shows the northward climatic (with the monthly temporal scales) meridional transport of temperature anomalies. To present an approximate physical description of this transport let us assume that one of the appropriate models for such a study can serve the diffusion process. According to Monin and Yaglom (1973, p. 580), the advection velocity practically coincides with the instantaneous flow velocity, and obviously, it cannot be greater than the flow velocity.

Table 7.8: Estimates (exceeding the 95% significance level) of the parameters of the eight-variate linear regression models for the latitude-temporal data on the 31-year interval (ε is the normalized standard error).

	u_1	u_2	u_3	u_4	u_5	u_6	u_7	u_8	ε
Observed data										
u_1	—	1.69	−1.47	0.83	−0.49	0.19	−0.21	—	.55
u_2	.35	—	1.03	−0.64	0.35	−0.17	0.14	—	.25
u_3	−.25	0.83	—	0.67	−0.37	0.17	—	—	.23
u_4	.25	−0.92	1.19	—	0.63	−0.32	0.18	—	.30
u_5	−.24	0.83	−1.08	1.03	—	0.69	−0.50	—	.39
u_6	—	−0.47	0.57	−0.61	0.79	—	0.91	−.32	.42
u_7	—	0.27	−0.24	0.24	−0.40	0.63	—	.56	.35
u_8	—	−0.13	—	—	0.18	−0.46	1.14	—	.50
Simulated data										
u_1	—	0.26	−0.13	—	—	—	—	—	.97
u_2	.21	—	0.39	—	—	—	—	—	.88
u_3	—	0.37	—	0.32	−0.20	—	—	—	.85
u_4	—	—	0.36	—	0.32	—	—	—	.90
u_5	—	—	−0.21	0.30	—	0.31	—	—	.87
u_6	—	—	—	—	0.34	—	0.21	—	.91
u_7	—	—	—	—	—	0.22	—	.23	.94
u_8	—	—	—	—	—	—	0.24	—	.96

Before the corresponding methodology is applied, the following remarks must be made. The character of the temperature advection under study is determined by many factors; for example, the direction and strength of the wind. If, for example, one is interested in the mean advection velocity of the synoptic scales fluctuations associated with the eastward wind, one must collect and analyze temperature observations that correspond to the kind of wind chosen. Therefore, careful planning is needed to determine what kind of advection must be evaluated. Given this situation, it was decided first to evaluate (where possible) the *mean* northward advection velocity for each latitude band, and then to find its mean value for the *entire* Northern Hemisphere.

One of the ways to explore the climatic diffusion of the temperature fluctuations and to estimate the advection velocity is to find the relationship between the AR models and a diffusive description of the temperature fluctuations (Polyak et al., 1994). Spatial averaging

7.5 The Diffusion Process

of data reduced dimensionality and brought the temperature fields to the coordinate system (x, t). Although this is the only case we will consider here, our reasoning will hold for a field of any dimension.

Therefore, one can approximate each of the latitude-temporal fields by the following diffusion equation:

$$\frac{\partial u}{\partial t} = s\frac{\partial^2 u}{\partial x^2} - v\frac{\partial u}{\partial x} - qu + f(x, t). \tag{7.2}$$

However, a multivariate model is the system of difference equations corresponding to a system of ordinary differential equations. In order to exploit this relationship, the diffusion equation must be approximated by a system of ordinary differential equations. The *straight − linemethod* (Berezin and Zhidkov, 1965) achieves such an approximation. If

$$x = x_k = kh, \ (k = 0, 1, 2, ...), \tag{7.3}$$

then replacement of the derivatives $\partial^2 u/\partial x^2$ and $\partial u/\partial x$ by the simplest finite differences gives

$$\begin{aligned}\frac{du_k(t)}{dt} &= \frac{s}{h^2}[u_{k+1}(t) - 2u_k(t) + u_{k-1}(t)] \\ &\quad - \frac{v}{2h}[u_{k+1}(t) - u_{k-1}(t)] - qu_k(t) + f_k(t).\end{aligned} \tag{7.4}$$

This system of ordinary differential equations approximates the diffusion equation and corresponds to the autoregressive equations of the multivariate model.

Any other form of the difference representation for the derivatives $\partial^2 u/\partial x^2$ and $\partial u/\partial x$ leads to a different, but linear, system of differential equations. The greater the order of the finite differences used, the more accurate an approximation can be achieved.

Thus, estimating the AR model parameters offers one of the possible solutions for the inverse problem of these differential equations.

The relationship between the parameters of the autoregressive model and the coefficients of the diffusion equation can be established by replacing the derivative du/dt in (7.4) by the simplest difference. We have

$$u_{k(t+1)} = au_{(k-1)t} + bu_{kt} + cu_{(k+1)t} + lf, \tag{7.5}$$

where l is the time step, and the parameters a, b, c are interconnected with l, h, q, s, and v by the following relationships:

$$s = \frac{h^2}{2l}(a + c), \tag{7.6}$$

$$v = \frac{h}{l}(a - c), \qquad (7.7)$$

$$q = \frac{1}{l}(1 - a - b - c). \qquad (7.8)$$

The autoregressive parameters (a, b, s) present appropriate samples for the subsequent estimation of the coefficients of the diffusion equation. This estimation makes sense when there is a real diffusive process in a given x direction, as is the case in the above spatial diffusion of temperature.

Let us associate the finite difference autoregressive equations (Table 7.7) for the obtained eight-variate AR models with the coefficients of the diffusion equation through formulas (7.5–7.8). For example, for the seventh time series (24° to 13° latitude band) in Table 7.7, one has $a = 0.13, b = 0.45$, and $c = 0.37$. By substituting these values in the equations (7.6–7.8) and having in mind that $l = 1$ month and $h \approx 1260$ km, the estimates

$$s \approx 630^2 \text{ km}^2/\text{month} \approx 391 \text{ m}^2/\text{sec};$$

$$v \approx -302 \text{ km}/\text{month} \approx -0.12 \text{ m}/\text{sec};$$

$$q \approx 0.05 \ 1/\text{month} \approx 0.00000002 \ 1/\text{sec};$$

are obtained (only two significant digits are presented in the final coefficient estimates). The negative value of v means southward advection.

The results for each of the AR equations in Tables 7.7 are given in Table 7.9. In the calculations by formulas (7.6–7.8), both the statistically significant and insignificant estimates (which are not shown in Table 7.7) of the AR model parameters were used. The means in Table 7.9 were obtained using the means $\bar{a}, \bar{b}$, and $\bar{c}$ of the AR parameter estimates which were found by averaging the corresponding values in Tables 7.7 on the main diagonal and on the two adjacent diagonals parallel to the main diagonal.

The estimates in Table 7.9 reflect the distinction in the Northern Hemisphere's three major atmospheric meridional circulation regions (Palmen and Newton, 1969; Oort and Rasmusson, 1971; Oort, 1983). For these three regions, the sign of the estimated values of the advection velocities are different, thereby reflecting the changes of the advection directions on the boundaries between the regions. The very crude estimation of the boundary between the tropics and the midlatitude circulation zone is 30–35° for the observed data and

7.5 The Diffusion Process

Table 7.9: Diffusion equation coefficients (s is in m^2/sec; v is in m/sec; q is in sec^{-1}) of the observed and simulated data for different latitude bands.

Time series number	Observed data			Simulated data		
	s	v	q	s	v	q
2	-362^2	0.12	0.0000004	262^2	0.00	0.0000002
3	310^2	-0.09	0.0000002	-101^2	0.05	0.0000003
4	240^2	0.18	0.0000002	154^2	0.10	0.0000002
5	295^2	0.05	0.0000002	280^2	0.03	0.0000002
6	-44^2	-0.12	0.0000002	79^2	0.03	0.0000002
7	391^2	-0.12	0.0000000	171^2	-0.01	0.0000002
mean	209^2	0.00	0.0000002	181^2	0.04	0.0000002

24–30° for the simulated data; between the arctic and the midlatitude circulation zone is 60–70° for the observed data and 70–80° for the simulated data.

Let us compare our estimates with the corresponding results obtained by Oort and Rasmussen (1971, pp. 71 and 228) using wind observations. Their approximate latitude points of the wind sign change (boundary values for the midlatitude region of the meridional atmospheric circulation) are 31° and 67° with the wind values from -0.9 to 0 m/sec for the tropics, 0 to 0.5 m/sec for the midlatitude, and 0 to -0.2 for the arctic. Roughly speaking, therefore, the advection velocity estimates in Table 7.9 for both kinds of data are approximately within the limits of accuracy of the Oort and Rasmussen (1971) results.

Relating to the turbulent diffusion coefficient s, Monin and Yaglom (1973) assumed that its value has to be of the order about 1 to 10 m^2/sec. Our results for both kinds of data are significantly greater, which means that the theoretical and experimental study of turbulence is a problem of great significance.

Generally, the numerical values of the advection velocities (as well as other coefficients of the diffusion equation) of different latitude bands for both types of data differ markedly. The estimate (-0.01 m/sec) of v in the tropics for the simulated data is many times less in absolute value than the corresponding estimates for the observed data, which means that the observed and simulated meridional atmospheric circulations in the tropics are very different. The reason for this difference needs to be given special consideration, which can be accomplished by averaging the data within latitude bands more narrow than $11.25°$ and, perhaps, with a time step (of data) of less than one month.

By accepting the above mean values in Table 7.9 for the observed data as the diffusion equation coefficient estimates, one gets the following:

$$\frac{\partial u}{\partial t} \approx 209^2 \frac{\partial^2 u}{\partial x^2} + 0.00 \frac{\partial u}{\partial x} - 0.0000002u + f(x,t). \quad (7.9)$$

This equation describes the *mean* meridional diffusion of the temperature fluctuations for the entire Northern Hemisphere obtained using the 31-year interval (1959–1989) of observations.

The questionable point in the above analysis, especially in the midlatitudes for the observed data, is the order of the finite differences used. As the estimated parameters of the autoregressive equations show (see Table 7.7), each value of u_{kt+1} must be a linear combination of the three to four values of u in the moment of t, which is not the case for the simplest finite difference schemes used in formulas (7.4) and (7.5). But increasing the finite differences order in (7.4) and (7.5) requires a special development of the methodology as well as a corresponding analysis of the numerical integration schemes used by the Hamburg GCM.

It is difficult to conclude from the above results what estimates (observed or simulated) most accurately represent the mean boundary values between the Rossby three-cell meridional circulation zones (Palmen and Newton, 1969). However, the analysis of the temporal variations of these latitude boundaries (which can be completed by applying the above methodology) can serve as an important source of information about climate change, and the 100-year set of surface air temperature observations can be used in such a study as an appropriate data base.

Comparison of the results obtained for the observed and simulated data shows that in different latitude bands the monthly temperature fluctuation advection velocities, as well as other coefficients

7.6 Latitude-Temporal Fields

of the diffusion equation, are markedly different. The absolute values of the v estimates are greater for the observed data, so the real processes of the transport of the climatic fluctuation are more powerful than those described by the model.

The eight zonal time series (observed or simulated) can be analyzed jointly as a two-dimensional latitude-temporal field of the normalized temperature anomalies of the NH.

But the two-dimensional spectral and correlation analysis methodology is valid only for the homogeneous random fields. The investigations that were carried out in the previous sections showed that neither of the two latitude-temporal fields above is homogeneous. Their statistical structure depends not only on the differences between coordinates of the grid points, but also on the values of these coordinates; that is, their statistical structure is latitude-dependent.

The application of the statistical methodology, developed for the homogeneous field (Chapter 3), to the nonhomogeneous field could bring about some very general characteristics which would not correspond to different, maybe approximately homogeneous, subdomains of the entire field. Therefore, such an approach can serve only as a very crude estimation of the real statistical properties, and in this case the spatial-temporal spectral and correlation characteristics could give only a rough, approximate description of the variability of such a latitude-temporal field.

Keeping the existence of the three different meridional atmospheric circulation regions of the NH in mind, one can try to overcome the inconvenience related to the nonhomogeneity by analyzing data on each such region; that is, by comparing the statistical structure of the corresponding observed and simulated fields for each region separately.

In addition to dividing the entire field into three different subfields, it is especially helpful to have high spatial resolution. To satisfy this requirement, we will consider the zonal temperature time series for different 5.625°-width latitude bands.

The first two subfields to be compared will correspond to the tropical region (0-28.125° latitude). The size of each subfield is 5[5.625° (or 630 km) width latitude bands]×372 months.

The main distinctive features of the two-dimensional spectral estimates (Figure 7.4) are the low frequency maxima which are centered in the origin. The non−low frequency part of the simulated

292   7 The GCM Validation

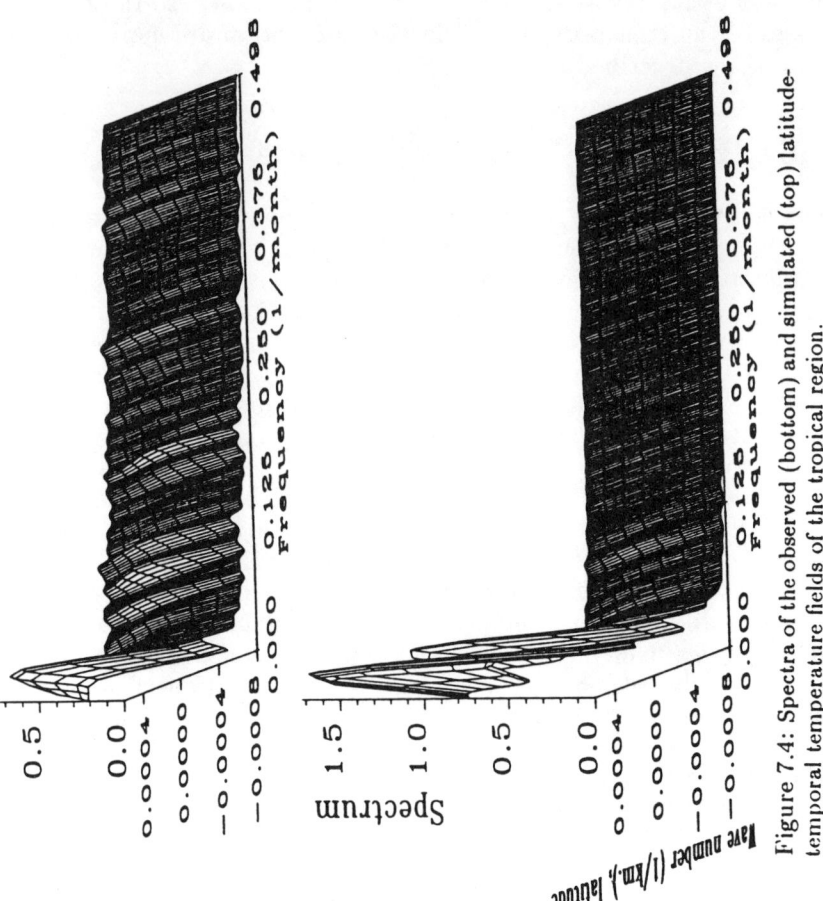

Figure 7.4: Spectra of the observed (bottom) and simulated (top) latitude-temporal temperature fields of the tropical region.

7.6 Latitude-Temporal Fields

spectrum looks more irregular than that of the observed one. The observed spectrum is smoother because the number of feedbacks and interdependences in nature are significantly greater than what are included in the model; therefore, the power of separate randomly distributed pikes in the observed periodogram must be smaller, whereas the smoothing procedure leads to more uniform estimates in this case.

The spectral maximum for the observed data is significantly more powerful than that for the simulated data. This indicates a very low level (the distance between the spectral surface and the horizontal plane) of the white noise component for the observed data. Therefore, it is possible that the observed long-period climate fluctuations are caused by some natural forces that are not identically described by the model. The low-frequency spectral maximum of the surface air temperature, which is well-known from the analysis of the historical records, reflects the possibility of the large-scale climate fluctuations. (But for the historical records this maximum is not very powerful because most of the meteorological stations with the long observational intervals are located in the middle latitudes.)

The existence of such a maximum (in spite of its small value) for the simulated data means that, generally speaking, the GCM rendered some climate features more or less realistically. Of course, we are far from attempting to identify the physical nature of the maxima for both types of data: they must have a different power as well as, for example, spectral maxima obtained using two sets of temperature data given on two different observational intervals. But the difference in the power, which is concentrated in the low-frequency part of the spectra in Figure 7.4, is too large to be attributed only to the sample variability.

The approximate values of the standard deviations of the spectral estimates were obtained using the residuals: periodogram − spectral estimate. The values of the standard deviations vary significantly for different regions of the frequency−wave-number domain. For example, for the entire domain of the spectral estimates of the observed data (Figure 7.4, bottom), this standard deviation is about 0.06, while for its low-frequency part it is about 0.20. Analogous values for the simulated data (Figure 7.4, top) are about 0.02 and 0.10. These estimates show that the low-frequency spectral maxima of both types of data are statistically significant at the 95% level.

The distinction in the distributions of the observed and simulated data becomes especially clear when one compares the correlation functions (see Figure 7.5 and Table 7.10). Such a comparison reveals that the numerical values of the estimates are absolutely different.

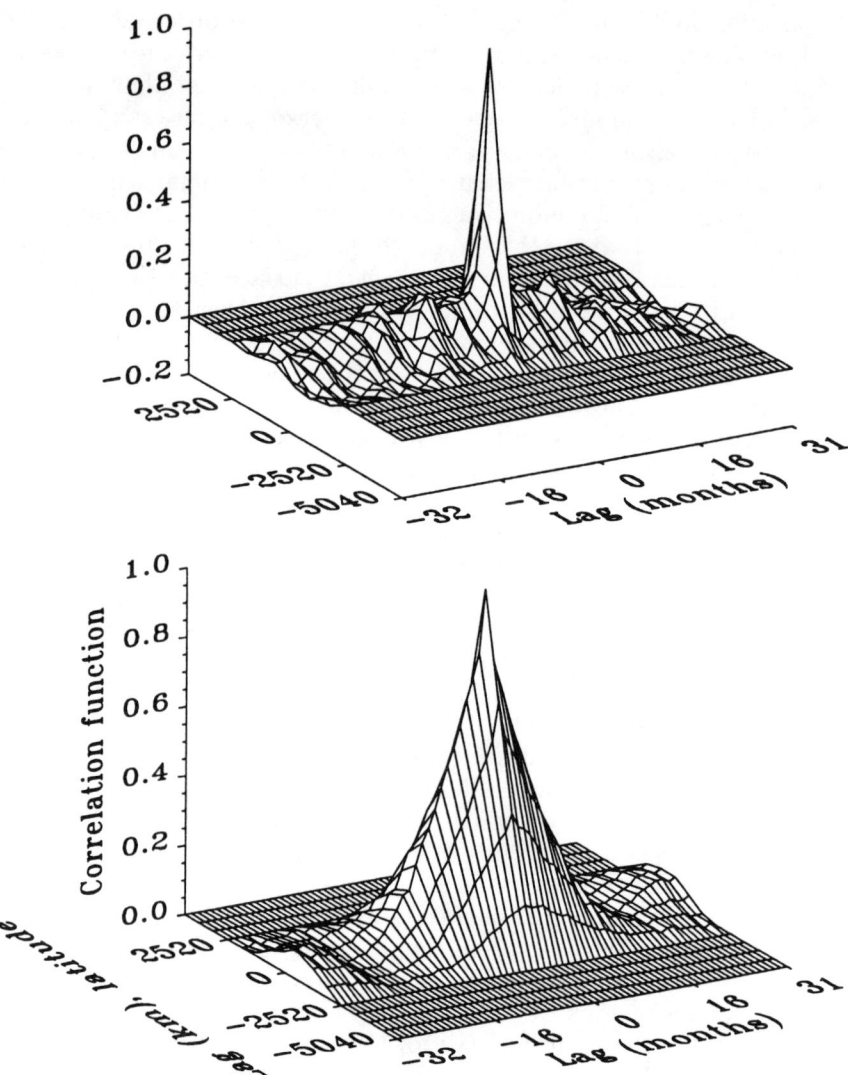

Figure 7.5: Correlation functions of the observed (bottom) and simulated (top) latitude-temporal temperature fields of the tropical region.

7.6 Latitude-Temporal Fields

Table 7.10: Correlations (see also Figures 7.5 and 7.7) for the first two spatial and temporal lags and corresponding statistics N.

Lags of the correlations	Observed data	Simulated data	N
Latitudes 90.00°–78.75°			
1 month, 0 km	0.13	0.16	−0.4
2 months, 0 km	0.04	0.08	−1.6
0 month, 630 km	0.45	0.32	2.1
0 month, 1260 km	0.06	0.06	0
Latitudes 78.75°–33.75°			
1 month, 0 km	0.26	0.29	1
2 months, 0 km	0.20	0.09	4
0 month, 630 km	0.79	0.58	15
0 month, 1260 km	0.46	0.14	12
Latitudes 28.125°–0°			
1 month, 0 km	0.82	0.45	26
2 months, 0 km	0.75	0.24	28
0 month, 630 km	0.70	0.46	13
0 month, 1260 km	0.40	0.06	12

For example, for the one-month temporal lag (and the zero spatial lag) the compared correlations are 0.82 and 0.45, and the N value is approximately 26. For the 630 km spatial lag (and the zero temporal lag) the correlations are 0.70 and 0.46, and $N \approx 13$. The differences of the correlations are statistically significant at any standard level. The spatial and temporal dependences of the observed data are high, which was more difficult to foresee.

The estimates of the correlation functions confirm the inferences derived from the spectral analysis: There is a significant qualitative and quantitative distinction in the power of the white noise and the low-frequency components in the observed and simulated climate variability for the tropical region, or the two samples analyzed are from different distributions.

The second pair of the subfields to be considered corresponds to the midlatitude region ($33.75-78.75°$ latitude). The size of each field is 8[$5.625°$ (or 630 km) width latitude bands]$\times 372$ months.

The most general feature of the frequency–wave-number spectra of these subfields (Figure 7.6) is the multiple crests spread along the frequency axis. This means that the temporal variability is reflected by the fluctuations with all possible periods from two months to many years. It seems that (in the case of simulated data) the small crests for different frequencies result from the large number of nonlinear transformations of the amplitude modulation type produced in the process of the numerical integrations of the hydrodynamic equations, while for the observed data such crests result from multiple interactions and feedbacks of many processes of nature. It is possible to eliminate all these irregularities by increasing the spectral window width. However, to keep the statistical methodology identical with that used in the tropical region, this was not done.

The spatial variability is realized mainly by the large spatial waves (the spatial part of the spectra has maxima corresponding to the zero wave number). The concentration of the power in the low-frequency part of the observed spectrum is also greater than that for the simulated data, but the maxima themselves are significantly smaller than those for the tropical region.

Spatial-temporal correlation functions (see Figure 7.7 and Table 7.10) slightly resemble each other especially in their temporal parts, but the differences in the spatial correlations are still large, and, of course, statistically significant (Table 7.10). The overall comparison of the two correlation's fields for the midlatitude suggests that the corresponding samples are from different distributions.

The volume of the available observations (as well as the size of the area) of the arctic region ($78.75-90.00°$ latitude) is too small to

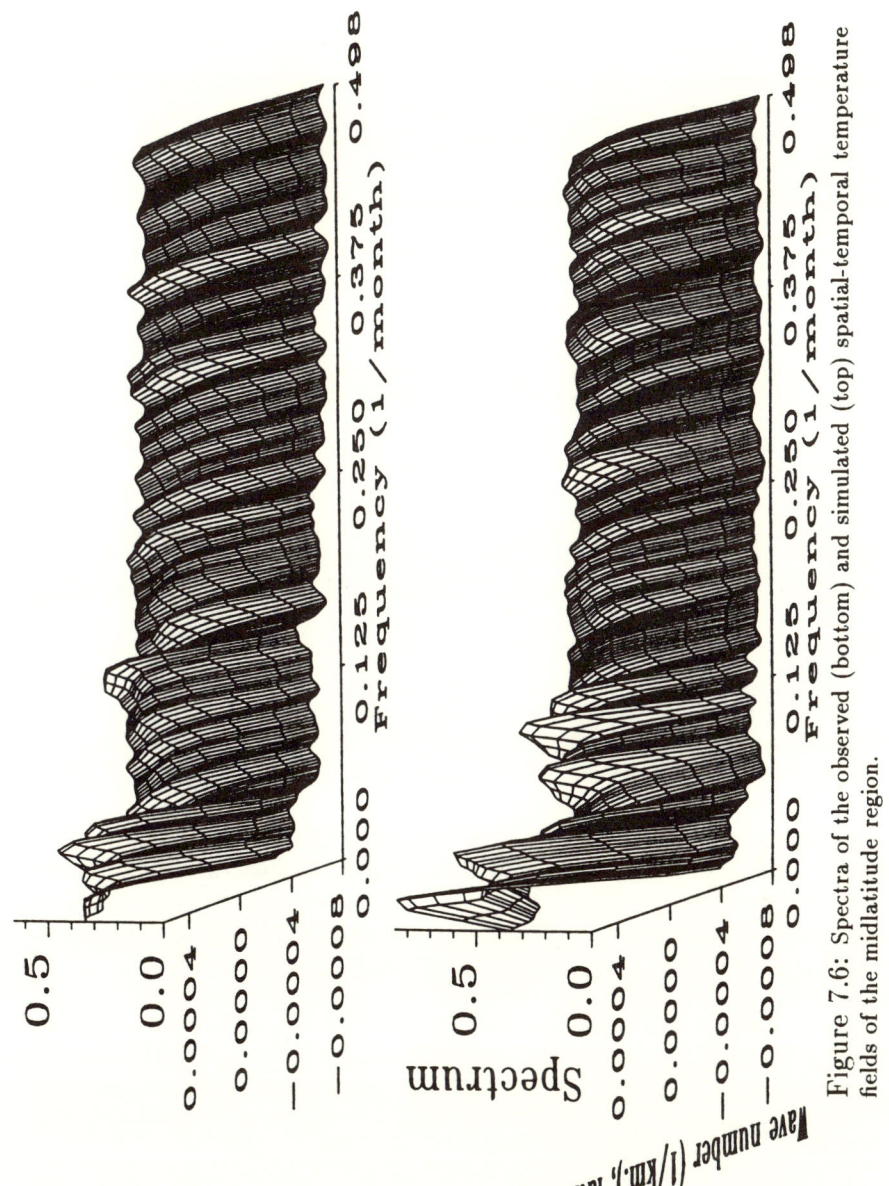

Figure 7.6: Spectra of the observed (bottom) and simulated (top) spatial-temporal temperature fields of the midlatitude region.

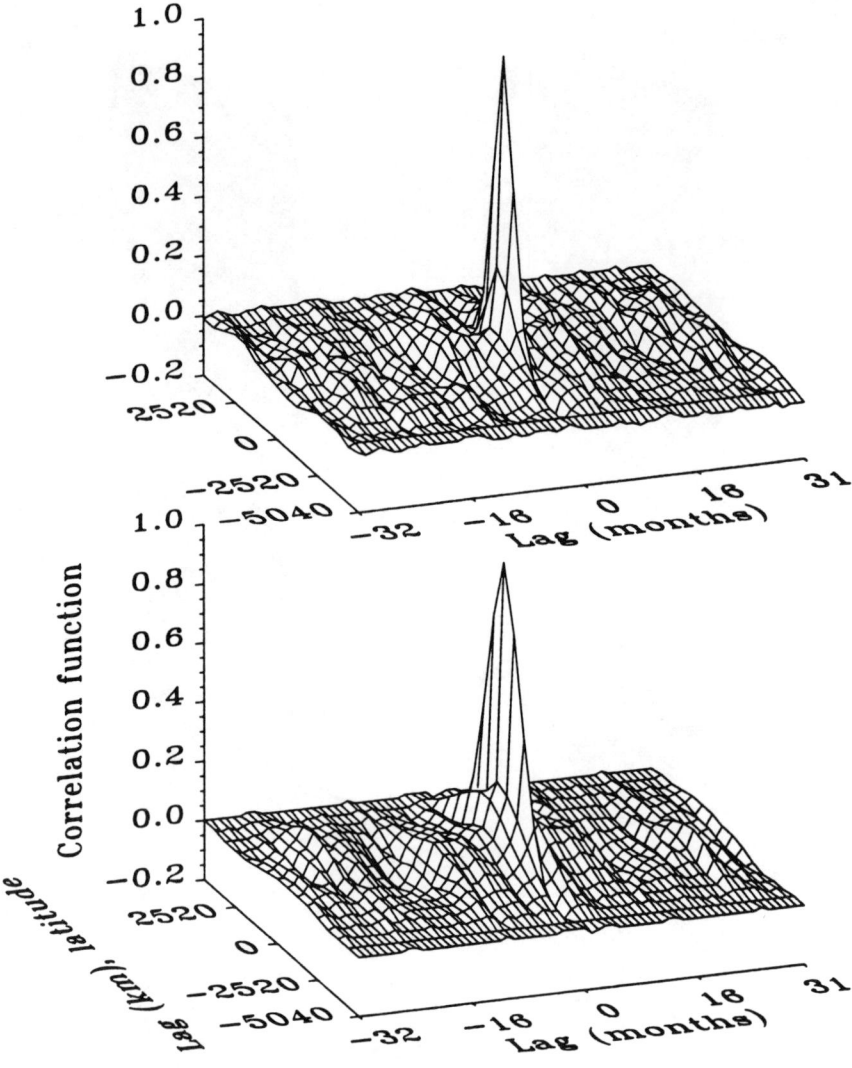

Figure 7.7: Correlation functions of the observed (bottom) and simulated (top) spatial-temporal temperature fields of the midlatitude region.

be appropriately represented by the two-dimensional field, but some conception about statistical structures of its observed and simulated temperatures can be gathered from the estimates of the temporal and spatial correlations given in Table 7.10. The temporal variability of both types of data is almost identical and are very close to a white noise process. Their spatial variability is still sufficiently different.

In short, the analysis provided in this section is an approximation of different latitude-temporal data structures by the homogeneous fields. Yet even though the analysis is rough, it makes possible to find the sufficient distinction of the sampled distributions. This distinction shows that there is much to be done before the practical application of the GCM for the analysis of the real climate variability can be scientifically justified.

7.7 Conclusion

There were two kinds of results obtained. The first includes the features of the observed spatial and temporal climate variability (the second-moment climatology) which can be used to validate any GCM and for some other purposes. The second includes the comparison of these features with the analogous characteristics of the climate simulated by this particular version of the Hamburg GCM.

The most general conclusion that can be made about the variability of the real climate is that all of the kinds of second moments considered (spectra, correlations, parameters of the stochastic models) are latitude-dependent. There are three clearly distinguishable latitude regions (arctic, midlatitude, and tropics) within which such variability can be assumed as approximately homogeneous.

The comparative study above showed the general similarity of the spectral and correlation spatial-temporal structures of the observed and simulated second moments. Some estimates, such as linear trends and advection velocities, have approximately the same orders for both types of data. However, the character of the latitudinal distribution of the white noise component in the fluctuations is quite different. The latitudinal trend of the observed spectra and correlation functions has a very specific feature: the dramatic change of its structure on the boundary between the tropics and the middle latitudes. This change is in complete agreement with the theoretical understanding of the character of the meridional atmospheric circulation. In contrast, such a latitudinal trend of the simulated data is gradual and smooth.

The autoregressive model for the observed data shows that the temperature fluctuations are most strongly dependent upon the fluctuations in the southern regions, which is not the case for the simulated data. This abrupt change of the second moments on the boundary between midlatitude and tropics, the different character of the northward transport of heat (parameters of the ARMA models), the level of a white noise component, and some other features, show that *the observed meridional circulation differs from that in the GCM*. A relationship between the coefficients of the diffusion equation and the parameters of the stochastic model has been found that has led to the estimation and comparison of the advection velocity for the corresponding latitude bands. Such advection velocities are different for both types of data.

Comparison of the statistical structures of the two-dimensional spatial-temporal fields for tropics and midlatitudes revealed that their main distinction is in the redistribution of their low-frequency composition, where the power of the observed spectrum is significantly greater than the power of the simulated one. This implies a relatively lower level of the white noise component for the observed data and provides evidence that the possible causes of the observed long-period climate fluctuations can be explained by some natural phenomena similar to El Niño or increasing of CO_2 concentration.

The methodology developed may help in the verification and comparison of many observed and simulated processes and fields. Moreover, problems identical to those presented in this study arise in comparison of different GCM models or the results of different simulations by the same model with different dynamical components or boundary conditions. The problem lies in identifying appropriate priorities and in determining the scale of the spatial and temporal averaging of data, as well as in revealing the possible uses of different kinds of statistical information obtained.

On the whole, the comparison of the observed and simulated second moments identifies that many details of their structures differ markedly in quality and quantity; and the largest differences in the compared statistics, found for the tropical regions, can, possibly, serve as a measure of such a distinction. This fact also shows that physical modeling of the heat fluxes must be done more accurately.

Therefore, the problems of the GCM diagnoses and verifications can be stated and solved in terms of standard statistical inference procedures concerning the distribution parameters associated with observed and simulated atmospheric systems.

The statistical methodology developed here can also be used for generating multiple samples of any size with statistical properties

7.7 Conclusion 301

that are identical to the properties of the observations; for finding the inverse solution of some of the hydrodynamic equations; and for studying some other problems of meteorological and climatological data analysis and modeling. Of course, the GCM verification is a continuous and complicated process (because the physical and computational methodologies are constantly being perfected and because the volume of observations is constantly being increased), but eventually the number of compared variables will become greater.

8

Second Moments of Rain

The first part of this chapter presents a description of the GATE rain rate data (Polyak and North, 1995), its two-dimensional spectral and correlation characteristics, and multivariate models. Such descriptions have made it possible to show the concentration of significant power along the frequency axis in the spatial-temporal spectra; to detect a diurnal cycle (a range of variation of which is about 3.4 to 5.4 mm/hr); to study the anisotropy (as the result of the distinction between the north-south and east-west transport of rain) of spatial rain rate fields; to evaluate the scales of the distinction between second-moment estimates associated with ground and satellite samples; to determine the appropriate spatial and temporal scales of the simple linear stochastic models fitted to averaged rain rate fields; and to evaluate the mean advection velocity of the rain rate fluctuations.

The second part of this chapter (adapted from Polyak et al., 1994) is mainly devoted to the diffusion of rainfall (from PRE-STORM experiment) by associating the multivariate autoregressive model parameters and the diffusion equation coefficients. This analysis led to the use of rain data to estimate rain advection velocity as well as other coefficients of the diffusion equation of the corresponding field.

The results obtained can be used in the ground truth problem for TRMM (Tropical Rainfall Measuring Mission) satellite observations, for comparison with corresponding estimates of other sources of data (TOGA-COARE, or simulated by physical models), for generating multiple rain samples of any size, and in some other areas of rain data analysis and modeling.

8.1 GATE Observations

8.1.1 Objectives

For many years, the GATE data base has served as the richest and most accurate source of rain observations. Dozens of articles presenting the results of the GATE rain rate data analysis and modeling have been published, and more continue to be released. Recently, a new, valuable set of rain data was produced as a result of the TOGA-COARE experiment. In a few years, it will be possible to obtain satellite (TRMM) rain information, and a rain statistical description will be needed in the analysis of the observations obtained on an irregular spatial and temporal grid. When satellite measurements over a certain region take place only twice a day, there may be significant losses of information, leading to inaccuracies in the climatological averages of rain. For example, if a rain time series has a diurnal cycle (hidden periodicity), the specific moments (of the day) in which the two daily readings are taken determine a bias in the spatial and temporal averages of the satellite data. A standard statistical technique for detecting such hidden periodicity is spectral and correlation analysis. Knowledge of a diurnal cycle of a non-zero rain rate is important in operational practice for retrieving data from the records of brightness temperature when one needs to know a range of variations of a rain rate.

These developments will lead to a broadening of the scope of comparative statistical description and modeling of rain in order to evaluate the scales of possible distinctions in rain climate characteristics for different geographical regions of the earth.

For a long time, the univariate autoregressive model has been one of the most widespread methodologies for modeling rainfall time series. The multivariate methodology applied here is statistically and physically richer than the univariate approach. Moreover, multivariate modeling enables us to study the diffusion of rain because there is a simple interconnection between diffusive and multivariate AR presentations of multiple rain time series.

Second-moment estimates provide the most appropriate way for comparing the climatic features of rain because the theory and practice of statistical inferences and hypotheses testing have already been developed for such moments.

On the whole, the objectives of this study are as follows:

1. To study the features of rain variability in the spaces of the temporal-spatial lags and of the frequency-wave numbers and to evaluate a possible distinction in estimates corresponding to radar and satellite observations.

2. To estimate the diurnal cycle of a non-zero rain rate.

3. To build multivariate AR models and to study the diffusion of rain.

4. To document the initial data structures and to demonstrate the real volume of the non-zero observations used for obtaining different statistics.

Multiple analogous studies (Bell and Reid, 1993; Chiu et al., 1990; Crane, 1990; Graves et al., 1993; Hudlow, 1977, 1979; Kedem et al., 1990; McConnell and North, 1987; Nakamoto et al., 1990; North and Nakamoto, 1989; Simpson et al., 1988; and so on) with the same data set were carried out. But our approach is different in presenting a detailed pictorial description of data as well as in its application of standard statistical technique of two-dimensional and multivariate analysis to spatially and temporally averaged fields. Such averaging of the original three-dimensional array helps to reduce the noise component and to show the second-moment characteristics of rain more distinctly. Whenever possible, we also tried to find the simplest stochastic model of one or two parameters, because the comparison of a small number of parameter estimates is the most effective way to solve the ground truth problem as well as to provide a comparative climatological description of rain.

8.1.2 Data Documentation

The GATE data base, created and discussed in detail by Arkell and Hudlow (1977) and by Patterson et al. (1979), offers the most complete archive of rain observations. The two 19-day intervals (June 28 to July 16 of GATE1 and July 28 to August 15 of GATE2) of rain rate observations used in this study were collected over a 400 km diameter hexagonal area in the Atlantic Ocean, centered on $8°30'N, 23°30'W$ off the west coast of Africa. The data were obtained by averaging the 15 minute rain rate (mm h^{-1}) radar and surface ship measurements over 4×4 km pixels. Therefore, the temporal step of the data is 15 minutes and the spatial step in both directions is 4 km.

Table 8.1 documents the number of observed spatial fields (100×100 size) within each hour for the time interval of the GATE1 experiment. Because there are only 1,716 terms of the two-dimensional time series in the GATE1 archives, the omitted or missed fields were replaced by zero fields. After such replacement, the initial structure obtained is the following matrix time series:

$$R_{ij}(t), i = 1, ..., 100; \quad j = 1, ..., 100; \quad t = 1, ..., 1824, \qquad (8.1)$$

8.1 GATE Observations

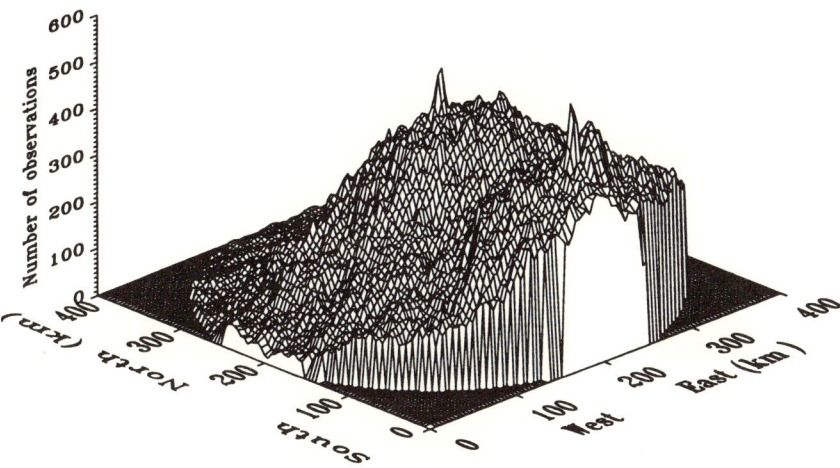

Figure 8.1: Number of non-zero observations of the GATE1 rain rate data base.

where i, j, and t are the subscripts in the directions of north-south (NS), west-east (WE), and time, respectively.

To obtain the general representation of the GATE1 array, the number of the non-zero observations for each spatial point (i, j) was counted and presented in Figure 8.1 (its contour lines are given in Figure 8.13).

This figure shows that the number of readings with rain for different points of the ocean varies from zero to 600; in other words, the proportion of readings with rain varies from zero to about one-third of the total number of terms. The farther out in the Northern part of the Atlantic Ocean, the fewer were the number of observations with rain. In the background of this significant north-south trend, there are two maxima of the amount of non-zero observations corresponding exactly to the locations of the two ships (the *Oceanographer* at the center and the *Researcher* at the most southern point of the GATE area) whose data were used for creating of the data base. Such a dependence (which does not really exist) between the rainfall amount and the ship locations clearly shows that not all of the rains in the GATE spatial area were recorded for the observational interval and that it is possible that not all of the non-zero readings were really rain.

Analysis of data averaged over all spatial points i and j for each moment t of the readings shows that the spatial mean rain rate varied from zero to about 13 mm/hr and that there were only a few 15-

minute intervals when rain was not recorded. Some of the general statistical characteristics of the GATE1 data appear in Table 8.2.

This study deals with the spectral and correlation analysis of three types of two-dimensional rain rate fields obtained by averaging data sequentially along axes (subscripts) i, j, and t (8.1). Averaging, as well as estimating the sample standard deviations, was done by adding only the non-zero observations and dividing the sum by their number. Then, the anomalies were divided by the corresponding estimates of standard deviations. The zeroes were left as zeroes after normalization. The elements of such two-dimensional fields of normalized anomalies will be denoted as u_{ij} or u_{it}.

The standard methodology with two-dimensional Tukey spectral window was employed. The values of the periodograms and the spectral estimates were multiplied by some factor to be conveniently presented by illustrations. The factor used is 10,000.

8.1.3 Spatial Field

Averaging data (8.1) along the t-axis over the entire observational interval for each point (i, j) of the surface gives the square spatial field of 100×100 size with a 4 km step along each axis. This field reveals that the range of variations of the temporal means is 0–20 mm/hr. Because the sample mean of the entire data is about 4.4 mm/hr, the sample distribution of the observations must be positively skewed.

The spatial spectrum estimates (Figure 8.2) show the concentration of almost all power in the vicinity of zero wave numbers with a significant peak centered in the origin.

Therefore, this spatial field has a two-dimensional trend possibly caused in part by the decreasing trend of non-zero observation number with distance from the ship (see Figure 8.1) and in part by the natural latitudinal variations of rain. Most of the spectral estimates are zeroes. The estimates also show that the shortest wavelengths with non-zero power are about 32 km. According to the Nyquist theorem, this means that it is useless to choose a spatial resolution of the observations that is less than 16 km for such a spectral analysis.

Thus, the first simple inference to be drawn from the spectrum above is that the optimal sampling of the rain rate observations must have spatial steps of about 16 km, which, as compared to the GATE1 data sampling, would reduce to 4×4 times a volume of the analyzed data without losing any information.

The two-dimensional spatial correlation function is given in Figure 8.3, its contour lines in Figure 8.4.

Table 8.1: Count of the GATE1 observed spatial fields within each hour.

Hour	\multicolumn{19}{c}{Day}

Hour	179	180	181	182	183	184	185	186	187	188	189	190	191	192	193	194	195	196	197
0	3	4	3	4	4	4	4	4	4	4	4	4	4	4	4	4	4	4	4
1	4	4	4	4	4	4	4	4	4	4	4	4	4	3	4	3	4	4	4
2	4	4	4	4	4	4	4	4	4	4	4	4	4	3	4	4	4	4	4
3	4	4	4	4	4	4	4	4	4	4	4	4	3	3	4	3	4	4	4
4	4	4	4	4	4	4	4	4	4	4	4	4	4	4	4	4	4	4	4
5	4	4	4	4	4	4	4	4	4	4	4	4	4	4	3	4	4	4	4
6	4	3	4	1	4	4	4	4	4	4	4	4	4	4	3	4	4	4	4
7	4	4	4	0	4	4	4	4	4	4	4	4	4	4	3	4	4	4	4
8	4	4	4	0	4	4	4	4	4	4	4	4	4	4	4	4	4	4	4
9	4	4	4	0	4	4	4	4	4	4	4	4	4	4	4	4	4	4	4
10	4	4	4	0	4	4	4	4	4	3	4	4	4	4	4	4	4	4	4
11	4	4	3	0	4	4	4	4	4	4	3	4	4	4	4	4	4	4	4
12	3	0	1	3	4	2	3	4	3	4	2	4	4	4	4	4	4	4	4
13	4	1	2	4	4	4	4	4	4	4	4	4	4	4	4	4	4	4	4
14	4	4	4	3	3	4	4	4	4	4	4	4	4	4	4	4	4	4	4
15	4	4	4	4	4	4	4	4	4	4	4	4	4	4	4	4	4	4	4
16	3	4	4	3	4	4	4	4	4	4	4	4	3	4	4	4	4	4	4
17	4	4	2	4	4	4	4	1	4	4	1	3	2	3	4	4	4	4	4
18	0	4	4	4	4	4	4	4	4	4	2	4	4	4	4	4	4	4	0
19	3	4	4	4	4	4	4	4	4	4	4	4	4	4	4	4	4	4	0
20	4	4	4	4	4	4	4	4	4	4	4	4	4	4	4	4	4	4	0
21	4	4	4	4	4	4	4	4	4	4	4	4	4	4	4	4	4	4	0
22	4	4	4	4	4	4	4	4	4	4	4	4	4	4	4	4	4	4	0
23	4	3	4	4	4	4	4	4	4	4	4	4	4	4	4	4	4	4	0

Table 8.2: Some general characteristics of the entire GATE1 data and 12-hour time-step sample.

Characteristics	GATE1 data	12-hour sample
Grand mean (mm h^{-1})	4.419	4.200
Standard deviation	7.092	6.932
Number of non-zero observations	1466271	24201
Percentage of non-zero observations	9	8

The isolines reflect the predominance of the transport of rain rate fluctuations along the west-east line that match the illustration (Hudlow, 1979) of the surface streamlines of the mean wind directions in the time and area of the GATE1. Of course, these isolines cannot point out the direction (west or east) of this transport. Therefore, the spatial dependence of rain rate fluctuations clearly show that the synoptic transport of rain along the parallels is reflected in the second moments and that the corresponding powerful propagating waves dominate over all other waves. Meridional propagation is significantly weaker. Although the prevalence of such transport of rain was expected, its domination of the two-dimensional second-moment statistics is not a trivial result.

The correlation function also shows that, for distances of about 50 km, the spatial correlation coefficient damps to values of about 0.3–0.38. Positive correlations, which can be traced up to distances of slightly more than 200 km into the parallel direction and up to distances of about 100 km into the meridional direction, show that the spatial rain rate field is anisotropic. The anisotropy appears especially clear for distances greater than 35 km, where the values of the correlations into the NS line damp to zero faster than they do along the WE line. However, for distances of less than 35 km, the anisotropy is hardly distinguishable.

This statement can be clearly illustrated by taking cross-sections of the above correlation function (Figure 8.3) at the two vertical

8.1 GATE Observations

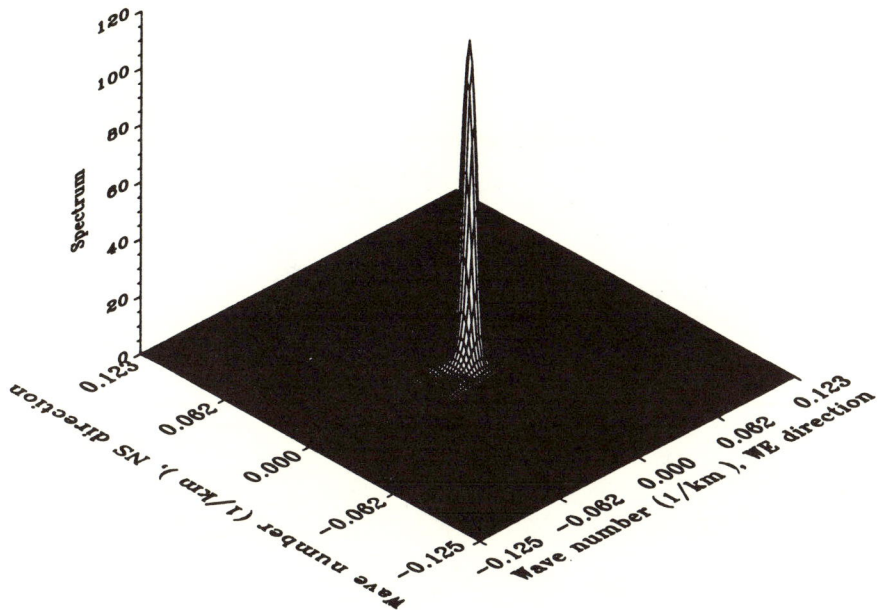

Figure 8.2: Spatial spectrum of the rain rate field obtained by averaging the GATE1 data over the time domain.

planes corresponding to the zero lags. The two resulting one-dimensional correlation functions (Figure 8.5) into the meridional and parallel directions, respectively, show that, for distances of about 35 km, their values are very close and can be approximated, for example, by the exponential correlation function $\rho(r) = 0.89^r$, where r is lag in km.

Therefore, for lags less than 35 km, the deviation of the statistical properties of the spatial rain rate field from the isotropic one is not significant. For lags greater than 35 km, the difference in the estimates of these one-dimensional correlation functions grows markedly, reflecting the real distinction of the statistical character of the WE and NS transports of rain rate fluctuations.

In order to find a better stochastic model of the spatial field considered, the approximation of the above one-dimensional correlation functions by the correlation functions of the second- and third-order autoregressive processes was done. Because the results are not much better than those for the first-order model with the correlation function $\rho(r) = 0.89^r$, they are not shown.

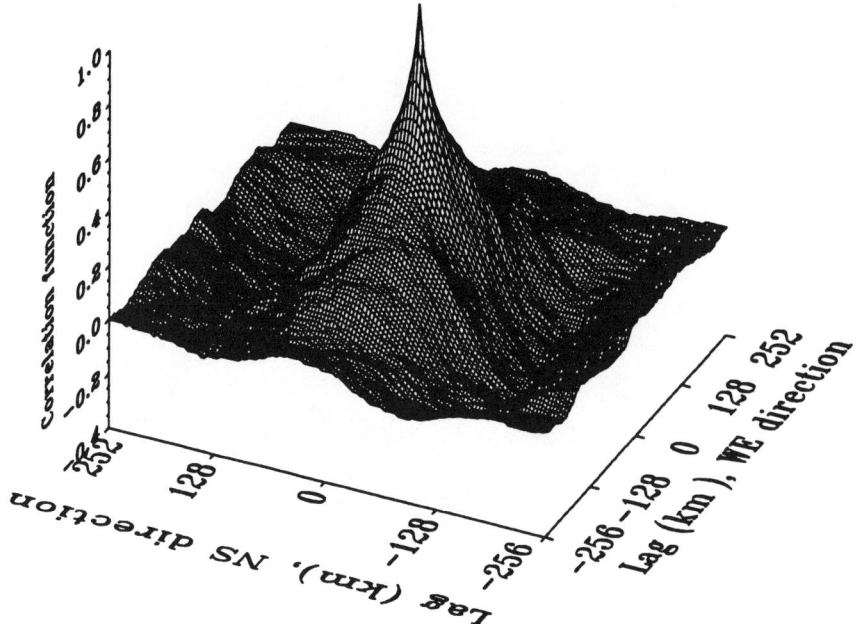

Figure 8.3: Spatial correlation function of the rain rate field obtained by averaging the GATE1 data over the time domain.

One other attempt at one-parameter modeling by the Laplace equation in accordance with the methodology of a paper by Jones and Vecchia (1993) should be mentioned (see Subsection 3.8.5). The two correlation functions of the solution of the Laplace equation (3.100), together with the above one-dimensional estimates, are given in Figure 8.6.

One can see that the correlation function of this model proves relatively close to the estimates only for small distances. On the whole, as follows from comparison of Figures 8.5 and 8.6, for distances of less than 35 km the concave shape of the correlation function of the first-order AR process is slightly more appropriate than the correlation function of the solution of the Laplace equation.

According to Jones and Vecchia (1993), some anisotropic fields can be transformed to the isotropic form by rotating the coordinates by an appropriate angle α and by scaling the two coordinate axes differently (see Subsection 3.8.5). However, careful consideration of the isolines in Figure 8.4 reveals that the anisotropic character of the rain rate field is very complicated and that it is impossible to transform it into an isotropic one by rotating and scaling coordinates.

8.1 GATE Observations

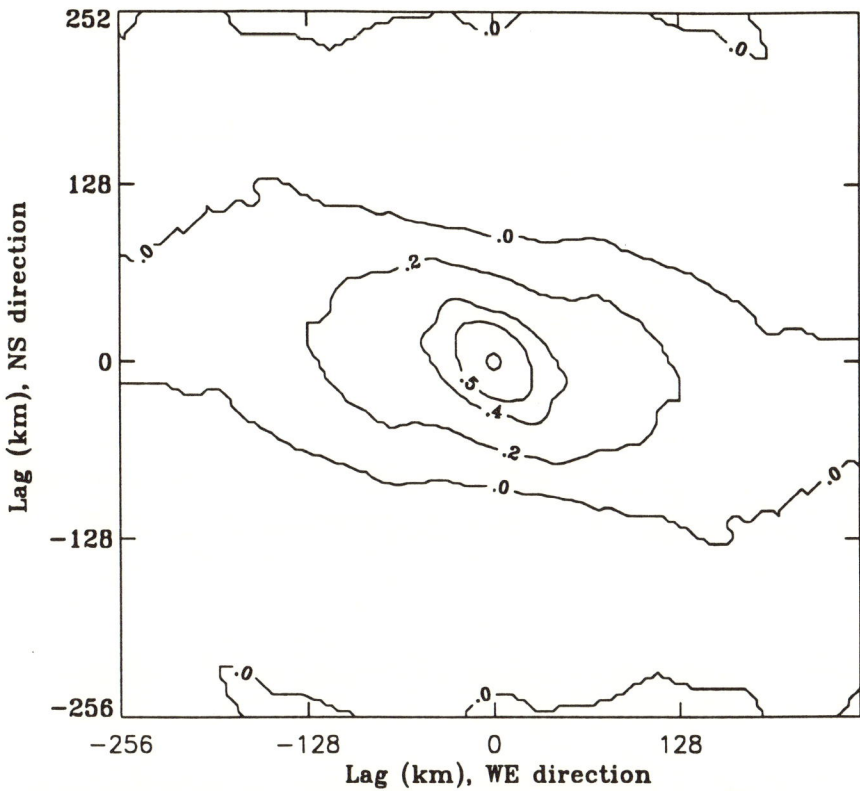

Figure 8.4: Contour lines of the spatial correlation function in Figure 8.3.

One can show that the angle α and the scaling parameter are not constant and that they differ for different points of the field. For example, the α is almost zero for distances of less than 35 km and grows significantly with increasing lag.

But both models considered are very simple (one parameter), so for small distances the estimates of these parameters can be used in comparing characteristics of rain obtained by different kinds of measurements.

8.1.4 Ground Truth Problem

Any measurements of rain present different samples from the unknown distribution of rain considered as a multidimensional random function. The finite and discrete character of the samples limits the

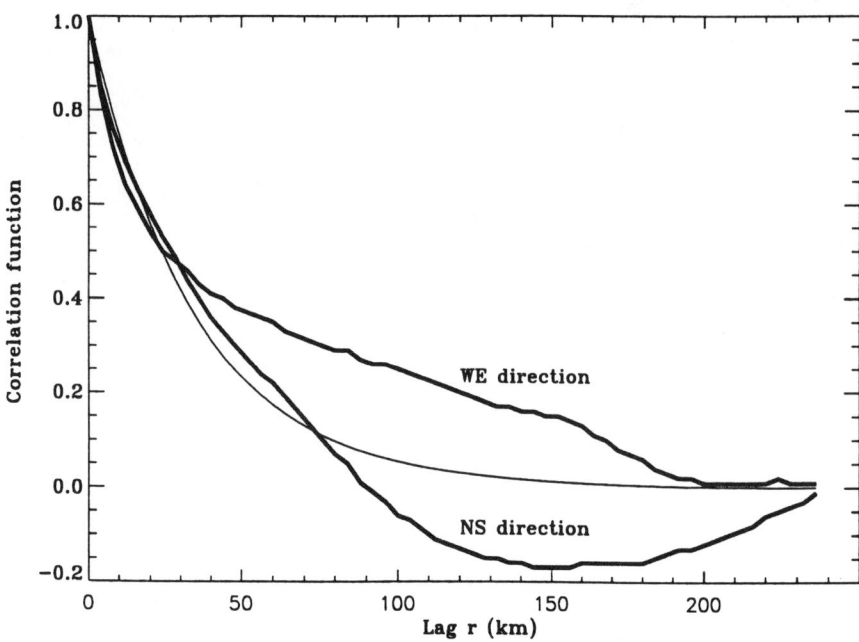

Figure 8.5: The GATE1 one-dimensional correlation functions into the NS and WE directions and the function $\rho(r) = 0.89^r$.

possibility of an accurate statistical description of rain in the spatial and temporal domains. Specifically, according to the Nyquist theorem, the time step of the observations limits the high frequencies that can be studied, while the sample size limits the low-frequency resolution.

There are many different possibilities for transforming the GATE1 observations into different time series and fields by spatial and temporal averaging and/or sampling. For a comparative study of the dependence of variability on the time step and the volume of data, a sample (which imitates satellite data) of spatial fields with a time step of 12 hours was selected from (8.1) and considered separately. The chosen fields correspond to exactly midnight and noon.

The counts, analogous to those given in Figure 8.1, of the non-zero observations in this sample for each spatial point of GATE1 area show that the real number of readings with rain varies from zero to 15 and that there is very little data for the northern region. Some of the statistical characteristics of the entire GATE1 data and the

8.1 GATE Observations

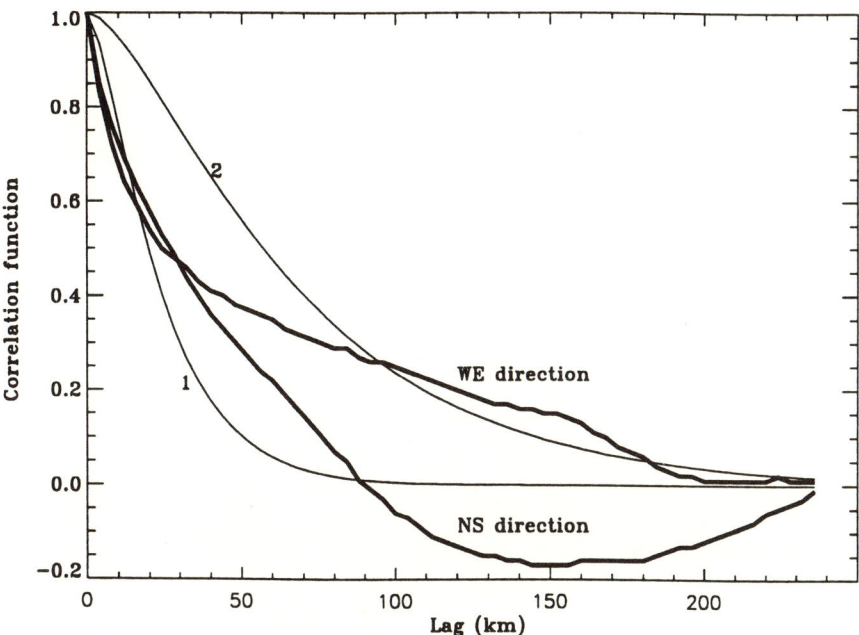

Figure 8.6: The GATE1 one-dimensional correlation functions into the NS and WE directions and the correlation functions of the solution of the Laplace equation, where $a = 0.0041$ (1) and 0.0005 (2).

12-hour time step sample are presented in Table 8.2. The estimates of their grand means, given in the first row of Table 8.2, are relatively close (4.4 mm/hr and 4.2 mm/hr), and it is easy to show that the 99% confidence interval for the mean of the distribution based on the entire GATE1 data is (4.404 to 4.434), while that for the 12-hour sample is (4.085 to 4.315). The small difference (0.2 mm/hr) of the above two means can be interpreted as bias of the 12-hour sample mean. The cause of these particular distinctions will be discussed shortly. Of course, accurate statistical inference is not completely appropriate in the above comparison because the samples are statistically dependent, but such inference will be valid when dealing with real satellite observations, which are statistically independent of ground data. The statistical significance of the distinctions in the standard deviation estimates (7.092 and 6.932) can also be tested in the case of independent samples.

Let us compare some of the results presented in Figures 8.2 and 8.3 with the analogous estimates for the field obtained by averaging

314   8 Second Moments of Rain

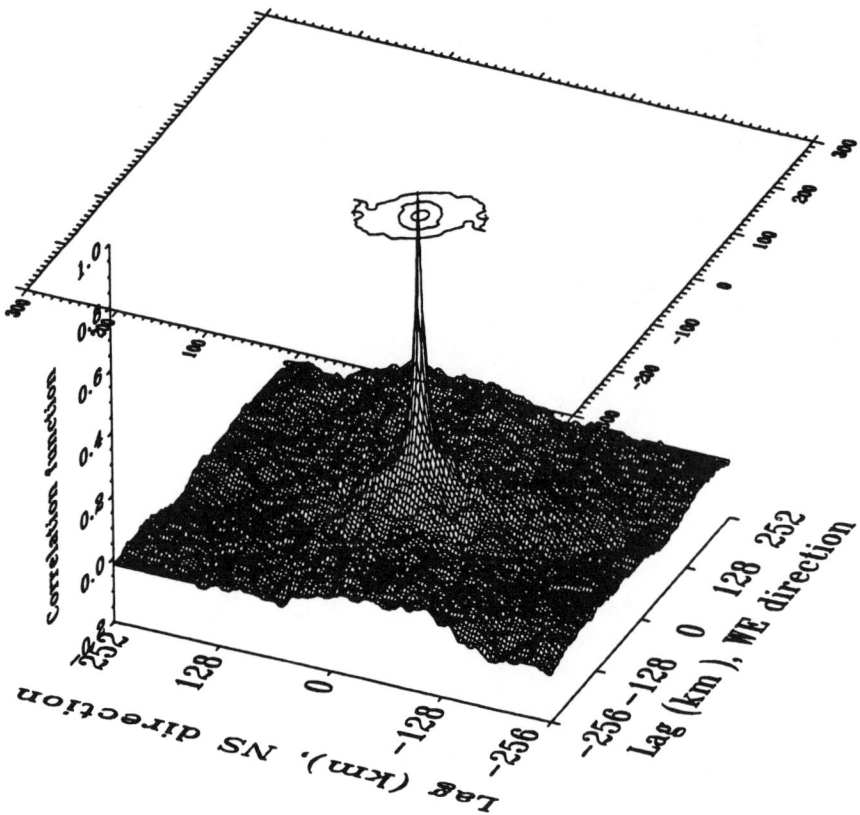

Figure 8.7: Spatial correlation function of the rain rate field obtained by the 12-hour time step sample averaging over the time domain.

the 12-hour time step sample over the time domain. The shape of the spectrum, which is not cited here, is similar to the spectrum in Figure 8.2, but the maximum in the origin is about three times smaller than that for the entire data.

The comparison of the correlation functions (Figure 8.7 with Figure 8.3) shows that the spatial statistical dependence of the averages of the 12-hour sample is significantly smaller, as expected, or that damping is faster. Positive correlations can only be traced up to distances of less than 100 km.

For example, for lags equal to 4 km, correlations in Figure 8.3 are equal to about 0.83–0.85, while for the 12-hour sample such correlations are only about 0.64–0.56. The values are different, but if the samples were independent, it would be possible to test a statistical hypothesis and make inferences about differences in the correlations of the distributions.

8.1 GATE Observations

Another feature of the 12-hour sample that should be mentioned is that instead of seeing the prevalence of the transport along the parallels we see the predominance of southeastward propagation. This distinction could possibly be connected with the diurnal cycle of the wind direction in this area in the time of the experiment. The corresponding one-dimensional correlation functions (which are not cited here), along the WE and NS lines, are close to each other, and they can be approximated by the exponential curve 0.7^r. But this model is valid only for distances of about 20 km. Modeling by the Laplace equation (see Subsection 3.8.5) is also appropriate only for the same distances, but the accuracy in this case is slightly lower than that of the above exponential approximation.

Concluding the comparative consideration of the statistics of the two rain rate spatial fields makes it possible to assert that the values of their spectral and correlation estimates differ markedly. Therefore, the 12-hour time step sample, an imitator of the satellite data, is not statistically representative for the rain rate field in the sense that its second-moment spatial statistics differ significantly from the corresponding estimates, which were obtained using the entire data set.

8.1.5 Spatial-Temporal Fields

8.1.5.1 Latitude-Temporal Field

Averaging data set (8.1) along the north-south direction over the subscript i gives the two-dimensional spatial-temporal rain rate field, which can be considered as a latitude-temporal field. The size of the field is 100×1824.

The central part of the spectrum of the normalized values of this field (Figure 8.8) shows a concentration of almost all power in the narrow wave number band along the frequency axis, which means that the observed variations of data take place mainly in time.

The primary feature of the spectrum is a significant crest over the frequency axis with a wide maximum that peaks in the origin. The maximum reveals that the data has a spatial-temporal trend, which can be associated with the trend of the non-zero observation number along the west-east direction and, possibly, with the seasonal cycle. On the background of the main crest, there are several smaller ones, the most noticeable and interesting of which is a lowering crest corresponding to a period of about 24 hours. It is spread in the domain of the positive (negative) frequencies and wave numbers quadrant along almost all wave numbers with a non-zero power. Moreover, it

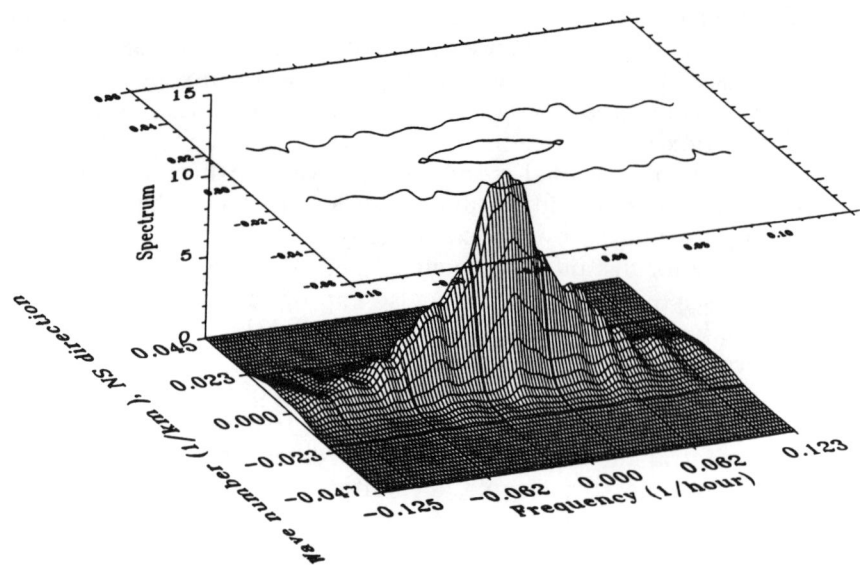

Figure 8.8: Central part of the spatial-temporal spectrum of the rain rate field obtained by averaging the GATE1 data along the NS direction.

can be seen that this diurnal maximum is more prominent for the non-zero wave numbers than for the zero wave number. This feature explains the cause of difficulties in detecting the diurnal cycle using spectral analysis of univariate time series. The standard deviations of the non-zero spectral estimates in Figure 8.8 vary from point to point with approximate values in the interval (0.03–0.5); therefore, the spectral maximum of a diurnal cycle is statistically significant.

The two-dimensional spatial-temporal correlation function (Figure 8.9) reveals that the temporal correlation for the 15-min lag is equal to 0.77 and that the spatial correlation for the 4-km lag along the west-east line is 0.81.

8.1.5.2 Longitude-Temporal Field

Averaging data set (8.1) along the west-east direction over the subscript j gives the two-dimensional spatial-temporal rain rate field, which can be considered as a longitude-temporal field of the same size as the previous one. The spectrum and correlation function (which are not cited here) shapes of this field reveal their similarity with those given in Figures 8.8 and 8.9. But it can be mentioned that

8.1 GATE Observations

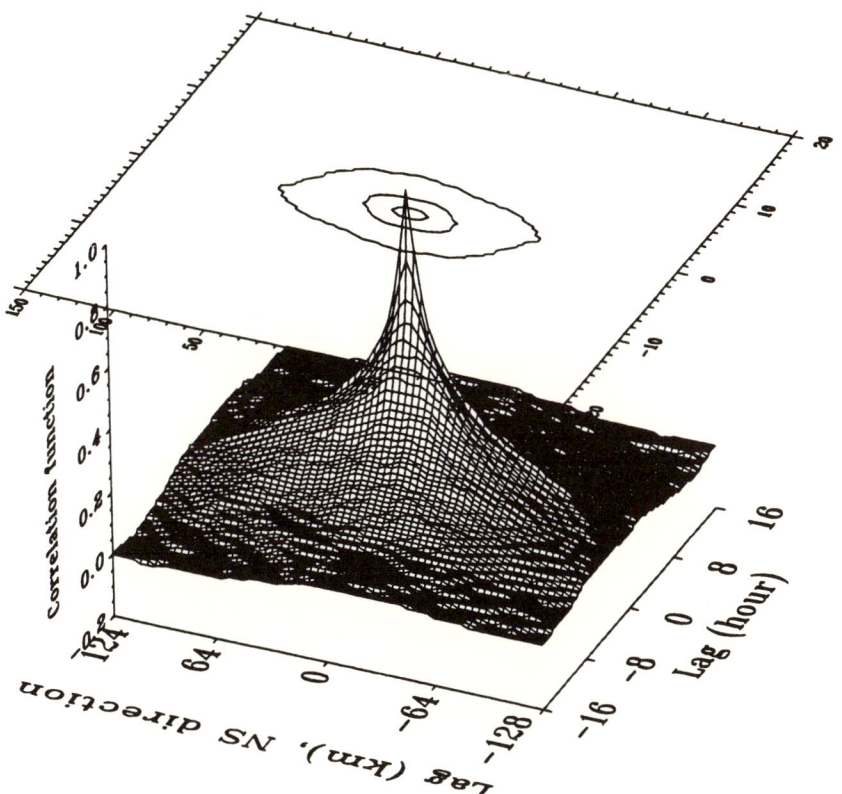

Figure 8.9: Spatial-temporal correlation function of the rain rate field obtained by averaging the GATE1 data along the NS direction (levels are 0.2, 0.4, and 0.6).

the peak of the spectrum in the origin is more powerful than that for the previous latitude-temporal field. The most plausible reason for this is the latitudinal trend of the amount of data. A small lowering crest corresponding to the diurnal frequency (0.042) can be seen here as well, although the transport of rain rate fluctuations is occurring primarily along the WE line.

For a more descriptive comparison, the univariate autocorrelation functions of the cross-sections by the plane with the zero spatial lags of the above two spatial-temporal correlation functions were considered separately. It was found that these autocorrelation functions can be approximated by a first-order autoregressive model [with correlation function $\rho(\tau) = 0.35^\tau, \tau$ is hours] only for lags within the interval of one hour. An attempt to fit the second- and third-order

AR models was not successful. In order to accurately approximate the temporal correlation functions on the 10 to 15 hour lag interval, the order of the AR model must be greater than 10. The most significant features of the two spatial-temporal spectra are the lowering crests corresponding to the diurnal cycle frequency. This means that the two-dimensional spectral analysis shows the existence of a diurnal cycle in the GATE1 rain rate observations with a small amplitude.

To gain insight into both the extent to which the results reflect reality and the reliance upon applied methodology, one must realize that the periodograms of spatial-temporal rain rate fields have a very complicated random character with many discrete lines and peaks for randomly spread frequencies and wave numbers. The smoothing procedure leads to the formation of one powerful peak (in the origin) and a few other small, but noticeable, spectral crests, one of which can be interpreted as a reflection of the diurnal cycle. The most likely cause for the maximum being at the origin, as already mentioned, is the trend of data imposed by the nonuniform number of observations (see Figure 8.1), the seasonal cycle, and the natural latitudinal variability. The most likely cause of other tiny maxima and crests are the sample variability.

8.1.6 Diurnal Cycle

Now, we can look at the diurnal cycle estimates obtained by spatial and temporal averaging over all non-zero observations corresponding to each of the 96 fixed moments of the day. The estimated curve (Figure 8.10) has a relatively simple sine wave shape with minimum values during the night and maximum values during the afternoon. The range of variation (3.4 to 5.4 mm/hr) is small compared with the standard deviation (about 7 mm/hr) of observations; this is the cause of difficulties in its detection in the spectral estimates.

The diurnal cycle estimate accuracy is relatively high; their standard deviations are less than 0.1 mm/hr.

The comparison and verification of the GATE1 diurnal cycle estimates with the results derived from other sources of data must be done very carefully because the diurnal cycle can have a different shape in different regions and its seasonal and latitudinal trends may obscure other patterns. But it is interesting that, in spite of the obvious fact that the estimates in Figure 8.10 reflect the features of the GATE1 sample, the rain rate diurnal cycle shape very closely follows the surface temperature diurnal cycle shape (see, for example, Schmugge, 1977).

8.1 GATE Observations

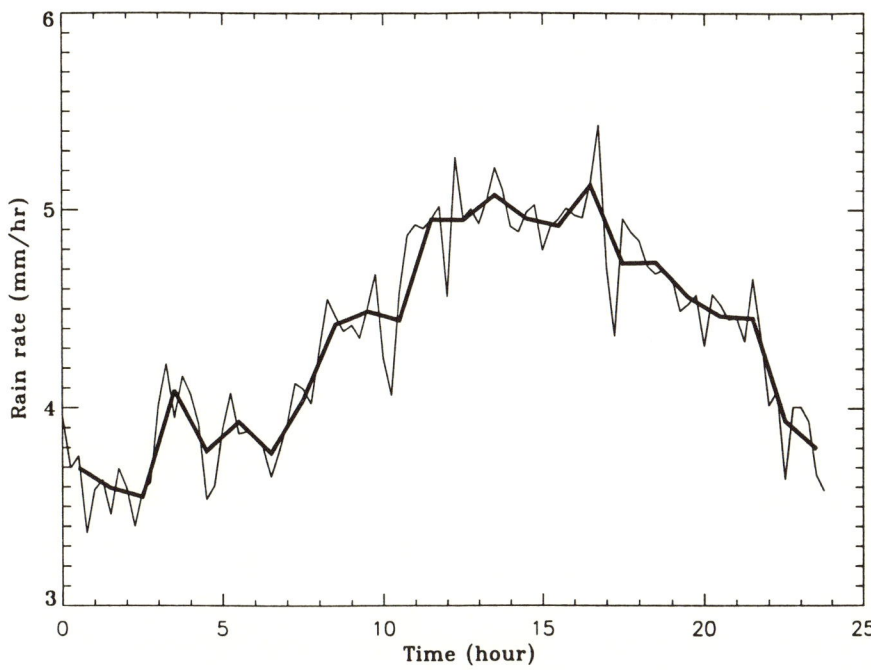

Figure 8.10: Estimates of the diurnal cycle of the GATE1 rain rate data and their averages within each hour.

The practical importance of taking the diurnal cycle into account when solving different engineering problems can be shown in the framework of the ground truth problem described above. Thus, averaging the two separate estimates, corresponding to midnight and noon (the two moments of the day for which the readings of the 12-hour sample were taken) on a curve in Figure 8.10 gives exactly the 12-hour sample mean estimate of 4.2 mm/hr (see Table 8.2).

Averaging all 96 estimates in Figure 8.10 gives the entire GATE1 sample mean of 4.419 mm/hr. The difference between these two estimates is equal to the above-mentioned bias, and it is completely conditioned by the presence of the diurnal cycle in the GATE1 data set. It is clear that such a bias can have nothing in common with the distinction in the probability distributions of errors of different kinds of rain measurements.

Let us associate the rain rate diurnal cycle with the diurnal cycle of number of non-zero observations. Such a diurnal cycle (Figure

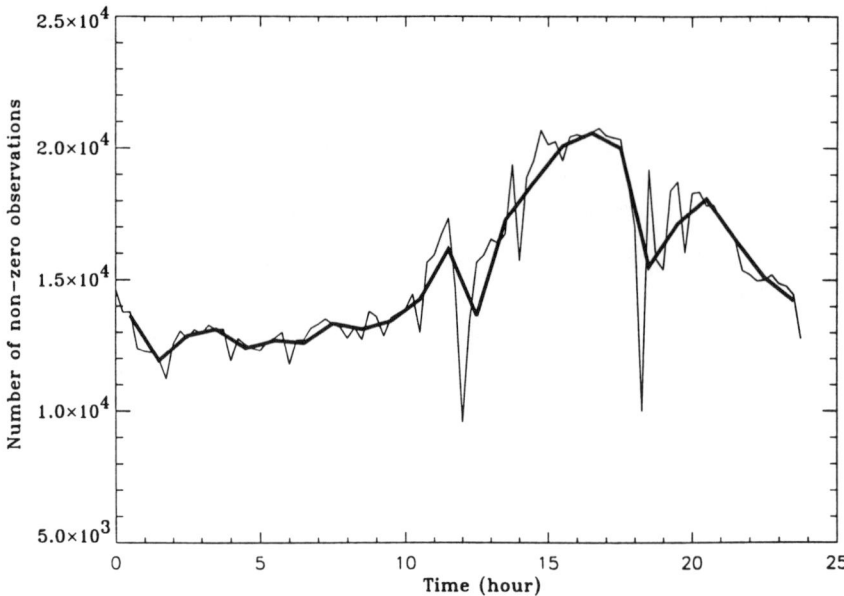

Figure 8.11: Diurnal cycle of the number of non-zero observations of GATE1 data and the averages within each hour.

8.11) shows the reliability of the results as well as the approximate character of the diurnal cycle of rainfall because, on average, the greater the number of non-zero observations, the greater the amount of precipitation. With the estimates of Figure 8.10 in mind, it seems natural to expect that the maximum of the number of non-zero observations must correspond to the afternoon hours. Of course, relatively small lunch and dinner minima on the curve in Figure 8.11 have nothing in common with rain statistics, but they do reveal the remarkably consistent character of data from the GATE1 experiment.

8.1.7 GATE2 Statistics

The source of GATE2 rain rate observations is the radar and surface measurements that took place from July 28 to August 15 of 1974 at the same region of the Atlantic Ocean. Table 8.3 gives a general representation about the volume of observations in the corresponding data base. There are almost three days with entirely missing data, and the total number of non-zero observations equals only two-thirds of the corresponding number of the GATE1 data.

8.1 GATE Observations

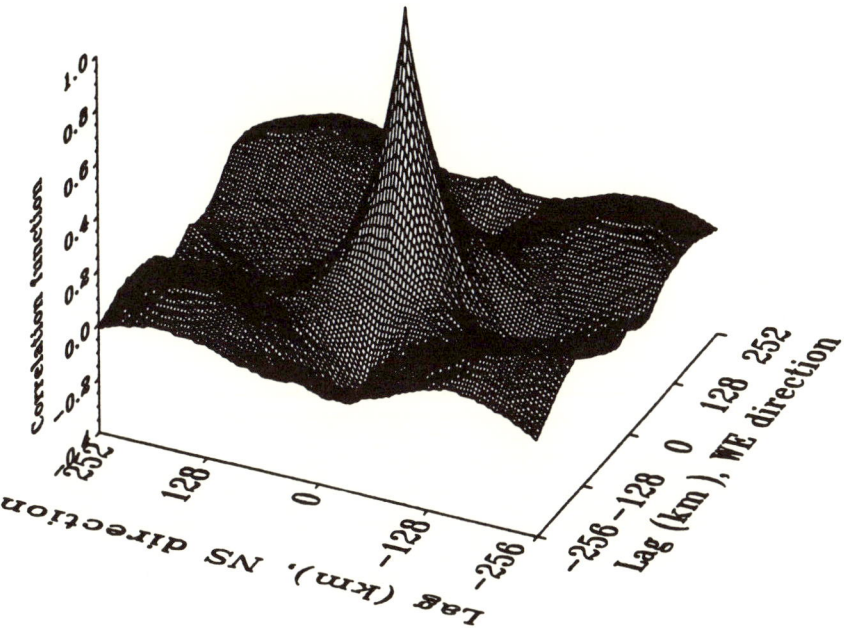

Figure 8.12: Spatial correlation function of the rain rate field obtained by averaging the GATE2 data over the time domain.

This inevitably leads to lower accuracy of the second-moment estimates. Nevertheless, some of them will be very briefly discussed. The shape of the spatial spectrum (which we do not show here) of the two-dimensional spatial GATE2 field is approximately the same as that of Figure 8.2, but the peak in the origin of the wave number coordinates is smaller.

Figure 8.12 shows the corresponding spatial correlation function, analogous to that given in Figure 8.3. It is hard to believe that such a beautiful structure reflects the properties of the most variable random process of nature rather than the result of simulation by a purely artificial deterministic model. It leaves hope that an accurate description of rain rate in the form of an interpretable physical model will be found. The anisotropy of the GATE2 spatial rain rate field carries a different character than that of the GATE1 data; specifically, the predominant transport of rain occurs along the NW–SE line rather than along the parallels, as it did for the GATE1 data. Moreover, the values of the correlations in both directions for the

Table 8.3: Count of the observed GATE2 spatial fields within each hour.

Hour	209	210	211	212	213	214	215	216	217	218	219	220	221	222	223	224	225	226	227
0	4	3	3	0	0	0	4	4	4	3	2	4	4	3	4	4	4	4	4
1	4	4	4	0	0	0	4	4	4	4	4	4	4	4	4	4	4	4	4
2	4	4	4	0	0	0	4	4	4	4	4	4	4	4	4	4	4	4	4
3	4	4	4	0	0	0	4	4	4	4	4	4	4	4	4	4	4	4	4
4	4	4	4	0	0	0	4	4	4	4	4	4	4	4	4	4	4	4	4
5	4	4	4	0	0	0	4	4	4	4	4	4	4	4	4	4	4	4	4
6	3	4	4	0	0	0	4	4	4	4	4	4	4	4	4	4	4	4	4
7	4	4	4	0	0	0	4	4	4	4	4	4	4	4	4	4	4	4	4
8	4	4	4	0	0	0	4	4	4	4	4	4	4	4	4	4	4	4	4
9	4	4	4	0	0	0	4	4	4	4	4	4	3	4	4	4	4	4	4
10	4	4	4	0	0	0	4	4	4	4	4	4	4	4	4	4	4	4	4
11	4	3	4	0	0	0	4	3	4	4	4	4	4	4	3	4	4	4	4
12	4	4	4	0	0	0	4	4	4	4	4	4	4	4	4	4	4	4	4
13	4	4	4	0	0	0	4	4	4	4	4	4	4	4	4	4	4	4	4
14	4	4	4	0	0	0	4	4	4	4	4	4	4	4	4	4	4	4	4
15	4	4	4	0	0	0	4	3	4	4	4	4	4	4	4	4	3	4	4
16	4	4	4	0	0	0	4	4	4	3	3	4	4	4	4	4	4	4	4
17	3	4	3	0	0	1	4	3	4	2	2	4	4	4	4	4	4	4	4
18	4	4	4	0	0	3	4	4	4	2	3	4	4	4	4	4	4	3	4
19	4	4	4	0	0	4	4	4	3	4	4	4	4	4	4	3	4	4	4
20	4	4	3	0	0	4	4	4	4	4	4	4	4	4	4	4	4	4	4
21	4	4	1	0	0	4	4	4	3	4	4	4	4	4	4	4	4	4	0
22	4	4	0	0	0	4	4	4	3	4	4	4	4	4	4	4	4	4	0

8.1 GATE Observations

GATE2 spatial field are slightly greater than those for the GATE1 spatial field (Figure 8.3).

In general, the shapes of the spatial-temporal spectral and correlation characteristics of the GATE2 data (the illustrations of which are not provided here) are similar to those of the GATE1 data in Figures 8.8 and 8.9. However, the details are different. First, instead of having a spectral maximum for the 0.042 frequency value (the 24-hour period), there are a couple of small peaks for the slightly lower and slightly greater frequencies. Therefore, it makes no sense to evaluate the corresponding diurnal cycle of rain rate for the GATE2 data.

Comparing the diurnal cycles of the number of non-zero observations for both phases of the experiment shows that the volume of GATE2 data equals approximately 60% of the volume of GATE1 data. The cause of this is not simply the three days with entirely missing data. The lunch and dinner minima of the GATE2 data are wide, and the corresponding losses of information are significant for only the 19-day interval of observations, especially for evaluation of the temporal characteristics. That is one of the reasons a diurnal maximum was not found in the corresponding spectral estimates.

8.1.8 Multivariate Models

Fitting linear stochastic processes to the multiple rain time series leads to the creation of a stochastic rain model, which has independent meaning as one possible approach to its description.

In the case of rain observations, the goal of modeling is not, of course, forecasting in the sense of predicting the onset or the end of rain. Our intention is to build rain rate models and to use them for diagnostic analysis of parameter matrices in terms of spatial-temporal interactions and feedbacks of observed processes.

Physical interpretation of parameters of multivariate stochastic models of spatially and temporally averaged data can provide the basis for a statistical description of a transport of rain fluctuations. To solve this problem, one must take into account a high level of white noise associated with fluctuations measured at separate points and attempt to reduce it. The signal detection may be enhanced through spatial and temporal averaging of data with a consecutive fitting of multivariate autoregressive models to the resulting time series; so averaging must be done to present the rain rate fields employed in the form of multiple time series. A stochastic model of rain rate fluctuations will generally be sensitive to the scales of spatial and

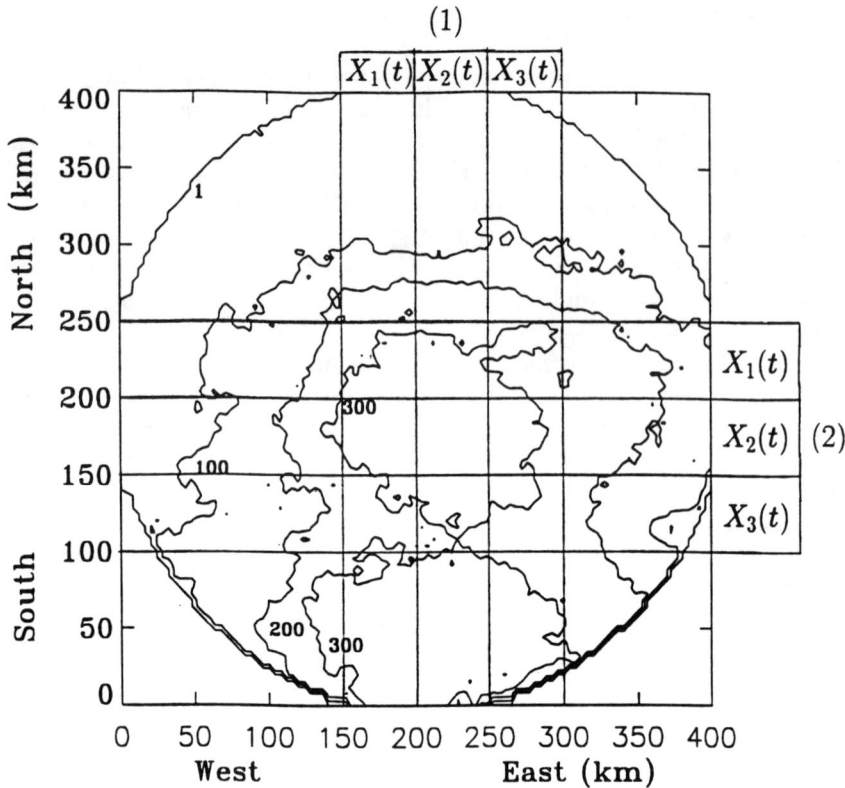

Figure 8.13: Contour lines of the number of non-zero observations of the GATE1 rain rate data base and the specific spatial bands of averaging for multivariate modeling.

temporal averaging. In the process of this work, several multivariate rain rate models corresponding to different scales of spatial and temporal averaging were constructed and analyzed, but only the two simplest models are presented here.

The data was averaged temporally within an hour (the temporal step employed is 60 minutes). Because of the distinctions in the statistical characteristics of the rain rate field in the NS and WE directions, the two different spatial averaging strategies, illustrated in Figure 8.13, were chosen.

The first strategy (Figure 8.13, 1), averaging within three 50-km width longitude bands with a maximum amount of data, is intended to describe west-east interdependence. The second strategy (Figure

8.1 GATE Observations

Table 8.4: Matrix correlation function for the first two lags of the three-variate latitude-temporal model (one-hour time step).

	u_1	u_2	u_3
$\tau = 0$			
u_1	1.00	0.40	0.23
u_2	0.40	1.00	0.64
u_3	0.23	0.64	1.00
$\tau = 1$			
u_1	0.65	0.45	0.23
u_2	0.34	0.72	0.54
u_3	0.26	0.64	0.77

8.13, 2), averaging within three 50-km width latitude bands with a maximum amount of data, is intended to explore the extent of meridional dependence of rain rate perturbations. Therefore, the time step between the two terms u_{t+1} and u_t of the time series studied is one hour, and the space distance between two adjacent time series is 50 km. Two three-variate models were constructed for each spatial and temporal averaging procedure shown in Figure 8.13.

Here are some of the results. The matrix correlation functions for the first two lags and the matrices of the estimated parameters of these two AR models, as well as the parameters of the corresponding multivariate linear regression models, are given in Tables 8.4, 8.5, 8.6, and 8.7. The statistical dependences presented by the matrices in Tables 8.4 and 8.6 show that the spatial correlation reaches a value of 0.65 (Table 8.4) into the WE directions, and only 0.39 (Table 8.6) into the NS directions. The temporal and spatial-temporal correlations of these data reach a value of 0.77 (Table 8.4) into the WE directions and of about 0.71 (Table 8.6) into the NS directions. These values are statistically significant (at any standard level of significance) and allow us to build a stochastic model.

Table 8.5: Estimates (exceeding the 99% significance level) of the parameters of the three-variate autoregressive and linear regression models for the time series of the first scheme of averaging (Figure 8.13, 1) and one-hour time step (ε is the standard error).

	Autoregressive model			
	u_{1t}	u_{2t}	u_{3t}	ε
u_{1t+1}	0.61	—	—	0.75
u_{2t+1}	0.20	0.44	0.31	0.63
u_{3t+1}	—	—	0.71	0.64
	Regression model			
	u_1	u_2	u_3	ε
u_1	—	0.42	—	0.92
u_2	0.26	—	0.58	0.72
u_3	—	0.65	—	0.77

Each row of Tables 8.5 and 8.7 with parameter estimates is associated with the corresponding autoregressive or regression equation. The values listed in column ε of these tables are the standard errors of the forecasts for one step ahead for the AR(1) model or the standard errors for the regression model.

Consider the parameters (Table 8.5) of the latitude-temporal AR model, which is fitted to the data, averaged as indicated in Figure 8.13, 1. With the results of the two-dimensional correlation analysis (Figure 8.4) in mind, one would expect to find a statistical dependence of the values at any moment upon the past rain rate fluctuations to the west and to the east. In the matrix of the model (Table 8.5) the parameters on the main diagonals are largest and there are only two statistically significant estimates, 0.20 and 0.31, out of the

8.1 GATE Observations

main diagonal. Therefore, the GATE1 rain rate fluctuations are mainly determined by their own past and partially determined by the history of the fluctuations of the closest region to the west and to the east. So, on average, the predominance of east-west transport of rain in the time of the GATE1 finds its reflection in the estimates of the parameters of the AR model when the appropriate spatial and temporal averaging are used to detect the signal.

The linear regression parameter estimates (Table 8.5) show the determining statistical dependence of the GATE1 data on the values in the two closest adjacent bands in any particular moment of time. As can be seen from comparison of the standard errors, the autoregressive model accuracy is slightly higher than the regression model accuracy.

Now the three-variate longitude-temporal model will be considered for the data obtained by the averaging procedure as shown in Figure 8.13, 2. The results (Table 8.7) are simple, and the accuracy of this three-variate model is lower [standard errors $\varepsilon(t+1)$ are greater] than for the latitude-temporal models. In this case, a nonzero estimate out of the main diagonal would be evidence of strong north-south transport of rain. However, the estimates in Table 8.7 do not support such a hypothesis. Only the estimates on the main diagonal are not zeroes. This means that the rain rate fluctuations into the NS directions in the time of GATE1 were mainly determined by its own past, or, in other words, the temporal statistical dependence between the mean rain rate fluctuations of the two adjacent 50-km width bands was not strong enough to cause the corresponding AR model parameter estimates to be statistically significant. So the transport of rain fluctuations observed was mainly realized along the parallels. The NS motions were significantly weaker.

Therefore, the approximation of the multiple time series by the multivariate AR model has the potential to detect the structure of second-moment rain rate fluctuations in spite of the significant variability and noise associated with the observations. The first autoregressive model shows that, on average, the rain rate fluctuations were most strongly dependent upon advection along the west-east line, since the computed matrix of the parameters has a special form with partially non-zero elements outside the main diagonal. It is interesting to estimate this advection velocity when one has only rain rate observations.

It is obvious that the advection of rain is determined by the direction and strength of the wind. Therefore, careful planning is needed for determining what kind of advection will be evaluated.

Table 8.6: Matrix correlation function for the first two lags of the three-variate longitude-temporal model (one-hour time step).

	u_1	u_2	u_3
$\tau = 0$			
u_1	1.00	0.22	0.27
u_2	0.22	1.00	0.39
u_3	0.27	0.39	1.00
$\tau = 1$			
u_1	0.56	0.22	0.29
u_2	0.18	0.44	0.37
u_3	0.25	0.26	0.71

The direction of the transport of rain in the time and area of the GATE1 was not stable and changed several times. As Hudlow (1979) has shown, the surface streamlines of the mean wind directions in the time and area of the GATE1 were curved from the northwest to the east. The purposeful selection of the time interval of the GATE1 observations with the fixed wind direction is complicated by the fact that we have too little data for the appropriate statistical analysis and modeling. In this situation, it was decided, first, to evaluate the *mean* advection velocity separately into the WE and NS directions for the *entire* observational interval of the GATE1, and then to obtain such velocities for separate days (whenever possible). Such a decision determined the strategy of data averaging (see Figure 8.13) for building multivariate autoregressive models and, finally, the values of the estimated mean advection velocities, as well.

8.1.9 Diffusion

One of the appropriate methodologies for analysis of rain advection is an approximation of the observed field by the diffusion equation

8.1 GATE Observations

Table 8.7: Estimates (exceeding the 99% significance level) of the parameters of the three-variate autoregressive and linear regression models for the time series of the second scheme of averaging (Figure 8.13, 2) and one-hour time step (ε is the standard error).

	Autoregressive model			
	u_{1t}	u_{2t}	u_{3t}	ε
u_{1t+1}	0.53	—	—	0.82
u_{2t+1}	—	0.38	—	0.89
u_{3t+1}	—	—	0.64	0.69
	Regression model			
	u_1	u_2	u_3	ε
u_1	—	—	0.21	0.96
u_2	—	—	0.35	0.91
u_3	0.19	0.35	—	0.90

(see, for example, Cahalan et al., 1982; North and Nakamoto, 1989; Polyak et al. 1994). This problem can be solved using the relationship between the AR model and a diffusion equation (see Section 7.5). The parameter estimates (Tables 8.5 and 8.7) give appropriate samples for subsequent estimation of the diffusion equation coefficients because the AR model presents the equations in finite deferences of the same kind as (7.5).

Let us consider the first three-variate AR model. The estimates in Table 8.5 make it possible to assume that $a \approx 0.20$, $b \approx 0.44$, and $c \approx 0.31$. By replacing these values in formulas (7.6)–(7.8) and by performing the computations, one obtains the values of s, v, and q.

The mean advection velocity v for the time of the GATE1 equals about -5.0 km/hr (-1.4 m/sec). A negative value means that the

Table 8.8: Mean advection velocity (km/hr) for different days of the GATE1.

Day	179	180	181	182	183	185	188	189	192	193	194	195
v_{WE}	−64	−8	−50	−5	−15	30	−13	0	—	−33	−13	−23
v_{NS}	0	0	−5	−30	−17	0	2	9	29	—	−50	−8

mean rain rate advection took place into the westward direction.

The diffusion coefficients are $s \approx 25^2$ km^2/hr (417^2 m^2/sec), and $q \approx 0.04$ 1/hr (0.00001 1/sec). It seems that these results are quite reasonable for such computations. Accepting these values as the estimates of the diffusion equation coefficients, one gets the following:

$$\frac{\partial u}{\partial t} \approx (25)^2 \frac{\partial^2 u}{\partial x^2} + 5 \frac{\partial u}{\partial x} - 0.04u + f(x,t). \qquad (8.2)$$

In using this approach, we attempted to evaluate the mean advection velocities into the WE direction for each day of the GATE1. This was possible only for 11 days, for which the volume of the non-zero data was large enough to estimate the corresponding correlations.

Of course, one can unite the observations for the adjacent days with little data, though not at this stage of analysis. The results (Table 8.8) show that in the time of GATE1, the mean daily advection velocity estimates into the WE direction v_{WE} fluctuate from 64 km/hr (18 m/sec) westward to 30 km/hr (8 m/sec) eastward with a predominance of the westward transport of rain.

As the longitude-temporal model (Table 8.7) with zero parameters outside the main diagonal makes clear (without any calculations), the *mean* advection velocity into the NS direction for the time interval of GATE1 was close to zero. The estimates of the mean daily advection velocities into the NS direction (Table 8.8) fluctuate from 36 km/hr (10 m/sec) southward to 50 km/hr (14 m/sec) northward.

We can see that the wind was unstable during the GATE1 experiment; and, as a result, the mean advection velocity in both directions is close to zero. Therefore, it is interesting to estimate the rain advection velocity for the observations with stable direction of wind, which can be accomplished by studying the PRE-STORM precipitation, as shown in the following sections of this chapter.

Notice here that, according to Monin and Yaglom (1973), the advection velocity is practically coincident with the instantaneous flow velocity, that is, with the wind velocity. By simultaneously analyzing the wind and precipitation satellite data, it is possible to verify and compare the results, if one realizes that the estimates of the spatial correlation function of rain (Figure 8.4) show the lines of transport of rain (or the isolines of the predominant wind) in the area, while the AR models and diffusive description give the direction and value of the mean wind velocity of the rain fluctuations.

Further development of this approach (by incorporating the second spatial coordinate into the diffusion equation) will promote the creation of a powerful methodology for describing the transport of rain in a real three-dimensional space. Based on these results, the problem of applying such a methodology to the mutual verification of the rain and wind satellite observations can be formulated.

8.2 PRE-STORM Precipitation

This section presents the results of the two-dimensional spectral and correlation analysis and multivariate modeling of PRE-STORM rainfall gauge data (Polyak et al., 1994).

Most of the previous analyses of rain have been based on the GATE data. There is a need to study other sources of rain observations to ascertain whether there is a difference between the conclusions drawn from GATE analyses and those based upon data from other areas such as subtropical or midlatitude land areas. One interesting experiment provides observations that can be used for such an analysis—the PRE-STORM (Preliminary Regional Experiment for STORM-Central) data from the Kansas/Oklahoma region. During the experiment the stable eastward or southeastward wind that enable us to investigate horizontal propagation of rain anomalies in these directions had been observed.

The intent of this study is, first, to develop a strategy for spatial averaging of rainfall observations, and second, to explore the implications of these results for the estimation of the diffusion equation coefficients.

8.2.1 Data

The PRE-STORM field experiment took place from May through June of 1985. Aggregated five-minute rainfall observations were obtained by the system of 42 Portable Automated Mesonet (PAM)

Table 8.9: Mean characteristics of PRE-STORM rainfall data.

Characteristic	May	June
Mean number of onsets of rain	35	63
Mean duration of rain	0.21hr	0.31hr
Mean interval between onsets of two consecutive rains	38hr	27hr
Mean relative frequency of observations with rain	0.004	0.008
Mean first autocorrelation coefficient (for 5 min lag)	0.52	0.66
Mean second autocorrelation coefficient (for 10 min lag)	0.43	0.44
First autocorrelation coefficient (for 5 min lag) of spatially averaged time series	0.69	0.81
Second autocorrelation coefficient (for 10 min lag) of spatially averaged time series	0.48	0.61

stations, most of which were situated in Kansas, and 42 Stationary Automated Mesonet (SAM) stations situated in Oklahoma. PRE-STORM rain observations were described in detail by Meitin and Cunning (1985) and analyzed by Graves et al. (1993). Consideration in the present study is limited to rainfall data collected at 40 PAM stations as shown in Figure 8.14(a). Missing and spurious data are replaced by zero.

Each of the time series has a very special property: More than 99% of their terms are zero. For a preliminary analysis, some statistics were evaluated separately for May and June data. The estimates for the point gauges of different stations vary significantly. Table 8.9 shows that the rains in June occurred, on average, almost two times more often than in May, and their mean duration was 1.5 times as long. A shortage of data in May is accompanied by higher variability, and it is expected that the accuracy of different June statistics must be higher than May statistics. As the values at the bottom of Table 8.9 show, spatial averaging over all points of rainfall observations leads to an increase of the first two autocorrelations.

8.2.2 Fields of Monthly Sums

Sums of rainfall data were found for each point gauge for May and June separately. The resulting two fields of monthly sums are considered as samples of the two-dimensional random field. The advantages

8.2 PRE-STORM Precipitation

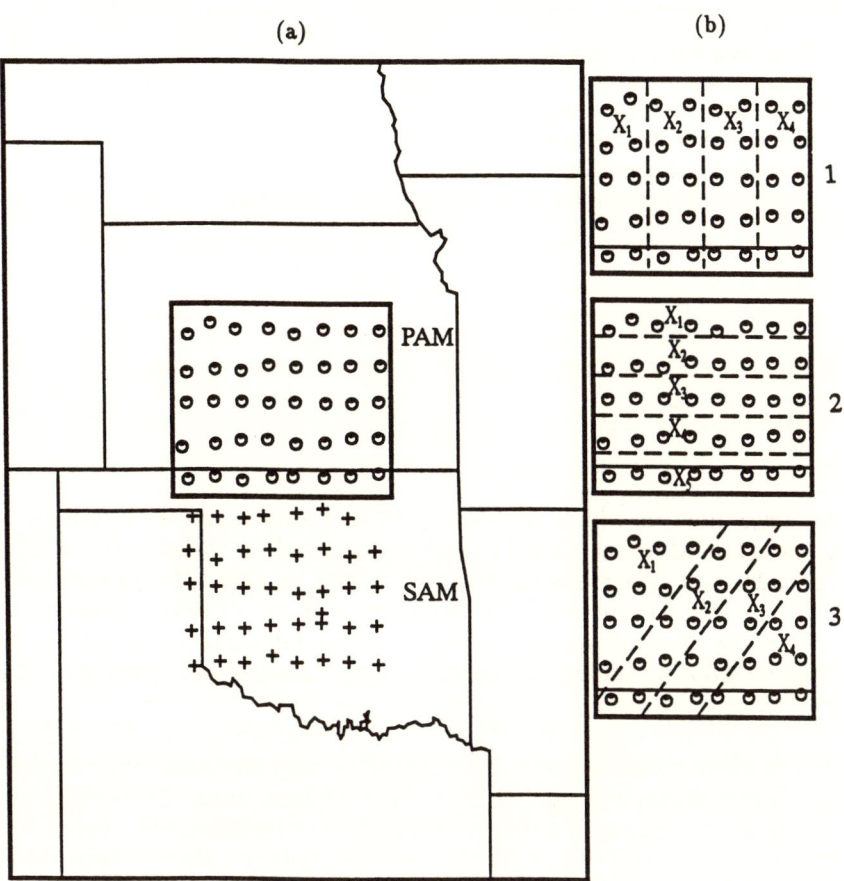

Figure 8.14: (a) PRE-STORM mesonet sites over the central part of the United States. The circles indicates the PAM stations, and the crosses indicates the SAM sites. (b) The specific regions for each spatial-averaging scheme for the PAM stations.

of such consideration are that there are no gaps in the monthly sums and that the standard statistical technique can be applied to their spectral analysis.

Figure 8.15 presents the periodograms of the fields. The values have a random character, but a noticeable concentration of a significant power along the east-west axis can be seen. The estimation of the spectra was done by applying a two-dimensional Tukey filter. Two-dimensional spectra (Figure 8.16) show the main features of the wave number composition of these fields, but their resolutions are too low to reveal the details of the spectral structure.

Only the general picture is clear: Powerful westerly spatial propagation waves dominate all other waves. Meridional propagations are significantly smaller.

Correlation functions (Figure 8.17) show that for distances of about 50 km the space correlation coefficient can reach values of about 0.35 to 0.45.

In spite of limitations in spatial resolution, the spatial dependence of monthly rainfall fluctuations is clear. Thus, the well-known synoptic eastward transport of rain clouds is reflected in the spectral representation of the second moments of monthly sums of rainfall.

This analysis also revealed that the rain data employed are representative samples of a multivariate random process.

8.2.3 Multivariate Models

Interpretations of the parameters of the multivariate stochastic models of spatially and temporally averaged data can provide the basis for a statistical description of a transport of rainfall fluctuations. As in the case of GATE data, the signal detection may be enhanced through spatial averaging and time aggregation of observational data with consecutive fitting of multivariate autoregressive models to the resulting time series.

Three different spatial averaging strategies are illustrated in Figure 8.14(b). The first strategy—averaging within four longitude bands—is intended for the detection of west-east interdependence. The second—averaging within five latitude bands—is intended to explore the extent of meridional propagations of rainfall perturbations across latitude zones. The third strategy is useful for the analysis of southeastward transport. Five (5, 15, 30, 45, and 60 min) temporal aggregation intervals were employed. Fifteen models were constructed from the combinations of spatial averaging and temporal aggregation assumptions. To reduce the seasonal contribution to the

8.2 PRE-STORM Precipitation

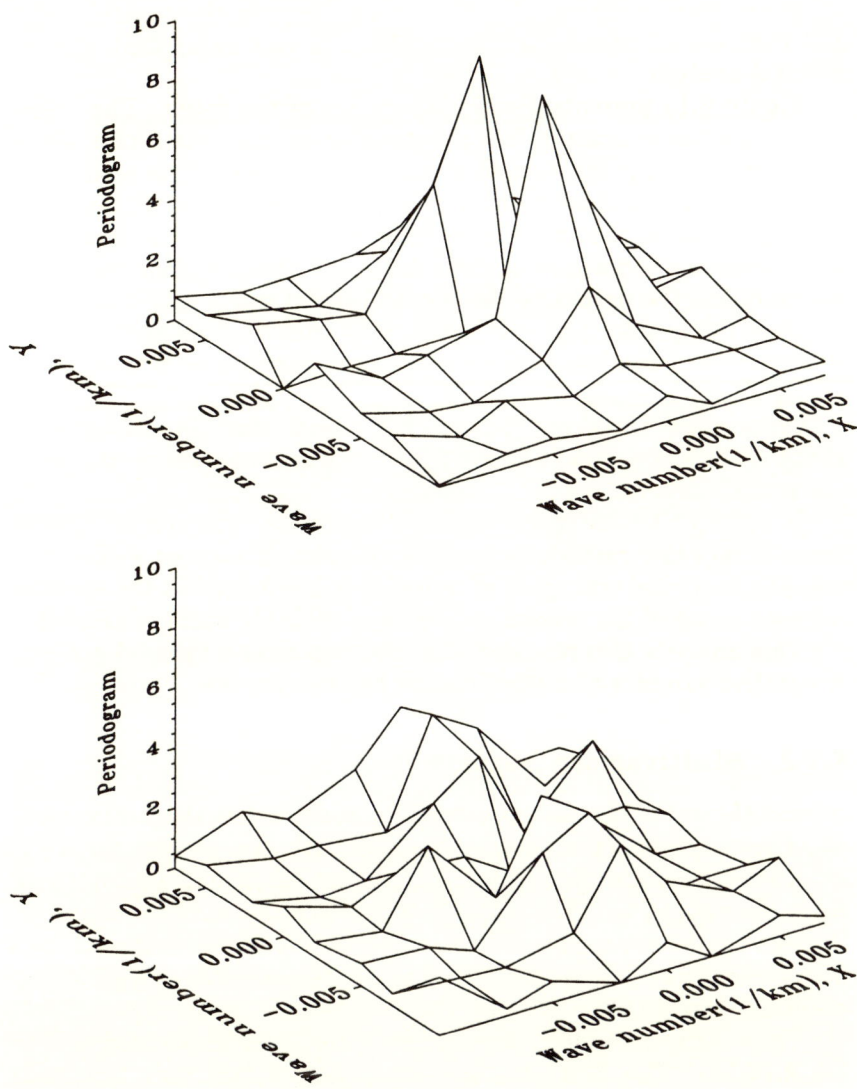

Figure 8.15: Two-dimensional periodograms of monthly sums of May (top) and June (bottom) rainfall fields of PAM stations (X is north-south direction; Y is west-east direction).

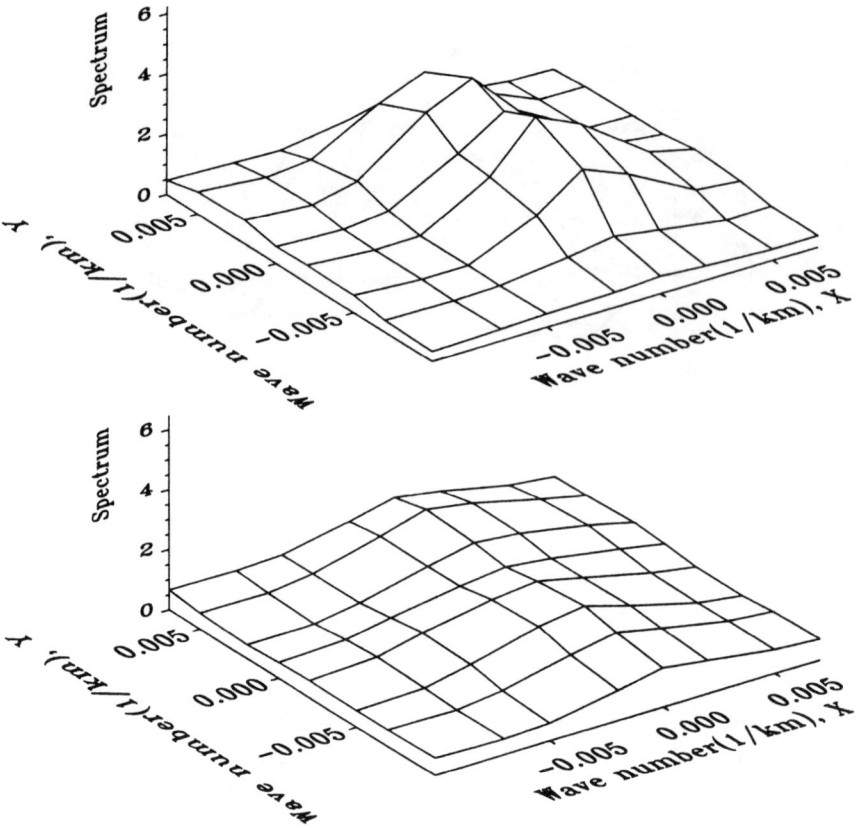

Figure 8.16: Estimates of two-dimensional spectra, computed by smoothing the periodograms shown in Figure 8.15.

rainfall perturbation analyzed, these models were built separately for May and June data. The most interesting results are presented here.

Each particular model is identified by number, month, and aggregation, where "number" corresponds to one of the three strategies in Figure 8.14(b), "month" is May (m) or June (j), and "aggregation" is the aggregation interval. The aggregation time determines the time step between the two terms u_{t+1} and u_t. Matrices of the estimated parameters of all the models are given in Tables 8.10, 8.11, and 8.12. Each row of the tables with parameter estimates is associated with an autoregressive equation.

Consider, for example, model (1,m,5), which is fitted to the data, as indicated in Table 8.10. The values listed in the columns ε are the

8.2 PRE-STORM Precipitation

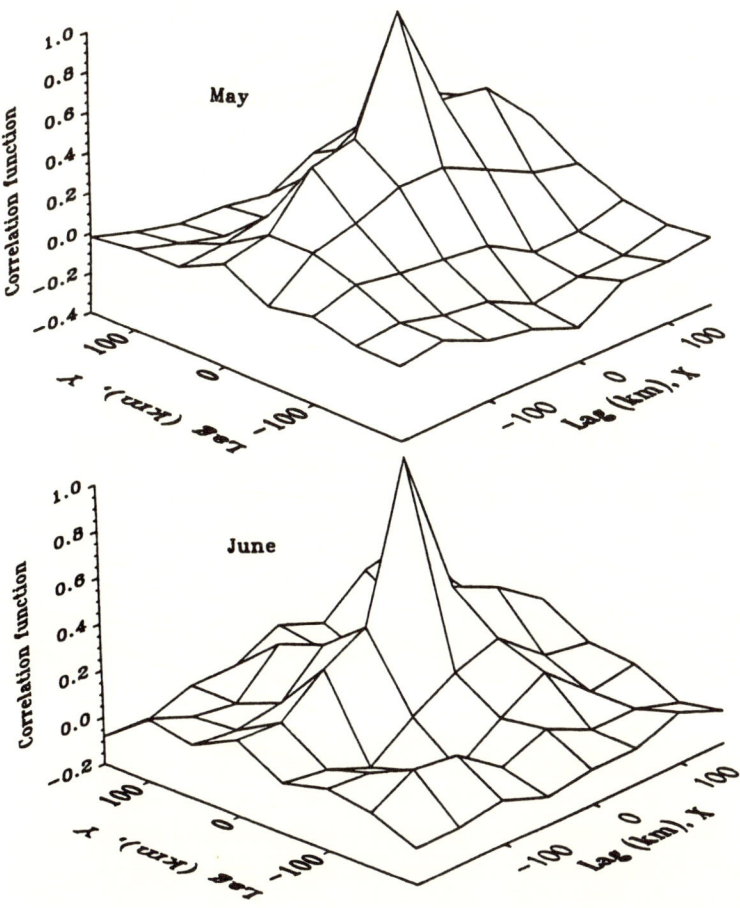

Figure 8.17: Estimates of two-dimensional correlation functions of monthly sums of May (top) and June (bottom) rainfall fields of PAM stations, obtained by Fourier transform of the spectrum fields shown in Figure 8.16 (X is north-south direction; Y is west-east direction).

standard deviations of the forecast errors for one step ahead. The autoregressive equations corresponding to model (1,m,5) are shown below.

$$\begin{aligned}
u_{1t+1} &\approx 0.63u_{1t} + 0.09u_{2t} + 0.04u_{3t} \\
u_{2t+1} &\approx 0.23u_{1t} + 0.50u_{2t} + 0.08u_{3t} \\
u_{3t+1} &\approx -0.08u_{1t} + 0.12u_{2t} + 0.56u_{3t} + 0.05u_{4t} \\
u_{4t+1} &\approx 0.06u_{4t}
\end{aligned} \tag{8.3}$$

We would expect to find a statistical dependence of the rainfall fluctuations (for each region of averaging) upon the past rainfall fluctuations of the western regions. In the matrices of parameters for the various models, this statistical dependence should manifest in the closeness to zero of the parameter estimates above the main diagonals of the matrices. For May data—when aggregation intervals are small (5 and 15 min)—this effect is clear but not strong. For temporal aggregations greater than 15 minutes, however, either parameter values above main diagonals are zeros or the number of statistically significant estimates (of above-main diagonals) is less than the number below. For the models of June data, the results are more dramatic, especially for the 30 and 45 minute aggregation intervals when all the estimates above the main diagonals are zeros. We now consider the five-variate models (Table 8.11) for zonal spatial averaging [see Figure 8.14(b2)]. In this case, a large number of small and zero estimates above main diagonals would be evidence that the rainfall fluctuations from the north influence the rainfall fluctuations of regions to the south.

However, the estimates in Table 8.11 do not suggest such an influence. On the whole, it seems that the numbers of zeros above and below main diagonals are approximately the same.

Models of the third type (Table 8.12)—based on diagonal band spatial averages [see Figure 8.14(b3)]—provide a clear quantitative picture of the strong dependence of rainfall fluctuations of any region upon the southeastward advection of rain clouds. All but one of the parameters above the main diagonals of June models and most such parameters of May models are zeros.

Rainfall fluctuations of each region with aggregation intervals larger than 15 minutes are partially determined by their own past (parameters on the main diagonals are the largest) and by the history of fluctuations of the closest region to the northwest (the values of the parameters of the lower second diagonals are not zeros). So, on average, southeastward transport of rain clouds in May and June of 1985 finds its reflection in rainfall statistics of the second moments when appropriate spatial averaging and temporal aggregation are used.

8.2 PRE-STORM Precipitation

Table 8.10: Estimates (exceeding the 99% significance level) of parameters of four-variate rainfall autoregressive models for spatial averaging scheme 1 [see Figure 8.14 (b1)].

May Parameters				ε	June Parameters				ε
(1,m,5)					(1,j,5)				
0.63	0.09	0.04	—	0.76	0.72	0.09	—	—	0.67
0.23	0.50	0.08	—	0.80	—	0.77	—	0.11	0.62
−0.08	0.12	0.56	0.05	0.79	0.05	0.05	0.75	—	0.64
—	—	—	0.60	0.80	—	—	0.07	0.67	0.73
(1,m,15)					(1,j,15)				
0.49	—	0.13	−0.06	0.86	0.65	0.08	—	—	0.74
—	0.64	−0.23	—	0.78	0.09	0.60	—	—	0.78
—	−0.06	0.50	0.14	0.85	—	0.14	0.66	—	0.71
—	—	—	0.58	0.81	0.10	−0.14	0.21	0.56	0.77
(1,m,30)					(1,j,30)				
0.37	0.12	—	—	0.90	0.66	—	—	—	0.73
0.23	0.41	−0.14	—	0.86	0.24	0.49	—	—	0.81
—	0.27	0.42	—	0.83	—	0.21	0.63	—	0.71
—	—	0.18	0.68	0.68	0.09	−0.15	0.28	0.58	0.71
(1,m,45)					(1,j,45)				
0.21	0.18	—	−0.15	0.94	0.59	—	—	—	0.79
0.26	0.38	—	—	0.86	0.32	0.57	—	—	0.71
—	0.33	0.47	—	0.79	—	0.35	0.60	—	0.64
0.10	−0.15	0.35	0.64	0.65	—	—	0.40	0.52	0.68
(1,m,60)					(1,j,60)				
0.31	0.12	—	—	0.93	0.50	—	—	—	0.85
0.41	0.33	−0.15	—	0.80	0.26	0.58	−0.26	0.13	0.75
—	0.43	0.59	−0.18	0.58	0.14	0.34	0.54	—	0.68
0.22	−0.17	0.45	0.41	0.74	—	—	0.42	0.43	0.68

Table 8.11: Estimates (exceeding the 99% significance level) of parameters of five-variate rainfall autoregressive models for spatial averaging scheme 2 [see Figure 8.14 (b2)].

		May			
		Parameters			ε
		(2,m,5)			
0.64	−0.06	0.04	—	—	0.77
—	0.48	—	−0.08	0.12	0.83
—	0.09	0.55	0.04	—	0.82
0.04	−0.10	−0.04	0.54	0.08	0.82
—	0.16	0.04	−0.07	0.55	0.76
		(2,m,15)			
0.39	—	—	0.16	—	0.90
0.15	0.51	—	—	—	0.82
—	0.32	0.28	—	−0.10	0.91
0.06	0.15	—	0.50	−0.16	0.87
—	0.14	0.11	0.09	0.46	0.80
		(2,m,30)			
0.34	—	0.10	0.28	—	0.86
0.30	0.34	—	—	0.11	0.82
—	0.67	0.17	0.14	−0.25	0.80
—	0.13	0.16	0.38	−0.12	0.90
0.17	0.24	0.12	0.15	0.19	0.85
		(2,m,45)			
0.27	—	0.12	0.21	0.15	0.88
0.44	0.41	—	—	—	0.80
—	0.98	0.15	—	−0.59	0.75
—	−0.21	0.35	0.20	0.14	0.90
0.34	—	—	0.16	0.40	0.83
		(2,m,60)			
0.23	—	0.13	—	0.21	0.88
0.45	0.48	−0.14	0.12	−0.13	0.78
0.23	0.47	0.23	—	−0.14	0.78
—	0.35	0.32	0.49	−0.47	0.78
0.35	—	—	0.15	0.27	0.85

(Continued)

8.2 PRE-STORM Precipitation

Table 8.11 (Cont.)

June					
Parameters					ε
(2,j,5)					
0.60	0.11	0.07	−0.09	−0.07	0.74
—	0.72	0.07	—	—	0.66
—	0.08	0.71	—	—	0.68
−0.11	—	0.11	0.63	—	0.74
—	—	0.08	—	0.68	0.71
(2,j,15)					
0.37	0.19	0.09	—	—	0.86
0.16	0.45	0.17	—	—	0.77
—	0.08	0.53	—	—	0.81
−0.14	–	0.26	0.39	0.06	0.83
−0.09	0.08	—	—	0.44	0.89
(2,j,30)					
0.20	0.27	0.18	—	0.11	0.86
0.36	0.27	0.36	—	—	0.65
—	0.18	0.33	0.18	—	0.83
—	—	0.17	0.39	—	0.86
—	—	—	—	0.32	0.94
(2,j,45)					
0.46	0.15	—	—	0.15	0.79
0.33	0.34	0.25	—	0.14	0.65
—	—	0.47	0.24	—	0.72
−0.13	0.15	0.20	0.36	—	0.85
—	—	—	—	0.43	0.88
(2,j,60)					
0.38	—	0.20	—	0.16	0.80
0.37	0.15	0.42	−0.15	0.12	0.67
—	—	0.34	0.30	—	0.76
—	—	0.24	0.45	—	0.79
−0.16	—	—	0.21	0.39	0.88

Consider the model (3, j, 30) of data spatially averaged over ten-gauge regions, as shown in Figure 8.14(b3), and aggregated over 30 minute time intervals (6 sequential points). The autocorrelation estimates for the first two lags are represented in Table 8.13. Statistically significant estimates of space-time correlations reach values of 0.33 and 0.39, for example, and allow us to build a stochastic model.

The constructed models describe the behavior of rainfall fluctuations with different aggregations. They show that in May and

Table 8.12: Estimates (exceeding the 99% significance level) of parameters of four-variate rainfall autoregressive models for spatial averaging scheme 3 [see Figure 8.14 (b3)].

May					June				
Parameters				ε	Parameters				ε
(3,m,5)					(3,j,5)				
0.68	0.05	—	−0.05	0.73	0.77	—	−0.03	—	0.63
—	0.58	—	—	0.81	0.06	0.80	—	—	0.59
—	—	0.65	—	0.76	—	—	0.77	—	0.63
−0.12	0.05	0.04	0.63	0.76	—	−0.06	—	0.69	0.71
(3,m,15)					(3,j,15)				
0.57	—	−0.11	—	0.81	0.61	—	—	—	0.78
—	0.45	0.06	—	0.88	0.16	0.64	—	—	0.71
—	0.07	0.49	0.06	0.86	−0.08	0.12	0.62	—	0.75
−0.14	—	0.14	0.52	0.81	—	−0.08	0.08	0.56	0.81
(3,m,30)					(3,j,30)				
0.46	—	—	—	0.88	0.55	—	—	—	0.85
0.22	0.43	0.12	—	0.85	0.27	0.59	—	—	0.69
0.23	0.19	0.44	—	0.82	—	0.25	0.49	—	0.81
—	—	0.22	0.59	0.73	0.12	−0.11	0.17	0.52	0.80
(3,m,45)					(3,j,45)				
0.25	—	—	−0.12	0.96	0.51	—	—	—	0.84
0.41	0.55	—	—	0.71	0.35	0.52	—	—	0.69
0.30	0.27	0.43	—	0.78	—	0.38	0.55	—	0.67
—	—	0.24	0.62	0.69	—	−0.14	0.34	0.52	0.76
(3,m,60)					(3,j,60)				
0.28	—	—	—	0.96	0.32	—	—	—	0.93
0.55	0.35	—	—	0.74	0.44	0.43	—	—	0.71
0.30	0.29	0.47	—	0.75	—	0.44	0.42	—	0.74
0.20	—	0.40	0.34	0.79	—	—	0.43	0.49	0.71

8.2 PRE-STORM Precipitation

Table 8.13: Matrix autocorrelation function for the first two lags (0 and 30 min) of the model (3, j, 30).

	u_1	u_2	u_3	u_4
$\tau = 0$				
u_1	1.00	0.29	−0.11	0.20
u_2	0.29	1.00	0.17	−0.07
u_3	−0.11	0.17	1.00	0.08
u_4	0.20	−0.07	0.08	1.00
$\tau = 1$ (30 min)				
u_1	0.52	0.42	0.00	0.18
u_2	0.11	0.68	0.33	−0.09
u_3	−0.08	0.08	0.54	0.18
u_4	0.07	−0.04	0.01	0.56

especially in June, when it rained more often, such fluctuations were most strongly dependent upon advection from the northwest. The computed parameter matrices of multivariate autoregressive models for longitudinal and diagonal band averaging (with temporal aggregations of a half hour or more) have a clear and stable triangular form with zero elements above main diagonals. Therefore, as in the case of GATE data, appropriate spatial averaging and the approximation of multiple time series by multivariate autoregressive models has the potential to detect the structure of second-moment rainfall fluctuations.

8.2.4 Diffusion

With sets of the AR parameter estimates, we can compute (exactly the same way as in Section 7.5) the diffusion equation coefficients (for temporal aggregations of a half hour or more and for eastward and southeastward horizontal atmospheric motions).

Consider the numerical scheme (7.2) to (7.8). We assume that $h \approx 100$ km (the approximate distance between the centers of the regions of averaging) and $c = 0$. The value of b was taken as the mean of the four estimates on the main diagonals of the parameter matrices (Tables 8.10 and 8.12) of corresponding models for the regions of averaging under schemes 1 and 3 [Fig. 8.14(b)]. The value of a is the mean of the three estimates on the diagonals below the main diagonals of the same models. The results of such computations are shown in Table 8.14. The advection velocity v varies between 34 and 48 km/hr with a mean of 44 km/hr (about 12 m/s). Such a result, as it seems, is a reasonable approximation of the mean velocity of the synoptic fronts for the considered region of the United States. Also note that the value of the diffusion coefficient $d = \sqrt{s}$ varies from 41 to 49 with mean of 47. If the mean values in Table 8.14 are accepted as the estimates of the coefficients of the diffusion equation, we get the following:

$$\frac{\partial u}{\partial t} \approx 47^2 \frac{\partial^2 u}{\partial x^2} - 44 \frac{\partial u}{\partial x} - 0.30u + f(x,t). \qquad (8.4)$$

Therefore, the mean values of the diffusion equation coefficients can be determined from appropriate modeling of averaged and aggregated systematic rainfall data when the wind direction is stable.

8.3 Final Remarks

The formalization of the rainfall statistical description can be achieved by applying standard techniques of multivariate modeling and multidimensional spectral and correlation analysis to data averaged spatially and temporally. The analysis can be briefly summarized as follows.

1. The fitted simple linear stochastic rain field models (as well as two-dimensional spectral and correlation estimates) can be used in the ground truth problem and in the comparative climatological studies connected with statistical inferences concerning the rain distribution parameters.

2. The structure of the multivariate AR model parameter matrix, as well as that of the two-dimensional spectrum and correlation function, paints a clear and stable picture of the predominance of the propagation of the GATE1 rain rate fluctuations along the parallels (as well as the eastward and southeastward wave propagation of the PRE-STORM rainfall fluctuations).

8.3 Final Remarks

Table 8.14: Estimates of the coefficients of the diffusion equation for different months, regions, and aggregation intervals l(hours).

l	a	b	$\sqrt{s}$	v	q
Region 1, May					
0.50	0.23	0.47	48	46	0.60
0.75	0.31	0.42	45	41	0.36
1.00	0.43	0.41	46	43	0.16
Region 1, June					
0.50	0.24	0.59	49	48	0.34
0.75	0.36	0.57	49	48	0.09
1.00	0.34	0.51	41	34	0.15
Region 3, May					
0.50	0.21	0.48	46	42	0.62
0.75	0.31	0.46	45	41	0.31
1.00	0.41	0.36	45	41	0.23
Region 3, June					
0.50	0.23	0.54	48	46	0.46
0.75	0.36	0.54	49	48	0.13
1.00	0.44	0.42	47	44	0.14
Mean	0.32	0.48	47	44	0.30

3. The comparison of the two-dimensional spectral and correlation functions of the entire GATE1 data and the 12-hour sample shows that data obtained from only one satellite is insufficient for making accurate climatological generalizations concerning the second moments of rain rate.

4. The diurnal cycle of rain is one of the causes for the possible bias in the mean rain statistics obtained with limited satellite data, and it must be taken into account in any comparison of ground and satellite observations. The diurnal cycle of the GATE1 rain data carries a relatively simple sinus wave shape, with minimum values occurring during the night and maximum values during the afternoon. The range of variation (3.4 to 5.4 mm/hr) is relatively small, compared with the standard deviation (about 7 mm/hr) of observations.

5. The complicated character of the anisotropy of the spatial rain fields precludes the possibility of transforming them into an isotropic form by the standard procedures of scaling and rotating coordinates. This is one of the reasons that any rain model that can accurately describe both large-scale spatial and temporal motions and the diurnal cycle—and also distinguish between the WE and NS transports—will necessarily be extremely complicated. No simple (one or two parameter) stochastic rain model that can describe the above-mentioned features of rain exists at present.

6. The numerical approximation of the diffusion equation that generates the fluctuation field by multivariate AR process has led to the solution of the corresponding inverse problem and obtained the mean advection velocity of the observed rain fields (for example, about 12 m/s for the PRE-STORM precipitation).

7. Based on the estimates of the advection velocity and of the spatial correlation function, the methodology of mutual verification of rain and wind satellite observations can be developed.

8. Comparing the diverse statistical estimates of this section with the very simple (white-noise-type) model for the precipitation climate time series in Chapter 6, one can understand the output of large-scale temporal averaging of meteorological observations.

8.3 Final Remarks

Therefore, the results obtained can be used for a comparative climatological study of rain in the statistical inference procedures about parameters of the distribution of observations from different sources of measurements and for solving some theoretical and engineering problems.

References

Anderson, T.W., 1971: *Statistical Analysis of Time Series.* Wiley, New York.

Arkell, R., and M. Hudlow, 1977: GATE International Meteorological Radar Atlas. U.S. Dept. of Commerce, NOAA Publication, Washington D.C.

Bayley G.V., and J.W. Hammersley, 1946: The "Effective" Number of Independent Observations in an Autocorrelated Time Series. *J. Roy. Stat. Soc.*, **8**, 29–40.

Bell, T.L., and N. Reid, 1993: Detecting the Diurnal Cycle of Rainfall Using Satellite Observations. *J. Applied Meteorology*, **32**, 311–322.

Berezin, I.S., and N.P. Zhidkov, 1965: *Numerical Calculations.* Pergamon, Oxford.

Bider, M., M. Schüepp, and H. Rudolff, 1959: Die Reduktion der 200 jàhrigen Basler Temperaturreihe. Arch. Meteorol., Geophis. u. Bioklimatol. Ser. B, Bd 9, H 3–4.

Bider, M., and M. Schüepp, 1961: Luftdruckreihen der letzten zwei Jàhrhunderte von Basel und Genf. Arch. Meteorol., Geophis. u. Bioklimatol. Ser. B, Bd 11, H 1.

Blackman, R.B., and J.W. Tukey, 1958: *The Measurement of Power Spectra from the Point of View of Communications Engineering.* Dover, New York.

Bolshev, L.N., and N.B. Smirnov, 1965: *Statistical Tables.* Nauka, Moscow. (In Russian.)

Box, G.E., and G.M. Jenkins, 1976: *Time Series Analysis: Forecasting and Control.* Holden-Day, San Francisco.

Budyko, M.I., 1974: *Climate Change.* Hydrometeoizdat, 1974. (In Russian.)

Cahalan, R.F., D.A. Short, and G.R. North, 1982: Cloud Fluctuation Statistics. *Monthly Weather Review*, **110**, 26–43.

Chiu, L.S., G.R. North, A. Short, and A. McConnel, 1990: Rain Estimation from Satellites: Effect of Finite Field of View. *J. Geophys. Res.*, **95**, 2177–2185.

Christakos, G., 1992: *Random Field Models in Earth Sciences*. Academic Press, San Diego.

Cooley, J.W., and J.W. Tukey, 1965: An Algorithm for the Machine Calculation of Complex Fourier Series. *Mathem. Comp.* **19** (90).

Craddock, J.M., and B.G. Wales-Smith, 1977: Monthly Rainfall Totals Representing the East Midlands for the Years 1726–1975. *Meteorol. Mag.* 106 (1257).

Crane, R.K., 1990: Space-Time Structure of Rain Rate Fields. *J. Geophys. Res.*, **95**, 2011–2020.

Cressie, N., 1991: *Statistics for Spatial Data*. John Wiley, New York.

Cubasch, U., K. Hasselmann, H. Hock, E. Maier-Reimer, U. Mikolajewicz, B.D. Santer, and R. Sausen, 1992: Time-Dependent Greenhouse Warming Computations with a Coupled Ocean-Atmosphere Model. *Climate Dynamics*, **8**: 55–69.

Daley, R., 1991: *Atmospheric Data Analysis*. Cambridge University Press.

Eikhoff, P.E. (editor), 1983: *Modern Methods of Systems' Identifications*. Mir, Moscow. (In Russian.)

Einstein, A., 1986: Method of Determining of the Observed Statistical Values Related to the Random Fluctuations. *Izvestiya Academy of Science, Physics of Atmosphere and Ocean*, **22** (1), 99–100. (In Russian.)

Epstein, E.S., 1985: *Statistical Inference and Prediction in Climatology: a Bayesian Approach*. American Meteorological Society, Boston.

Essenwanger, O., 1989: *Applied Statistics in Atmospheric Sciences*. Elsevier, Amsterdam.

Flon, G., 1977: History and Climate Intransitivity. In: *Physical Foundation of Climate Theory and its Modeling*. Hydrometeoizdat, Leningrad. 114–124. (In Russian.)

Gandin, L.S., 1965: Objective Analysis of Meteorological Fields. Program for Scientific Translations. Jerusalem, Israel.

Gandin, L.S., R.L. Kagan, 1976: *Statistical Methods of Meteorological Data Interpretation*. Hydrometeoizdat, Leningrad. (In Russian.)

Graves, C.E, J.B. Valdes, S.S.P. Shen, and G.R. North, 1993: Evaluation of Sampling Errors of Precipitation from Spaceborne and Ground Sensors. *J. Applied Meteorology*, **32** (2), 374–385.

Gruza, G.V. (editor), 1987: *Climate Monitoring Data*. Goskomhydrometizdat, Moscow. (In Russian.)

Guyon, X., 1982: Parameter Estimation for a Stationary Process on a d-Dimensional Lattice. *Biometrika,* **69**, 95–105.

Hamming, R.W., 1973: *Numerical Methods for Scientists and Engineers*. McGraw-Hill, New York.

Hannan, E.J., 1970: *Multiple Time Series*. Wiley, New York.

Hasselmann, K., 1976: Stochastic Climate Models. *Tellus*, **28** (6), 473–485.

Hasselmann, K., 1988: PIPs and POPs: The Reduction of Complex Dynamical Systems using Principal Interaction and Oscillation Patterns. *J. Geophys. Res.*, **93**, 11015–11021.

Hays, J.D., J. Imbrie, and N.J. Shackleton, 1976: Variation in the Earth's Orbit: Pacemaker of Ice Ages. *Science*, **194** (4270).

Hudlow, M.D., 1977: Precipitation Climatology for the Three Phases of GATE. *Preprints Second Conf. Hydrometeorology*, Toronto, Amer. Meteor. Soc., 290–297.

Hudlow, M.D., 1979: Mean Rainfall Pattern for the Three Phases of GATE. *J. Appl. Meteor.*, **18**, 1656–1668.

Jenkins, J., and D. Watts, 1968: *Spectral Analysis and its Applications*. Holden-Day, San Francisco.

Jones, P.D., S.C.B. Raper, R.S. Bradly, H.F. Diaz, P.M. Kelly, and T.M.L. Wigley, 1986: Northern Hemisphere Surface Air Temperature Variations, 1851-1984. *J. Clim. Appl. Meteorol.*, **25**, 161–179.

Jones, R.H., and A.V. Vecchia, 1993: Fitting Continuous ARMA Models to Unequally Spaced Spatial Data. *J. American Statistical Association*, **88** (423), 947–954.

Kagan, R.L., 1979: *Averaging of the Meteorological Fields*. Hydrometeoizdat, Leningrad. (In Russian.)

Kashyap, R.L., and A. Rao, 1976: *Dynamic Stochastic Models from Empirical Data*. Academic Press, New York.

Katz, R.W., 1992: The Role of Statistics in the Validation of General Circulation Models. *Climate Research*, **2**, 34–45.

Kendall, M., and A. Stuart, 1963: *The Advance Theory of Statistics: Distribution Theory*. Hafner, New York.

Kendall, M., and A. Stuart, 1967: *The Advance Theory of Statistics: Inference and Relationship*. Hafner, New York.

Lanczos, C., 1956: *Applied Analysis*. Prentice-Hall, Englewood Cliffs, N.J.

Landsberg, H.E., J.M. Mitchell, and H.L. Gruntcher, 1959: Power Spectrum Analysis of Climatological Data for Woodstock College, Maryland. *Month. Weath. Rev.,* **87** (8).

Lebrijn, A., 1954: The Climate of the Netherlands During the Last Two and a Half Centuries. *Koninklijk Nederlandsch Meteorologisch Institut,* **102** (49), 94–99.

Lepekhina, N.A., and E.I. Fedorchenko, 1972: Temporal Structure of Temperature Time Series. *Trudy Main Geophysical Observatory,* **286**. (In Russian.)

Loeve, M., 1960: *Probability Theory.* Van Nostrand, Princeton, N.J.

Lorenz, E.N., 1977: Climate Predictability. In: *Physical Foundation of Climate Theory and its Modeling.* Hydrometeoizdat, Leningrad. 137–141. (In Russian.)

Madden, R.A., 1977: Estimates of the Autocorrelation and Spectra of Seasonal Mean Temperatures over North America. *Mon. Wea. Rev.,* **105** (1), 9–18.

Manley, G., 1974: Central England Temperatures: Monthly Means 1659 to 1973. *Quart. J. Roy. Meteor. Soc.,* **100** (425), 389–405.

Matrosova, L.E., and I. I. Polyak, 1990: Presentation of the Upper Atmosphere Temperature in Stochastic Models. *Izv. Acad. Sci., U.S.S.R., Physics of Atmosphere and Ocean,* **26** (9), 925–934.

McConnell, A., and G.R. North, 1987: Sampling Errors in Satellite Estimates of Tropical Rain. *J. Geophys. Res.* **92** (D8), 9567–9570.

Meitin, J., and J. Cunning, 1985: The Oklahoma-Kansas Preliminary Regional Experiment for Storm-Central. Vol. I: Daily Operations Summary. NOAA Technical Memorandum ERL ESG-20.

Mitchell, J.M., 1976: An Overview of Climatic Variability and Its Causal Mechanisms. *Quart. Res.,* **6** (4).

Monin, A.S., 1969: *Weather Forecasts as a Physical Problem.* Nauka, Moscow. (In Russian.)

Monin, A.S., 1972: *The Earth's Rotation and Climate.* Hydrometeoizdat, Leningrad. (In Russian.)

Monin, A.S., 1977: *The Earth History.* Nauka, Moscow. (In Russian.)

Monin A.S., and I.L. Vulis, 1971: On the Spectra of Long-Period Oscillations of Geophysical Parameters. *Tellus,* **23** (4–5).

Monin, A.S., and A.M. Yaglom, 1973: *Statistical Fluid Mechanics.*

MIT Press, Cambridge.
Nakamoto, S., J.B. Valdes, and G.R. North, 1990: Frequency-Wavenumber Spectrum for GATE Phase I Rain Fields. *J. Appl. Meteor.*, **29**, (9), 842–850.
North, G.R., and S. Nakamoto, 1989: Formalism for Comparing Rain Estimation Designs. *J. Atmos. Ocean. Tech.*, **6**, 985–992.
North, G.R, and K.Y. Kim, 1991: Surface Temperature Fluctuations in Stochastic Climate Models. *J. Geophysical Research*, **96**, (D10), 18573–18580.
Oort, A.H., and Rasmusson, 1971: *Atmospheric Circulation Statistics.* NOAA Professional Paper No. 5, U.S. Govt. Printing Office, Washington, D.C.
Oort, A.H., 1978: Adequacy of the Radiosonde Networks for Global Circulation Study Tested Through Numerical Model Output. *Monthly Weather Review*, **106**, 107–115.
Oort, A.H., 1983: *Global Atmospheric Circulation Statistics, 1958-1973.* NOAA Professional Paper No. 14, U.S. Govt. Printing Office, Washington, D.C.
Palmen, E., and C.W. Newton, 1969: *Atmospheric Circulation Systems.* Academic Press, New York.
Panofsky, H.A., and G.W. Brier, 1958: *Some Applications of Statistics to Meteorology.* The Pennsylvania State University.
Parzen, E., 1966: Time Series Analysis for Models of Signal Plus White Noise. In *Spectral Analysis of Time Series*, Proceedings. J.Wiley and Sons, New York. pp. 233–258.
Parzen, E. (editor), 1984: Time Series Analysis of Irregularly Observed Data. Proceedings of a symposium held at Texas A&M University, College Station, Texas, February 10–13, 1983. Springer-Verlag, New York.
Patterson, V.L., M.D. Hudlow, P.J. Pytlowany, F.P. Richards, and J.D. Hoff, 1979: GATE Radar Rainfall Processing System, NOAA Tech. Memo. EDIS 26, Washington, D.C.
Polyak, I.I., 1975: *Numerical Methods for Data Analysis.* Hydrometeoizdat, Leningrad. (In Russian.)
Polyak, I.I., 1979: *Methods for Random Processes and Fields Analysis.* Hydrometeoizdat, Leningrad. (In Russian).
Polyak, I.I., 1989: *Multivariate Stochastic Models of Climate.* Hydrometeoizdat, Leningrad. (In Russian).
Polyak, I.I., 1992: Multivariate Stochastic Climate Models. 12th Conference on Probability and Statistics in the Atmospheric Sciences. Toronto. pp.273–276.

Polyak, I.I., Z.I. Pivovarova, and L.V. Sokolova, 1979: Long-Period Variations of the Solar Radiation. *Trudy of the Main Geophysical Observatory.* No. 428.

Polyak, I.I., G.R. North, and J. B. Valdes, 1994: Multivariate Space-Time Analysis of PRE-STORM Precipitation. *J. of Applied Meteorology.* **33** (9), 1079–1087.

Polyak, I.I., and G.R. North , 1995: The Second Moment Climatology of the GATE Rain Rate Data. *Bulletin of the American Meteorological Society,* **76** (4), 535–550.

Polyak, I.I., 1996: Observed versus Simulated Second-Moment Climate Statistics in the GCM Verification Problems. *J. Atmospheric Sciences,* **53** (5), 677–694.

Preisendorfer, R.W., and C.D. Mobley, 1988: *Principal Component Analysis in Meteorology and Oceanography,* Elsevier, New York.

Raibman, N., 1983: Methods of Non-Linear and Minimax Identification. In: *Modern Methods of System Identification.* Mir, Moscow. (In Russian.)

Rao, R., 1973: *Linear Statistical Inference and Its Applications.* Wiley, New York.

Roeckner E., 1989: *The Hamburg Version of the ECMWF Model (ECHAM).* G.J. Boer, editor. CAS/JSC Working Group on Numerical Experimentation, Res. Act. Atm. and Ocean Modeling, Rep. 13, WMO/TD-322, pp. 7.1–7.4.

Santer, B.D. and T.M.L. Wigley, 1990: Regional Validation of Means, Variances, and Spatial Patterns in General Circulation Model Control Runs. *J. Geophysical Res.,* **95**, 829–850.

Santer, B.D., W. Brüggemann, U. Cubasch, K. Hasselmann, H. Höck, E. Maier-Reimer, and U. Mikolajewicz, 1994: Signal-to-Noise Analysis of Time-Dependent Greenhouse Warming Experiments. *Climate Dynamics,* **9**, 267–285.

Schmugge, T., 1977: Remote Sensing of Surface Soil Moisture. Preprints Second Conf. Hydrometeorology, Toronto, Amer. Meteor. Soc., 304–310.

Simpson, J., R. Adler, and G.R. North, 1988: On Some Aspects of a Proposed Tropical Rainfall Measuring Mission (TRMM). *Bull. Amer. Meter. Soc.,* **69**, 278–295.

Thiebaux, H.J., 1994: *Statistical Data Analysis for Ocean and Atmospheric Sciences.* Academic Press, New York.

Tong, H., 1990: *Non-Linear Time Series. A Dynamical System Approach.* Oxford Science Publications, Oxford.

Tutubalin, V.N., 1972: *The Probability Theory.* Moscow University. (In Russian.)

Understanding Climate Change. A Program for Action., 1975: National Academy of Sciences. Washington, D.C.

van der Waerden, B.L., 1960: *Mathematical Statistics.* Moscow Publishing House of Foreign Literature. (In Russian.)

Wigley, T.M.L., and B. D. Santer, 1990: Statistical Comparison of the Spatial Fields in Model Validation, Perturbation, and Predictability Experiments. *J. Geophy. Res.*, **95**, 851–865.

Wilks, D.S., 1995: *Statistical Methods in the Atmospheric Sciences.* Academic Press, New York.

World Weather Records, 1975: U.S. Department of Commerce, NOAA, National Climatic Center, Asheville, N.C.

Yaglom, A.M., 1986: *Correlation Theory of Stationary and Related Random Functions.* Springer-Verlag, New York.

Index

advection velocity 308
    GATE rain rate 330
    PRE-STORM data 344
    temperature 288
algorithm for multivariate AR modeling 218
amplitude modulated time series 131
anisotropy
    climatological fields 160
    rain rate field 308
AR model and diffusion process 285
AR process 165
    multivariate 208
AR(1) process 167
    multivariate 212
AR(2) process 165
    multivariate 223
ARMA process 163
ARMA(1,1) process 183
atmospheric pressure 248
autocovariance function 111
    estimator 112
autocorrelation function 111
    atmosperic pressure 253
    precipitation 255
    temperature 251
autodispersion function 108
autoregressive parameters 163

Bernoulli polynomials 10
Bessel function 161
Blackman-Tukey methodology 115

central England temperature 193
circulation statistics 225
climate change 263

climate system identification 234
climatological fields 147
coefficients of polynomial 7
coherency function 118, 139–146
    estimator 121, 139–146
Cooley-Tukey methodology 115
correlated observations 61
correlation function
    latitude-temporal rain rate field 316
    latitude-temporal temperature fields 271
    point estimates 27, 32
    PRE-STORM rainfall fields 334
correlation matrix 62
correlation window 113
    two-dimensional 135
cospectrum 118
covariance
    function of point estimates 32
    matrix 61
covariances of point estimates 26
cross-covariance function 118, 139–146
    two-dimensional 132
cross-periodogram 119, 139-146
    two-dimensional 133
cross-spectral density 118, 139–146
    two-dimensional 132
cross statistical analysis 115
cumulative spectral function 116

derivative estimation 44, 49
derivatives of observed function 14
differentiation of fields 54
diffusion 287

diffusion equation coefficients estimation 288
diffusion of GATE rain rate 329
diffusion of PRE-STORM observations 344
diffusion of temperature 285
digital filter 26
discrete Fourier expansion 139–146
   two-dimensional 133
dispersion function 107
diurnal cycle of rain rate 318
double frequency spectral density 125

equivalence of filters 42

fast Fourier transform (FFT) 115
filters
   correlated observations 84
   differentiating 44, 54
   finite differences 91
   harmonic 37
   multi-dimensional 56
   parameters 26
   regressive 28
   smoothing 24
   two-dimensional 51
   width 26
   window 26
finite differences 92
   Markov process 96
   white noise 93
Fourier coefficients 21
Fourier set 21
Fourier transforms 111, 139–146
   two-dimensional 140
frequency characteristics 28, 37

GATE data 304
GATE1 data 304
GATE2 data 320
Gauss-Markov Theory 3
GCM diagnosis and verification 267
geopotential 147
ground truth problem 311

Hamburg GCM zonal temperature 268

harmonic filter 39
harmonic structure 28
harmonizable processes 125
historical records 239

identification 178
   ARMA processes 162
   climate system 234
   polynomial degree 20
instrumental variable 100
inverse normal matrix 11

Kolmogorov-Smirnov criterion 117

Laplace equation 160, 310
latitude-temporal
   rain rate field 315
   temperature field 291
least squares method 4, 61
   polynomial approximations 6
linear regression 99
   observed and simulated temperature 289
linear trend 70
   air temperature 196, 240, 245,
   atmospheric pressure 243
   precipitation 243, 247

MA process 179
matrix correlation function 209
   GATE data 324
   observed and simulated temperature 280–283
   PRE-STORM data 334
matrix covariance function 209
mean estimator 70, 75, 76
menorah equation 18
meridional circulation 288
models
   atmospheric circulation statistics 225
   GATE1 rain rate data 323
   observed and simulated temperatures 281
   PRE-STORM precipitation 336
   temperature time series 225–233

Index

moving average
   harmonic analysis 124
   parameters 163
   procedure 26
   processes 179
multidimensional field 138
   filter 58
multiple correlation coefficient 165
multivariate AR processes 208
   AR(1) process 209
   AR(2) process 221

nonergodic stationary process 127
nonlinear
   random processes 107
   model 205
nonoptimal estimation 74
nonstationary process 122
normal equations 4, 7, 9, 62
   matrix 4, 11
normalized standard deviation of forecast error 164, 167–183
normalized variances
   forecast error 164
   point estimates 5, 22, 63
   spectral density 111
numerical differentiation of field 54

Parseval identity 113, 139–146
permissible correlations 100, 104
   AR process 168
   ARMA(1,1) process 184
   instrumental variable 101
   linear regression 99
   MA process 179
   multivariate AR(1) process 213
periodogram 112, 139–146
   historical records 247, 251
   PRE-STORM rainfall fields 334
phase spectrum 118, 121
   estimator 121, 139–146
point estimates 5, 22, 62
polynomial coefficients 14
precipitation trend over territory of Russia 245
PRE-STORM data 332

process with stationary increments 198

quadratic form 62
quadrature spectrum 118
q-variate AR process 210
q-variate correlation function 209
q-variate white noise 210

regressive filters 29

sample
   cospectrum 119
   covariance function of two-dimensional field 133
   cross-covariance function 117
      two-dimensional 136
   quadrature spectrum 119
second-order derivative
   estimation 47
   of field 54
separable statistical dependence 83
signal-plus-white-noise processes 82, 190
signal ratio 84, 191
   climate time series 193
   surface monthly mean temperatures 196
smoothing
   correlated observations 87
   two-dimensional 51
   window 30
solar radiation budget 196
spatial averaging 79
spatial correlation function of GATE1 data 306
spatial spectrum of GATE1 data 306
spatial-temporal covariance 80
spectra
   altitude-temporal field 158
   historical records 259
   latitude-temporal climate fields 150
   latitude-temporal temperature fields 291
   monthly-annual field 154

PRE-STORM rainfall fields 334
  zonal temperature time series 271
spectral analysis of 500 mb surface geopotential field 147
spectral density 111, 132
spectral window 113
  two-dimensional 135
spectrum of latitude-temporal rain rate field 315
standardized autodispersion function 108
stationary process 111
statistics and climate change 263
streamflow time series 196
Student statistic 6
sum of signal and white noise 80
surface air temperature 267
surface air temperature spectra and autocorrelation functions 251
symmetrical grid points 7
symmetrical indexing 7

three Fourier transformation method 115
time series with missing data 129
Toeplitz matrix 75
Tukey filter 41

univariate AR processes 165
univariate ARMA processes 163
univariate modeling of historical records 259

validation of GCM 266
variance of forecast error 164
variance of point estimates 6, 15, 62, 89

window width 26

Yule-Worker
  equations 165
  multivariate equations 211